T0207109

Beginning IoT Projects

Breadboard-less Electronic Projects

Charles Bell

Apress®

Beginning IoT Projects: Breadboard-less Electronic Projects

Charles Bell
Warsaw, VA, USA

ISBN-13 (pbk): 978-1-4842-7233-6 ISBN-13 (electronic): 978-1-4842-7234-3
https://doi.org/10.1007/978-1-4842-7234-3

Managing Director, Apress Media LLC: Welmoed Spahr
Acquisitions Editor: Susan McDermott
Development Editor: James Markham
Coordinating Editor: Jessica Vakili

Distributed to the book trade worldwide by Springer Science+Business Media New York, 233 Spring Street, 6th Floor, New York, NY 10013. Phone 1-800-SPRINGER, fax (201) 348-4505, e-mail orders-ny@springer-sbm.com, or visit www.springeronline.com. Apress Media, LLC is a California LLC and the sole member (owner) is Springer Science + Business Media Finance Inc (SSBM Finance Inc). SSBM Finance Inc is a **Delaware** corporation.

For information on translations, please e-mail booktranslations@springernature.com; for reprint, paperback, or audio rights, please e-mail bookpermissions@springernature.com.

Apress titles may be purchased in bulk for academic, corporate, or promotional use. eBook versions and licenses are also available for most titles. For more information, reference our Print and eBook Bulk Sales web page at http://www.apress.com/bulk-sales.

Any source code or other supplementary material referenced by the author in this book is available to readers on GitHub via the book's product page, located at www.apress.com/978-1-4842-7233-6. For more detailed information, please visit http://www.apress.com/source-code.

Printed on acid-free paper

Table of Contents

About the Author

Charles Bell conducts research in emerging technologies. He is a member of the Oracle MySQL development team and is a principal developer for the MySQL cloud services team. He lives in a small town in rural Virginia with his loving wife. He received his Doctor of Philosophy in Engineering from Virginia Commonwealth University in 2005. Dr. Bell is an expert in the database field and has extensive knowledge and experience in software development and systems engineering. His research interests include 3D printers, microcontrollers, three-dimensional printing, database systems, software engineering, and sensor networks. He spends his limited free time as a practicing maker focusing on microcontroller projects and refinement of three-dimensional printers.

About the Technical Reviewer

Sai Yamanoor is an embedded systems engineer working for an industrial gases company in Buffalo, NY. His interests, deeply rooted in DIY and open source hardware, include developing gadgets that aid behavior modification. He has published two books with his brother, and in his spare time, he likes to build things that improve quality of life. You can find his project portfolio at `http://saiyamanoor.com`.

PART I

Getting Started with IoT

This part begins with an overview of the Internet of Things and then introduces the hardware platforms we will use in the book to learn how to build IoT projects. Platforms include the Arduino and Raspberry Pi. The part also includes tutorials on how to write the software to run on the platforms using the Arduino language for writing sketches and Python for programming on the Raspberry Pi.

CHAPTER 1

Introduction to the Internet of Things

Much has been written about the Internet of Things (IoT). Some sources are more about promoting IoT as their latest innovation (that costs more money); other sources seem to suggest IoT is something everyone needs or be left behind in the dust of antiquity.

Fortunately, books and similar media avoid the sales pitch to expand on the science and technology for implementing and managing the data for IoT, while other texts concentrate on the future or the inevitable evolution of our society as we become more connected to the world around us each and every day. However, you need not dive into such tomes or be able to recite rhetoric to get started with the IoT. In fact, through the efforts of many companies, you can explore the IoT without intensive training or expensive hardware and software.

However, most publications[1] assume the reader knows or wants to know how to connect discrete components together to build the hardware from scratch. That means if you'd like to learn more about building

[1] Including my own books!

C. Bell, *Beginning IoT Projects*, https://doi.org/10.1007/978-1-4842-7234-3_1

Internet of Things (IoT) solutions, but don't have the time or will to learn all of the nuances of electronics and discrete components, you've been left with little recourse.[2] That is, until now.

In this book, we will explore how to build IoT solutions using a series of basic projects without the need to learn the difference between a diode and a resistor. In fact, we will be using two modular platforms that you can connect to your host board without the need to wire anything together![3] This is accomplished by using a special adapter board/shield for your device that permits you to connect to modules that include sensors, displays, switches, and more!

However, before we get into the details of devices, boards, modules, etc., let's take a moment and learn what the Internet of Things is and what sort of solutions are classified as IoT projects.

What Is the Internet of Things?

So what is this IoT?[4] I'll begin by explaining what it isn't. The IoT is not a new device or proprietary software or some new piece of hardware. It is not a new marketing scheme to sell you more of what you already have by renaming it and pronouncing it "new."[5] While it is true that the IoT

[2] Well, there are IoT kits out there that you can use to build specific, simple projects, but not much in the way of help for those that want to take on new projects without spending a lot of time learning enough about electronics to implement the project.

[3] We will discuss advanced uses of the modular components and some simple wiring connections, but nothing that requires prior knowledge of electronics.

[4] `https://en.wikipedia.org/wiki/Internet_of_Things`

[5] For example, everything seems to be cloud this, cloud that when in reality nothing was changed.

employs technology and techniques that already exist, the way they are employed, coupled with the ability to access the solution from anywhere in the world, makes the IoT an exciting concept to explore. Now let's discuss what the IoT is.

The essence of the IoT is simply interconnected devices that generate and exchange data from observations, facts, and other data, making it available to anyone you'd like - only yourself or immediate family - or share it with the world. While there seem to be some marketing efforts attempting to make anything connected to the Internet an IoT solution or device (not unlike the shameless labeling of everything "cloud"), IoT solutions are designed to make our knowledge of the world around us timelier and more relevant by making it possible to get data about anything from anywhere at any time.

As you can imagine, if we were to connect every device around us to the Internet and make sensory data available for those devices, it is clear there would be potential for the number of IoT devices to exceed the human population of the planet[6] and for the data generated to rapidly exceed the capabilities of all but the most sophisticated database systems. These concepts are commonly known as addressability and big data, which are two of the most active and debated topics in IoT. But don't worry about these terms - I mention them here for completeness and possibly to pique your interest. You can read more about these issues at your leisure.

However, the IoT is all about understanding the world around us. That is, we can leverage the data to make our world and our understanding of it better.

[6] We aren't so far away from that now. Think about how many iWatches there are out there - yes, they're IoT devices too!

Before we proceed, let's review some terms that can help us understand the context and subject better. The following are the major terms used in this book:

- *IoT solution*: A complete project that implements the software and hardware to perform one or more tasks
- *IoT device*: The hardware (and associated software) that connects to the Internet sending data to one or more IoT services
- *IoT service*: A product or set of services in the cloud used to process IoT data
- *IoT vendor*: Those businesses that provide services for IoT solutions
- *IoT data*: Data generated from one or more IoT devices such as observations from one or more sensors
- *Knowledge*: The conclusions one can draw from the data once it has been made available in the IoT services (cloud) for review

Now that we know what the IoT is and some terms we use to describe it, let's dive further into what IoT means to us.

The Internet of Things and You

The best example of a sophisticated IoT device is the human body. It is a complex marvel of ingenious sensory apparatus that allow us to see, hear, taste, and even feel through touch anything we encounter or get near. Even our brains can store visual and auditory events recalling them at will. IoT solutions mimic many of these sensory capabilities and therefore can become an extension of our own abilities.

While that may sound a bit grandiose (and it is), IoT solutions can record observations in the form of data from one or more sensors. Sensors are devices that produce either analog or digital values. We can then use the data collected to draw conclusions about the subject matter. IoT devices can also retrieve information from one device and forward it to another, but let's keep it simple and focus on devices that detect things about the world around us and what that knowledge could do for us.

For example, an IoT device could be connected to a sensor to detect when a mailbox is opened. In this case, the knowledge we gain from a simple switch opening or closing (depending on how it is implemented and interpreted) may be used to predict when incoming mail has arrived or when outgoing mail has been picked up. I use the term predict because the sensor (switch) only tells us the door was opened or closed, not that anything was placed in or removed from the mailbox itself – that would require additional sensors.

When working with IoT projects that include sensors, you should always think about what conclusions you can draw from the data. Sometimes, like the switch in the mailbox, it can be only a few things, which is most often the case. By defining what we can perceive (learn) from the sensor data, we can better understand what our IoT project and its data can do for us.

A more sophisticated example is using a series of sensors to record atmospheric data such as temperature, humidity, barometric pressure, wind speed, ambient light, rainfall, and so forth, to monitor the weather and perform analysis on the data to predict trends in weather. That is, we can predict within a reasonable certainty that precipitation is in the area and to some extent its severity.

Now, add the ability to see this data not only in real time (as it occurs) but also remotely from anywhere in the world, and the solution becomes more than a simple weather station. It becomes a way to observe the weather about one place from anywhere in the world.

This example may be a bit commonplace since you can tune into any number of television, web, and radio broadcasts to hear the weather from anywhere in the world. But consider the implications of building such a solution in your home. Now you can see data about the weather at your own home from anywhere!

In the same way, but perhaps on a smaller scale, we can build solutions to monitor plants to help us understand how often they need water and other nutrients. Or perhaps we can monitor our pets while we are away at work. Further, we can record data about wildlife in our area to better understand our effect on nature.

IoT Is More Than Just Connected to the Internet

So, if a device is connected to the Internet, does that make it an IoT solution? That depends on whom you ask. Some will say the answer is yes. However, others (like me) contend that the answer is no unless there is some benefit from doing so.

For example, if you could connect your toaster to the Internet, what would be the benefit of doing so? What knowledge would you gain? It would be pointless (or at least extremely eccentric) to get a text on your phone from your toaster stating that your toast is ready given that it only takes a couple of minutes to complete. In this case, the answer is no. However, if you have someone - such as a child or perhaps an older adult - whom you would like to monitor, it may be helpful to be able to check to see how often and when they use a device like a toaster so that you can check on them.[7] That is, you can use the data to help you make decisions about their care and safety.

[7] Toasters and toaster ovens have appeared in the top five most dangerous appliances in the home. Scary.

Allow me to illustrate with another example. I was fortunate to participate in a design workshop held on the Microsoft campus in the late 1990s. During our tour of the campus, we were introduced to the world's first Internet-enabled refrigerator (also called a smart refrigerator).

There were sensors in the shelves to detect the weight of food. It was suggested that, with a little ingenuity, you could use the sensors to notify your grocer when your milk supply ran low, which would enable people to have their grocery shopping not only online but also automatic. This would have been great if you lived in a location where your grocer delivers, but not very helpful for those of us who live in rural areas.[8] While it wasn't touted an IoT device (the term was coined later), many felt the device illustrated what could be possible if devices were connected to the Internet.

Thus, being connected to the Internet doesn't make something IoT. Rather, IoT solutions must be those things that provide some meaning – however small that benefit is to someone or some other device or service. More importantly, IoT solutions allow us to sense the world around us and learn from those observations. The real tricky part is in how the data is collected, stored, and presented. We will see all of these in practice through examples in later chapters.

IoT solutions can also take advantage of companies that provide services that can help enhance or provide features that you can use in your IoT solutions. These features are commonly called IoT services and range from storage and presentation to infrastructure services, such as hosting.

---

[8] However, given the COVID-19 stay-at-home orders in many places, this idea may have come back into practicality.

IoT Services

Sadly, there are companies that tout having IoT products and services that are nothing more than marketing hype - much like what some companies have done by prepending "cloud" or appending "for the cloud" to the name. Fortunately, there are some good products and services being built especially for IoT. These range from data storage and hosting to specialized hardware and sophisticated data analysis and visualization.

Indeed, businesses are adding IoT services to their product offerings, and it isn't the usual suspects, such as the Internet giants. I have seen IoT solutions and services being offered by Cisco, AT&T, HP, and countless start-ups and smaller businesses.

You may be wondering what these services and products are and why someone would consider using them. That is, what is an IoT service, and why would you decide to buy it? The biggest concerns in the decision to buy a service are cost and time to market.

For example, if you want to use IoT in your organization but your developers do not have the resources or expertise and obtaining them will require more than the cost of the service, it may be more economical to purchase the service. However, you should also consider any additional software or hardware changes (sometimes called retooling) necessary in the decision. I once encountered a well-meaning and well-documented contracted service that permitted a product to go to market sooner than projected at a massive savings. Sadly, while the champions of that contract won awards for technical achievement, they failed to consider the fact that the systems had to be retooled to use the new service. More specifically, it took longer to adopt the new service than it would to write one from scratch. So instead of saving money, the organization spent nearly twice the original budget and was late to market. Clearly, you must consider all factors.

Similarly, if your time is short or you have hard deadlines to meet to make your solution production-ready, it may be quicker to purchase an IoT service rather than create or adapt your own. This may require spending a bit more, but in this case, the motivation is time and not (necessarily) cost. Of course, project planning is a balance of cost, time, and features.

So what are some of the IoT services available? The following lists a few that have emerged in the last few years. It is likely more will be offered as IoT solutions and services mature:

- *Enterprise IoT data hosting and presentation*: Services that allow your users to develop enterprise IoT solutions from connecting to managing and customizing data presentation in a friendly form, such as graphs, charts, and so forth.
- *IoT data storage*: Services that permit you to store your IoT data and get simple reports.
- *Networking*: Services that provide networking and similar communication protocols or platforms for IoT. Most specialize in machine-to-machine (M2M) services.
- *IoT hardware platforms*: Vendors that permit you to rapidly develop and prototype IoT devices using a hardware platform and a host of supported modules and tools for building devices ranging from a simple component to a complete device.

For the hobbyist or enthusiast, you may not need such sophistication. Rather, you may need only a place to store or display your data. In those cases, there are IoT vendors that provide such products (some free, from fee-based) using relatively simple-to-configure features. Two such examples include Microsoft Azure (`https://portal.azure.com`) and ThingSpeak for IoT Projects (`https://thingspeak.com/`). We will see ThingSpeak in action later on in book.

Now that you know more about what IoT is, let's look at a few examples of IoT solutions to get a better idea of what IoT solutions can do and how they are employed.

A Brief Look at IoT Solutions

Recall an IoT solution is simply a set of devices designed to produce, consume, or present data about some event or series of events or observations. This can include devices that generate data, such as a sensor, devices that combine data to deduce something, devices or services designed to tabulate and store the data, and devices or systems designed to present the data. Any or all of these may be connected to the Internet.

IoT solutions may include one or all of these qualities, whether it is combined into a single device such as a web camera; used as a sensor package and monitoring unit, such as a weather station; or used as a complex system of dedicated sensors, aggregators, data storage, and presentation, such as a complete home automation system. Figure 1-1 shows a futuristic picture of all devices - everywhere - connected to the Internet through databases, data collectors or integrators, display services, or other devices.

Figure 1-1. *The future of IoT – all devices, everywhere*[9]

Let's take a look at some example IoT solutions. The IoT solutions described in this section are a mix of solutions that should give you an idea of the ranges of sizes and complexities of IoT solutions. I also point out how some of these solutions leverage services from IoT vendors.

Sensor Networks

Sensor networks are one of the most common forms of IoT solutions. Simply stated, sensor networks allow you to observe the world around you and make sense of it. Sensor networks could take the form of a pond monitoring system that alerts you to water level, water purity (contamination), or water temperature or detects predators or even turns on features automatically, such as lighting or fish feeders.

[9] https://pixabay.com/en/network-iot-internet-of-things-782707/

If you, or someone you know, have spent any time in a medical facility, it's likely that a sensor network was employed to monitor body functions, such as temperature, cardiac and respiratory rates, and even movement. Modern automobiles also contain sensor networks dedicated to monitoring the engine, climate, and, even in some cars, road conditions. For example, the lane warning feature uses sensors (typically a camera, microprocessor, and software) to detect when you drift too far toward lane or road demarcations.

Thus, sensor networks employ one or more sensors that take measurements (observations) about an event or state and communicate that data to another component or node in the network, which is then presented, in some form or another, for analysis. Let's take a look at an example of an important medical IoT solution.

Medical Applications

Medical applications – including health monitoring and fitness – are gaining a lot of attention as consumer products. These solutions cover a wide range of capabilities, such as the fitness features built into the Apple Watch to fitness bands that keep track of your workout and even medical applications that help you control life-threatening conditions. For example, there are solutions that can help you manage diabetes.

Diabetes is a disease that affects millions of people worldwide (`www.diabetes.org`). There are several forms, the most serious being type 1 (`www.diabetes.org/diabetes-basics/type-1/?loc=db-slabnav`). Those afflicted with type 1 diabetes do not produce enough (or any) insulin due to genetic deficiencies, birth defects, or injuries to the pancreas. Insulin is a hormone that the body uses to extract a simple sugar called glucose, which is created from sugars and starches, from blood for use in cells.

Thus, type 1 diabetics must monitor their blood glucose to ensure that they are using their medications (primarily insulin) properly and balanced with a healthy lifestyle and diet. If their blood glucose levels become too low or too high, they can suffer from a host of symptoms. Worse, extremely low blood glucose levels are very dangerous and can be fatal.

One of the newest versions of a blood glucose tester consists of a small sensor that is inserted in the body along with a monitor that connects to the sensor via Bluetooth. You wear the monitor on your body (or keep it within 20 feet at all times). The solution is marketed by Dexcom (`dexcom.com`) and is called a continuous glucose monitor (CGM) that permits the patient to share their data to others via their phone. Thus, the patient pairs their CGM with their phone and then shares the data over the Internet to others. This could be loved ones, those that help with their care, or even medical professionals. Figure 1-2 shows an example of the Dexcom CGM app and sensor. The monitor is on the left, and the sensor and transmitter are on the right. The sensor is the size of a small syringe needle and remains inserted in the body for up to a week.

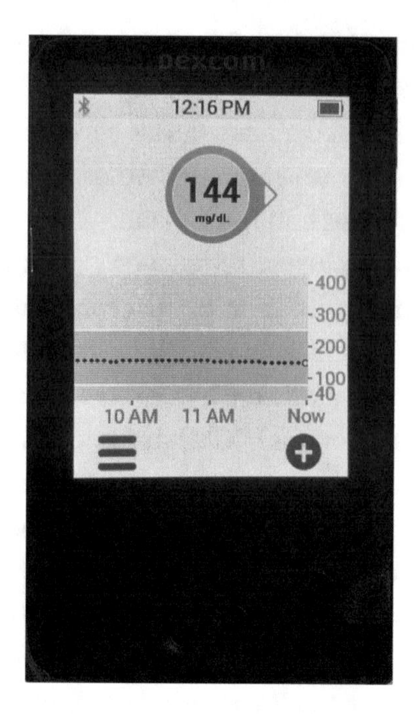

Figure 1-2. *Dexcom continuous glucose monitor with sensor*

WHAT ABOUT BLOOD GLUCOSE TESTERS (GLUCOMETERS)?

Until solutions like the Dexcom CGM came about, diabetics had to use a manual tester. Traditional blood glucose testers are single-use events that require the patient to prick their finger or arm and draw a small amount of blood onto a test strip. While this device has been used for many years, it is only recently that manufacturers have started making blood glucose testers with memory features and even connectivity to other devices, such as laptops or phones. The ultimate evolution of these devices is a solution like the Dexcom CGM, which is a medical IoT device that improves the quality of life for diabetics.

Dexcom also provides a free web-based reporting software called Clarity that is accessed from a special uploading application called the Clarity Uploader (see `http://dexcom.com/clarity` for more details)[10] to allow patients to see the data collected and generate a host of reports they can use to see their glucose levels over time. Reports include averages, patterns, daily trends, and more. They can even share their data with their doctor. Figure 1-3 shows an example of Dexcom Clarity with typical data loaded.

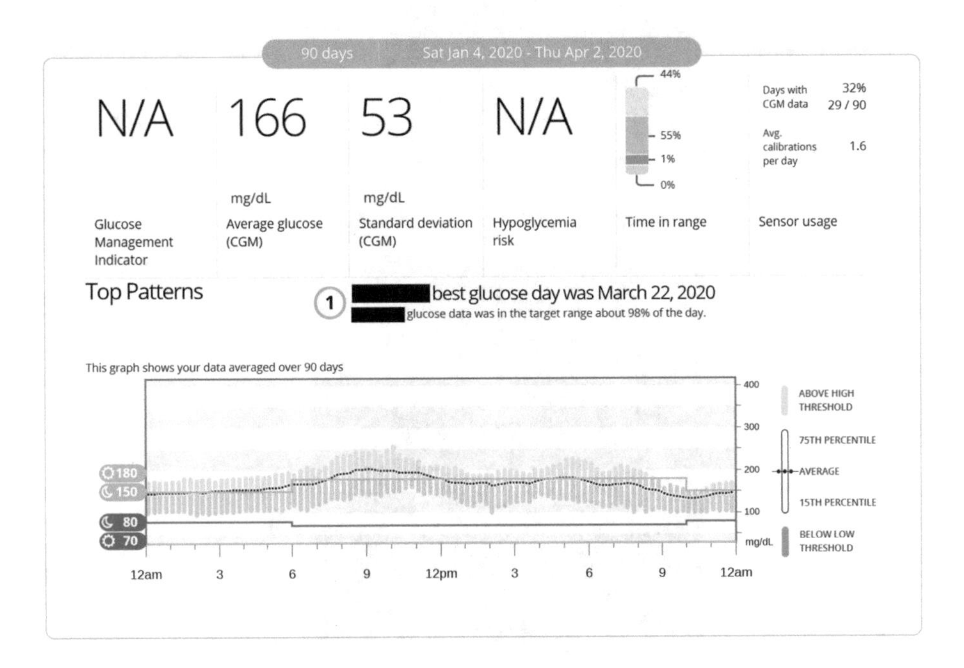

Figure 1-3. *Dexcom Clarity*

[10] Dexcom also provides a mobile version of Clarity for iOS or Android.

A feature called Dexcom Share permits the patient to make their data available to others via an app on their phone. That is, the patient's phone transmits data to the Dexcom cloud servers, which is then sent to anyone who has Dexcom Share iOS app and has been given permission to see the data. Figure 1-4 shows an example of the CGM report from the Dexcom Share iOS app, which allows you to check the blood glucose of a friend easily and quickly or loved one.

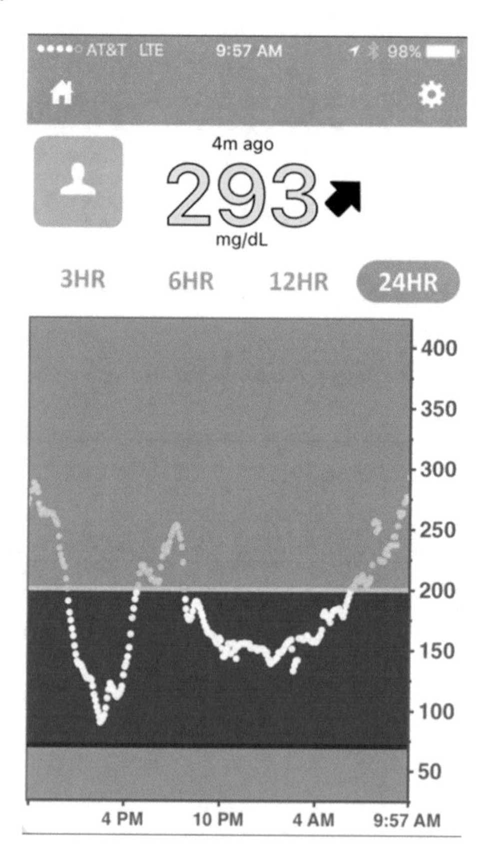

Figure 1-4. *Dexcom Share app report*

Not only does the app allow the visualization of the data, it can also relay alerts for low or high blood glucose levels, which has profound implications for patients who suffer from additional ailments or complications from diabetes. For example, if the patient's blood glucose level drops while they are alone, incapacitated, or unable to get treatment, loved ones with the Dexcom Share app can respond by checking on the patient and potentially avoiding a critical diabetic event.

While this solution is a single sensor connected to the Internet via a proprietary application, it is an excellent example of a medical IoT device that can enhance the lives of not only the patient but everyone who cares for them.

Combined with the programmable alerts, you and your loved ones can help manage the effects of diabetes. If you have a loved one who suffers from diabetes, a CGM is worth every penny for peace of mind alone. This is the true power of IoT materialized in a potentially life-saving solution.

Automotive IoT Solutions

Another personal IoT solution is the use of Internet-connected automotive features. One of the oldest products is called OnStar (`onstar.com`), which is available on most late-model and new General Motors (GM) vehicles. While OnStar predates the IoT evolution, it is a satellite-based service that has several levels and many fee-based options. It incorporates the Internet to permit communication with vehicle owners. Indeed, the newest GM vehicles come with a WiFi access point built into the car! Better still, there are some basic features that are free to GM owners that, in my opinion, are very valuable.

The free, basic features include regular maintenance reports sent to you via email and the ability to use an app on your phone to remotely unlock, lock, and start the car – all the features on your key fob. This is a really cool feature if you have ever locked your keys in your car! Figure 1-5 shows an example of the remote key fob app on iOS. Of course, there are even more features available for a fee, including navigation, telephone, WiFi, and on-call support.

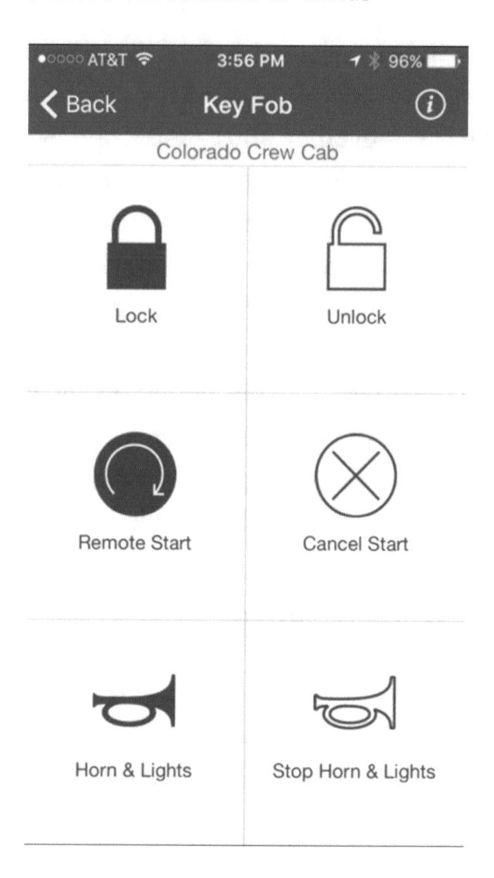

Figure 1-5. *OnStar app key fob feature*

The OnStar app works by connecting to the OnStar services in the cloud, requesting the feature (e.g., unlock) that is sent to the vehicle via the OnStar satellite network. So it is an excellent example of how IoT solutions use multiple communication protocols.

The feature I like most is the maintenance reports. You will receive an email with an overview of the maintenance status of your vehicle. The report includes such things as oil life, tire pressure, engine and transmission warnings, emissions, airbag, and more. Figure 1-6 shows an excerpt of a typical email that you receive.

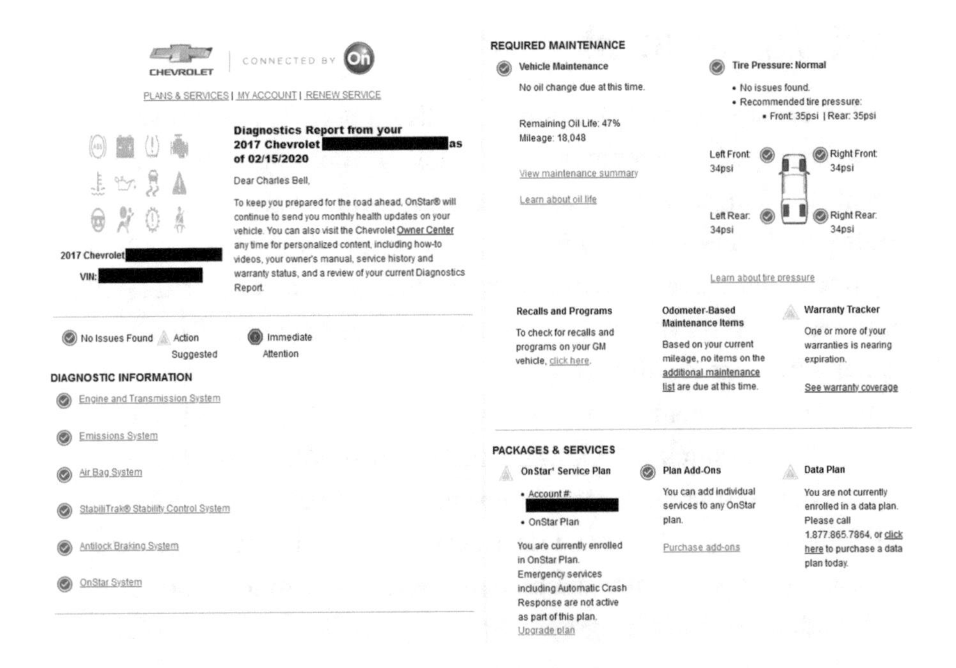

Figure 1-6. *OnStar maintenance report*

Notice the information displayed. This is no mere idiot light! Actual data is transmitted to OnStar from your vehicle. For example, the odometer reading and tire pressure data are taken directly from the vehicle's onboard data storage. That is, data from the sensors is read and interpreted and the report generated for you. This feature demonstrates how automatic compilation of data in an IoT solution can help us keep our vehicles in good mechanical condition with early warning of needed maintenance. This serves us best by helping us keep our vehicles in prime condition and thus in a state of high resell value.

I should note that GM is not the only automotive manufacturer offering such services. Many others are working on their own solutions, ranging from an OnStar-like feature set to solutions that focus on entertainment and connectivity.

Fleet Management

Another example of an IoT solution is a fleet management system.[11] While developed and deployed well before the coining of the phrase Internet of Things, fleet management systems allow businesses to monitor their cars, trucks, ships, and just about any mobile unit, to not only track their current location but also to use the location data (GPS coordinates taken over time) to plan more efficient routes, thereby reducing the cost of shipment.

Fleet management systems are not just for routing. Indeed, fleet management systems also allow businesses to monitor each unit to conduct diagnostics. For example, it is possible to know the amount of fuel in each truck; when its last maintenance was performed or, more importantly, when the next maintenance is due; and much more. The combination of vehicle geographic tracking and diagnostics is called telematics. Figure 1-7 shows a drawing of a fleet management system.

[11] https://en.wikipedia.org/wiki/Fleet_management

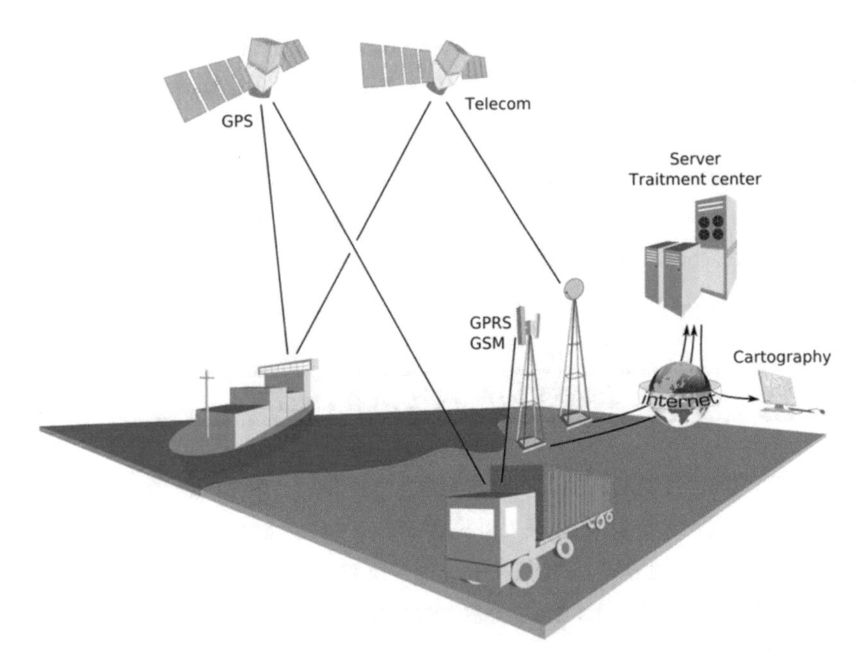

Figure 1-7. *Fleet management example*[12]

In Figure 1-7, you see the application of GPSs to track location as well as satellite communication to transmit additional data, such as diagnostics, payload states, and more. All these ultimately traverse the Internet, and the data becomes accessible by the business analysts.

You may think fleet management systems are only for large shipping companies, but with the proliferation of GPS modules and even the microcontroller market, anyone can create a fleet management system. That is, they do not cost millions of dollars to develop.

[12] Éric Chassaing – via CC BY-SA 3.0 (`http://creativecommons.org/licenses/by-sa/3.0/`).

For example, if you owned a bicycle delivery company, you could easily incorporate GPS modules with either cellular or wireless connectivity on each delivery person to track their location, average travel time, and more. More specifically, you can use such a solution to minimize delivery times by allowing packages to be handed off from one delivery person to another, rather than having them return to the depot each time they complete a set of deliveries.

CAMERA DRONES AND THE IOT

One possible use of the IoT is making data that drones generate available over the Internet. Some people feel that drones are an invasion of privacy. I agree in situations where they are misused or established laws are violated. Fortunately, most drone owners obey local laws, regulations, and property owners' wishes.[13]

However, there are many legitimate uses of drones, be they land, air, or sea based. For example, I can imagine home monitoring solutions where you can check on your home remotely by viewing data from fixed cameras, as well as data from mobile drones. I for one would love to see a solution that allowed me to program a predetermined sentry flight path to monitor my properties with a flying camera drone.

While some vendors have WiFi-enabled drones, there are not many consumer-grade options available that stream data in real time over the Internet. However, it is just a matter of time before we see solutions that include drones. Of course, the current controversy and the movement of the US government to register and track drones, along with increasing restrictions on their use, may limit the expansion of drones and IoT solutions that include drone-acquired data.

[13] Drones are increasingly under scrutiny, and the rules change often. If you have a drone and operate in the United States, be sure to check the following website for the latest rules: `https://registermyuas.faa.gov/`

While typically not considered for most home IoT projects, a discussion of IoT solutions would be incomplete without spotlighting security.

IoT and Security

The recent rash of massive data breaches proves that basic security simply is not good enough. We have seen everything from outright theft to exploitation of the data stolen from very well-known businesses, like popular brick-and-mortar retailers, convenience stores, and even some government agencies!

IoT solutions are not immune to security threats. Indeed, as IoT solutions become more and more integrated into our lives, so too will our personal data. Thus, security must be taken extremely seriously and built into the solution from the start.

This includes solutions that we develop ourselves. More specifically, if you design a weather station for your own use, you should take reasonable steps to ensure that the data is protected from both accidental and deliberate exploitation. You may think weather data is not a high risk, but consider the case where you include GPS coordinates for your sensors (a reasonable feature) so that people can see where this weather is being observed. If someone could see that information and determine the solution uses an Internet connection, it is possible they could gain physical access to the Internet device and possibly use it to further penetrate and exploit your systems. Thus, security is not just about the data; it should encompass all aspects of the solution – from data to software, to hardware, and to physical access.

There are four areas where you may want to consider spending extra care ensuring that your IoT solution is protected with good security. As you will see, this includes several things you should consider for your existing infrastructure, computers, and even safe computing habits. By leveraging all these areas, you will be building a layered approach to security, often called a defense-in-depth method.

Security Begins at Home

Before introducing an IoT solution to your home network, you should consider taking precautions to ensure that the machines on your home network are protected. Some of the best practices for securing your home networking include the following:

- *Use passwords.* This may seem like a simple thing, but always make sure that you use passwords on all your computers and devices. Also, adopt good password habits, such as requiring longer strings, mixed case, numbers, and symbols to ensure that the passwords are not easily guessed.[14]
- *Secure your WiFi.* If you have a WiFi network, make sure that you add a password and use the latest security protocols, such as WPA2, or, even better, the built-in secure setup features of some wireless routers.
- *Use a firewall.* You should also use a firewall to block all unused ports (TCP or UDP). For example, lock down all ports except those your solution uses, such as port 80 for HTML.

[14] You also need to balance complexity of passwords with your ability to remember them. If you have to write it down, you've just defeated your own security!

- *Restrict physical access.* Lock your doors! Even if your network has a great password and your computers use espionage quality encrypted biometric access, these things are meaningless if someone can gain access to your networking hardware directly. For IoT solutions, this means any external components should be installed in tamper-proof enclosures or locked away so that they cannot be discovered. This also includes any network wiring.

Secure Your Devices

As I mentioned, your IoT devices also need to be secured. The following are some practices to consider:

- *Use passwords.* Always add passwords to the user accounts you use on your IoT devices. This includes making sure that you rename any default passwords. For example, you may be tempted to consider an IoT device such as an Arduino, Raspberry Pi, or similar too small of a device to be a security concern, but if you consider that the Raspberry runs one of the most powerful operating systems available (forms of Linux), a Raspberry Pi could be a very powerful hacking tool if one were to gain access.
- *Keep your software up-to-date.* You should try to use the latest versions of any software that you use. This includes the operating system as well as any firmware that you may be running. Newer versions often have improved security or fewer security vulnerabilities.

- *If your software offers security features, use them.* If you have servers or services running on your devices and they offer features such as automatic lockout for missed passwords, turn them on. Not all software has these features, but if they are available, they can be a great way to defeat repeated attacks.

Use Encryption

This is one area that is often overlooked. You can further protect yourself and your data if you encrypt the data as it is stored and the communication mechanism as it is transmitted. If you encrypt your data, it cannot be easily deciphered, even if someone were to gain physical access to the storage device. Use the same care with your encryption keys and passcodes as you do your computer passwords.

Security Doesn't End at the Cloud

There are many considerations for connecting IoT devices to cloud services. Indeed, Microsoft has made it very easy to use cloud services with your IoT solutions. However, there are two important considerations for security and your IoT data:

- *Do you need the cloud?* The first thing you should consider is whether you need to put any of your data in the cloud. It is often the case that cloud services make it very easy to store and view your data, but is it really necessary to do so? For example, you may be eager to view logistical data for where your dog spends his time while you are at work, but who else would really care to view this data? In this case, storing the data in the cloud to make it available to everyone is not necessary.

- *Do not relax!* Many people seem to let their guard down when working with cloud services. For whatever reason, they consider the cloud more secure. The fact is it is not! In fact, you must apply the very same security best practices when working in the cloud that you do for your own network, computers, and security policies. Indeed, if anything, you need to be even more vigilant because cloud services are not in your control with respect to protecting against physical access (however remote and unlikely) nor are you guaranteed your data isn't on the same devices as tens, hundreds, or even thousands of other users' data.

Now that you have an idea of how you should include security in your projects, let's look at how Windows 10 has evolved into a modern platform that not only supports the usual productivity and gaming tasks but also helps us build IoT solutions.

Summary

The Internet of Things is an exciting new world for us all. Those of us young at heart but old enough to remember *The Jetsons* TV series recall seeing a taste of what is possible in the land of make believe. Self-aware robotic maids with personalities (attitude), talking toasters, flying cars that spring from briefcases, and robotic everything - television fantasy of decades ago is now coming true. We have wristwatches that double as phones and video players. We can unlock our cars from around the world, find out if our dog has gone outside, and even answer the door from across the city. All of this is possible and working today with the advent of the IoT.

In this chapter, we discovered what the IoT is and saw some examples of well-known IoT solutions.

In the next chapter, we will learn about the hardware platform that has become ubiquitous with learning hardware – the Arduino. We will discover more about the Arduino hardware and how to get started programming our first Arduino project as a building block for a simple IoT project.

CHAPTER 2

Introducing the Arduino

Since this is a beginner's book, you are likely just getting started working with hardware and IoT solutions, and you may not have encountered the world that is Arduino and microcontrollers. Arduino boards are small boards with components that support general-purpose input/output (GPIO) pins with a limited processor (called a microcontroller, not a CPU) and memory that permits you to write small programs to control the hardware. In essence, it is a hardware development platform.

There are many such boards and the Arduino is perhaps the most popular with a community that spans the globe providing a vast assortment of sample libraries, code, blogs, books, and documentation. This makes the Arduino one of the most popular choices for hardware development. Some may say it is even more popular than the Raspberry Pi.

In this chapter, you explore the Arduino platform with the goal of using the Arduino to build IoT devices. You see a list of the current Arduino boards along with a short tutorial on the Arduino development environment and explore sample projects to help get you started working with the Arduino.

C. Bell, *Beginning IoT Projects*, https://doi.org/10.1007/978-1-4842-7234-3_2

What Is an Arduino?

The Arduino is an open source hardware prototyping platform supported by an open source software environment. It was first introduced in 2005 and was designed with the goal of making the hardware and software easy to use and available to the widest audience possible. Thus, you do not have to be an electronics expert to use the Arduino. Yay!

The original target audience included artists and hobbyists who needed a microcontroller to make their designs and creations more interesting. However, given its ease of use and versatility, the Arduino has quickly become the choice for a wider audience and a wider variety of projects.

This means you can use the Arduino for all manner of projects from reacting to environmental conditions to controlling complex robotic functions. The Arduino has also made learning electronics easier through practical applications.

Another aspect that has helped the rapid adoption of the Arduino platform is the growing community of contributors to a wealth of information made available through the official Arduino website (`http://arduino.cc/en/`). When you visit the website, you find an excellent "getting started" tutorial as well as a list of helpful project ideas and a full reference guide to the C/C++ language for writing the code to control the Arduino (called a sketch).

Note Don't worry. The C++ programming concept (from the view of the main sketch, it resembles C, but includes many C++ concepts and features) is very easy to learn and does not require any training beyond the tutorial in this chapter.

Arduino also provides an integrated development environment called the Arduino IDE. The IDE runs on your computer (called the host), where you can write and compile sketches and then upload them to the Arduino via USB connections. The IDE is available for Linux, Mac, and Windows. It is designed around a text editor especially designed for writing code and a set of limited functions designed to support compilation and loading of sketches.

Sketches are written in a special format consisting of only two required methods - one that executes when the Arduino is reset or powered on and another that executes continuously. Thus, your initialization code goes in `setup()`, and your code to control the Arduino goes in `loop()`. The language is C-like (without all of the baggage typical in C compilers), and you may define your own variables and functions. For a complete guide to writing sketches, see `http://arduino.cc/en/Tutorial/Sketch`.

You can expand the functionality of sketches and provide for reuse by writing libraries that encapsulate certain features such as networking, using memory cards, connecting to databases, doing mathematics, and the like.

The Arduino supports a number of analog and digital pins that you can use to connect to various devices and components and interact with them. The mainstream boards have specific pin layouts, or headers, that allow the use of stackable expansion boards called shields. Shields let you add additional hardware capabilities such as Ethernet, Bluetooth, and XBee support to your Arduino. The physical layout of the Arduino and the shield allow you to stack shields. Thus, you can have an Ethernet shield as well as an XBee shield, because each uses different I/O pins. You learn the use of the pins and shields as you explore the application of Arduino to sensor networks.

The next sections examine the various Arduino boards and briefly describe their capabilities. I list the boards by when they became available, starting with the most recent models. Many more boards and variants are available, and a few new ones are likely to be out by the time this book is printed, but these are the ones that are typically used in beginning projects.

Arduino Hardware

There are a growing number of Arduino boards. Some are configured for special applications, while others are designed with different processors and memory configurations. There are boards that are considered official Arduino boards because they are branded and endorsed by Arduino.cc. Since the Arduino is open source, anyone can build and even sell Arduino-compatible boards (often called an Arduino clone). In this section, you examine some of the more popular Arduino branded boards.

The basic layout of an Arduino board consists of at least one USB connection, a power connector, a reset switch, LEDs for power and serial communication, and a standard spaced set of headers for attaching shields (boards that can be mounted adding hardware capabilities in a modular fashion).

The official boards sport a distinctive blue-colored PCB with white lettering. With the exception of one model, all the official boards can be mounted in a chassis (they have holes in the PCB for mounting screws). The exception is an Arduino designed for mounting on a breadboard.

Uno

The Uno board is the standard Arduino board that most new to the Arduino will choose. It features an ATmega328P processor; 14 digital I/O pins, of which 6 can be used as pulse width modulation (PWM)[1] output; and 6 analog input pins. The Uno board has 32KB of flash memory and 2KB of SRAM.

[1] https://en.wikipedia.org/wiki/Pulse-width_modulation

The Uno is available either as a surface-mount device (SMD) or a standard IC socket. The IC socket version allows you to exchange processors, should you desire to use an external IC programmer to build custom solutions. Details and a full datasheet are available at `https://store.arduino.cc/usa/arduino-uno-rev3`. It has a standard USB type B connector and supports all shields. Figure 2-1 shows the Arduino Uno board.

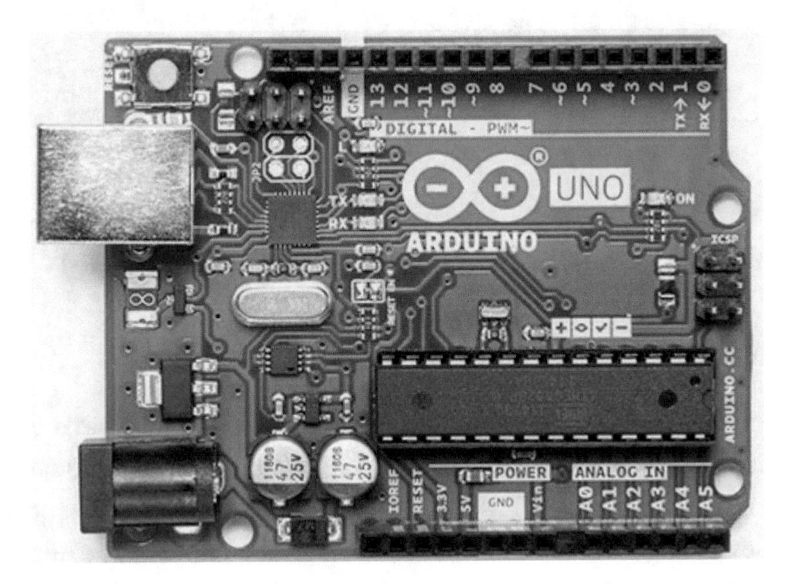

Figure 2-1. *Arduino Uno Rev3 (courtesy of Arduino.cc)*

There is also a version of this board that has a built-in WiFi chip making it ideal for IoT projects or situations where using a WiFi shield is problematic (lack of space, conflicts with other shields, etc.). While it is named the same, it differs from the standard Uno in several ways. Aside from the WiFi chip, it has a different processor and one less PWM pin. You can read more about the Uno WiFi board at `https://store.arduino.cc/usa/arduino-uno-wifi-rev2`. Figure 2-2 shows the Arduino Uno WiFi board.

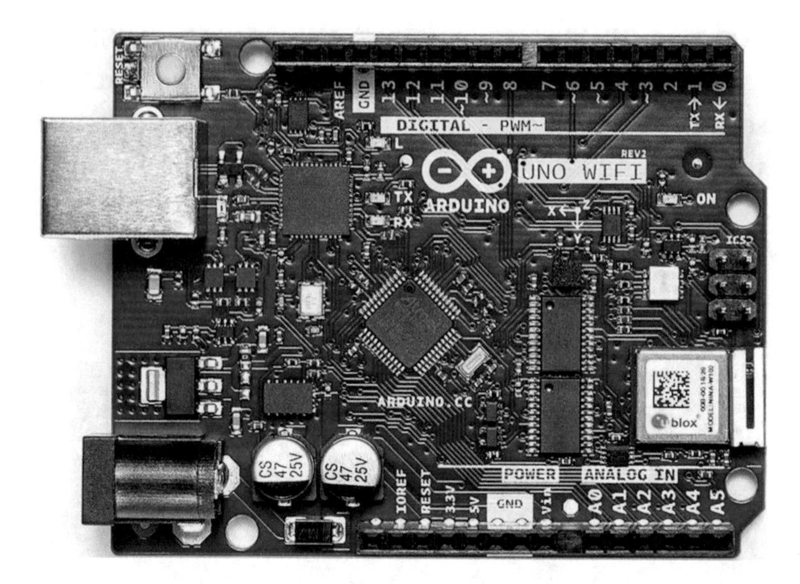

Figure 2-2. *Arduino Uno WiFi Rev2 (courtesy of Arduino.cc)*

Leonardo

The Leonardo board represents another of the standard boards in the Arduino platform. It is a little different in that, while it supports the standard header layout, it also has a USB controller that allows the board to appear as a USB device (e.g., mouse or keyboard) to the host computer. The board uses a newer ATmega32U4 processor with 20 digital I/O pins, of which 12 can be used as analog pins and 7 can be used as pulse width modulation (PWM) output. It has 32KB of flash memory and 2.5KB of SRAM.

The Leonardo has more digital pins than the Uno, but continues to support most shields. The USB connection uses a smaller USB connector. The board is also available with and without headers. Figure 2-3 depicts an official Leonardo board. Details and a full datasheet can be found at `https://store.arduino.cc/usa/leonardo`.

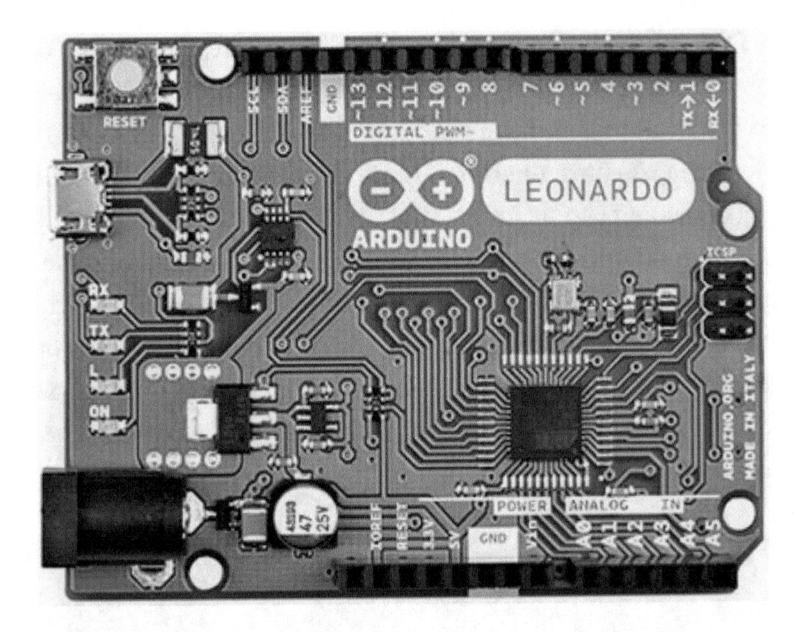

***Figure 2-3.** Arduino Leonardo (courtesy of Arduino.cc)*

Due

The Arduino Due is a larger, faster board based on the Atmel SAM3X8E ARM Cortex-M3 processor. The processor is a 32-bit processor, and the board supports a massive 54 digital I/O ports, of which 14 can be used for PWM output; 12 analog inputs; and 4 UART chips (serial ports), as well as 2 digital-to-analog (DA) and 2 two-wire interface (TWI) pins. The new processor offers several advantages:

- 32-bit registers
- DMA controller (allows CPU-independent memory tasks)
- 512KB flash memory
- 96KB SRAM
- 84MHz clock

The Due has the larger form factor (called the mega footprint) but still supports the use of standard shields as well as mega format shields. The new board has one distinct limitation: unlike other boards that can accept up to 5V on the I/O pins, the Due is limited to 3.3V on the I/O pins. Details and a full datasheet can be found at `https://store.arduino.cc/usa/due`.

The Arduino Due is intended to be used for projects that require more processing power, more memory, and more I/O pins. Despite the significant capabilities of the new board, it remains open source and comparable in price to its predecessors. Look to the Due for your projects that require the maximum hardware performance. Figure 2-4 shows an Arduino Due board.

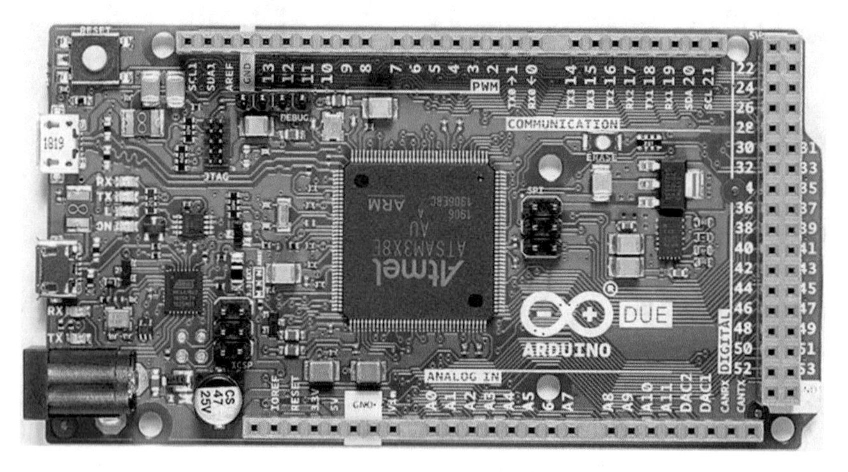

Figure 2-4. *Arduino Due (courtesy of Arduino.cc)*

Tip Notice how much larger the Due is than the Uno. If you choose to incorporate a Due, Mega, or similar board, you may have to set aside more room to mount the board.

Micro

The Arduino Micro is a special form of the Leonardo board and uses the same processor with 20 digital I/O pins, of which 12 can be used as analog pins and 7 can be used as PWM output. It has 32KB of flash memory and 2.5KB of SRAM. Details and a full datasheet can be found at `https://store.arduino.cc/usa/arduino-micro`.

The Micro was made for use on breadboards in the same way as the Mini but in a newer, updated form. But unlike the Mini, the Micro is a full-featured board complete with a USB connector. And like the Leonardo, it has built-in USB communication, allowing the board to connect to a computer as a mouse or keyboard. Figure 2-5 shows the Arduino Micro board.

Figure 2-5. *Arduino Micro (courtesy of Arduino.cc)*

Although branded as an official Arduino board, the Arduino Micro is produced in cooperation with Adafruit.

Nano

The Arduino Nano is an older form of the Arduino Micro. In this case, it is based on the functionality of the Duemilanove and has the ATmega328 processor (older models use the ATmega168) and 14 digital I/O pins, of which 6 can be used as PWM output, and 8 analog inputs. The Nano has 32KB of flash memory and uses a 16MHz clock. Details and a full datasheet can be found at `https://store.arduino.cc/usa/arduino-nano`.

Like the Micro, it has all the features needed for programming via a USB connection. Figure 2-6 shows an Arduino Nano board.

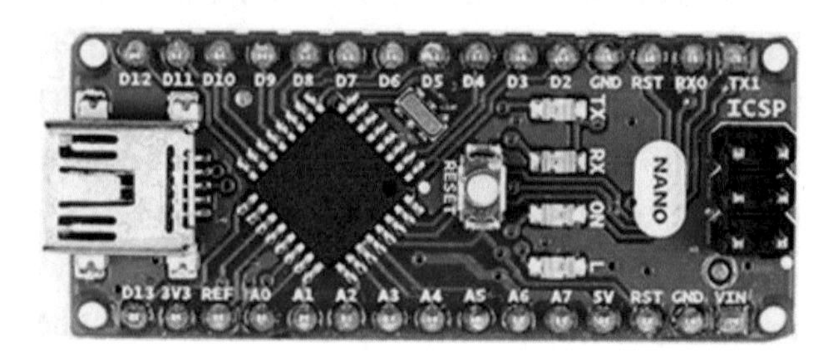

***Figure 2-6.** Arduino Nano (courtesy of Arduino.cc)*

MKR Series Boards

There is another form of Arduino called the MKR (for "maker") series. The MKR series includes a variety of boards based on the (now retired) Zero board that have various communication capabilities such as WiFi, LoRa, LoRaWAN, and GSM.

They are based on the Atmel ATSAMW25 SoC (System on Chip) and designed for IoT projects and devices. They also support cryptographic authentication. For those working on projects that require a battery port, the MKR series of boards include a lithium polymer (Li-Po) charging circuit for charging a Li-Po battery while running on external power. Details and a full datasheet can be found at `https://store.arduino.cc/usa/arduino-mkr1000`.

The boards do not use the same pin layout as the Uno-compatible shield-based boards (but you can get an adapter). Rather, they are designed like the Nano and Mini (but a bit larger) to minimize the size of the board to make it easier to incorporate into your projects. In fact, they are one of the boards of choice for Internet of Things (IoT) projects and make an excellent choice for sensor network projects. Since they are relatively new and some have specialized communication options, most new to Arduino would be better served starting with the Arduino boards that support Uno-compatible shields. Figure 2-7 shows a MKR1000 board.

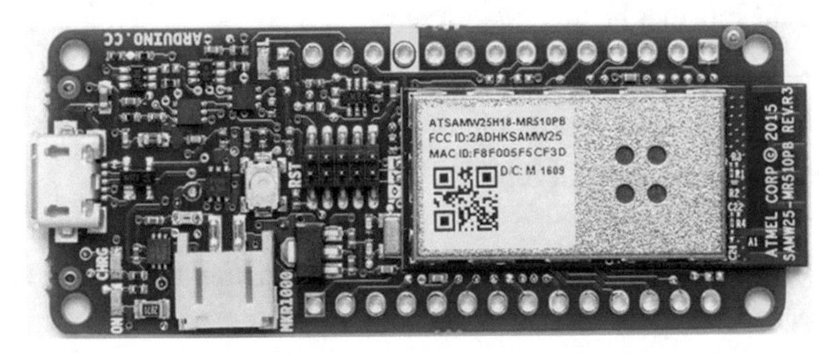

Figure 2-7. *MKR1000 (courtesy of Arduino.cc)*

A companion to the MKR series is the MKR2UNO adapter (`https://store.arduino.cc/usa/mkr2uno-adapter`) that allows you to use the MKR board with any Uno-compatible shield. Figure 2-8 shows the MKR2UNO adapter.

Figure 2-8. *MKR2UNO adapter (courtesy of Arduino.cc)*

Not only does the adapter provide support for shields, it also provides a standard Uno power connector making it easy to move your project from an older Uno board to the MKR series board. If you plan to use any shields, you should consider buying this adapter.

Caution The MKR boards run on 3.3V power and have a maximum input on the GPIO pins of 3.3V.

Now that we've seen a number of the Arduino branded boards, let's consider the cloned boards.

Arduino Clones

A growing number of Arduino boards are available from a large number of sources. Because the Arduino is open hardware, it is not unusual or the least bit illicit to find Arduino boards made by vendors from all over the world.

Although some would insist the only real Arduinos are those branded as such, the truth of the matter is that as long as the build quality is sound and the components are of high quality, the choice of using a branded vs. a copy, hence clone, is one of personal preference. I have sampled Arduino boards from a number of sources, and with few exceptions they all perform their intended functions superbly.

Except for the Arduino Mini, the Arduino clone boards have a greater variety of hardware configurations. Some Arduinos are designed for use in embedded systems or on breadboards, and some are designed for prototyping. I examine a few of the more popular clone boards in the following sections, but you are likely to encounter many variations in the wild.

Arduino Pro Mini

The Arduino Pro Mini is a board from SparkFun. It is based on the ATmega168 processor (older models use the ATmega168) and has 14 digital I/O pins, of which 6 can be used as PWM output, and 8 analog inputs. The Pro Mini has 16KB of flash memory and 1KB of SRAM, and it uses a 16MHz clock. Details and a full datasheet can be found at `www.sparkfun.com/products/11113`.

The Arduino Pro Mini is modeled on the Arduino Mini and is also intended for use on breadboards but does not come with headers. This makes the Arduino Pro Mini ideal for use in semipermanent installations where the pins can be soldered to the components or circuitry and space is a premium. Figure 2-9 shows an Arduino Pro Mini board. It really is that tiny.

Figure 2-9. *Arduino Pro Mini (courtesy of SparkFun)*

Also, the Pro Mini does not include a USB connector and therefore must be connected to and programmed with an FTDI cable or similar breakout board. It comes as either a 3.3V model with an 8MHz clock or a 5V model with a 16MHz clock.

Fio

The Arduino Fio is another board made by SparkFun. It was designed for use in wireless projects. It is based on the ATmega32U4 processor with 14 digital I/O pins, of which 6 can be used as PWM output, and 8 analog pins. Details and a full datasheet can be found at `www.sparkfun.com/products/11520`.

The Fio requires a 3.3V power supply, which allows for use with a lithium polymer (Li-Po) battery, which can be recharged via the USB connector on the board.

Its wireless pedigree can be seen in the XBee socket on the bottom of the board. Although the USB connection lets you recharge the battery, you must use an FTDI cable or breakout adapter to connect to and program the Fio. Similar to the Pro models, the Fio does not come with headers, allowing the board to be used in semipermanent installations where connections are soldered in place. Figure 2-10 shows an Arduino Fio board.

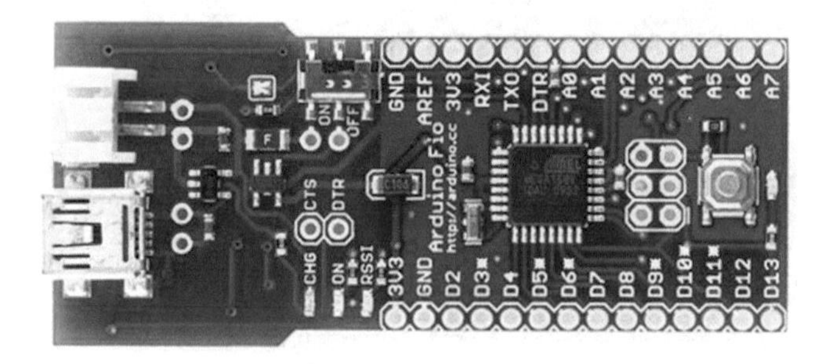

Figure 2-10. *Arduino Fio (courtesy of SparkFun)*

Seeeduino

The Seeeduino is an Arduino clone made by Seeed Studio (`www.seeedstudio.com`). It is based on the ATmega328P processor and has 14 digital I/O pins, of which 6 can be used as PWM output, and 8 analog pins. It has 32KB of flash memory and 2KB of SRAM. Details and a full datasheet can be found at `www.seeedstudio.com/Seeeduino-V4-2-p-2517.html`.

The board has a footprint similar to the Arduino Uno and supports all standard headers. It supports a number of enhancements such as I2C and serial Grove connectors and a mini USB connector, and it uses SMD components. It is also a striking red color with yellow headers. Figure 2-11 shows a Seeeduino board.

***Figure 2-11.** Seeeduino (courtesy of Seeed Studio)*

Tip Seeed Studio also makes several versions of this board. For more details, see `www.seeedstudio.com/seeeduino-boards-c-987.html`.

Metro from Adafruit

The Metro from Adafruit is a set of Arduino-compatible boards supporting a number of formats including several that support Arduino shields. The version I like as a balance of compatibility and cost is the Metro 328 (`www.adafruit.com/product/2488`). Figure 2-12 shows the Metro 328 board from Adafruit. The board uses the ATmega328P at 16MHz and a host of minor improvements to make the Metro an excellent alternative to an Arduino Uno. Check out the product page for more details.

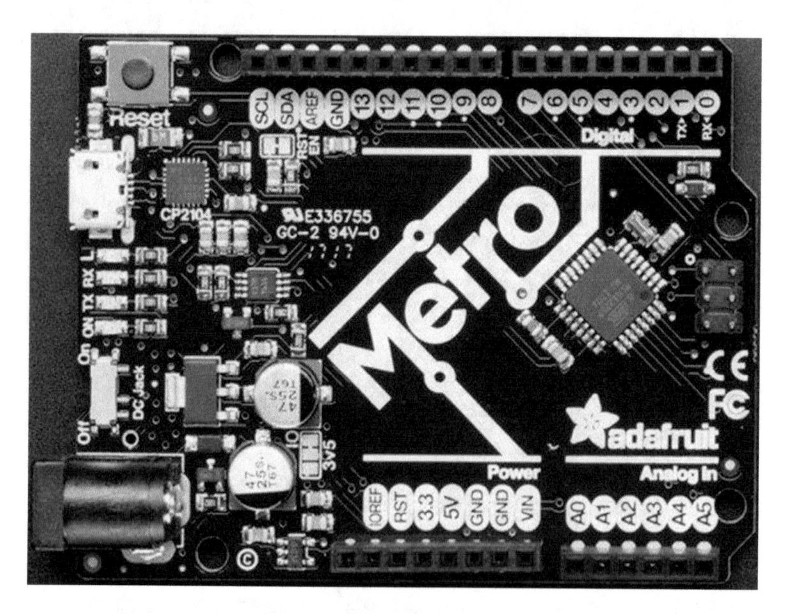

Figure 2-12. *Metro 328 (courtesy of Adafruit)*

Adafruit makes a number of versions of the Metro board including smaller form factors and some with advanced features. See `www.adafruit.com/category/834` for more details.

Espressif Boards

Made popular by their relatively low cost are a series of boards that, while they are not technically Arduino clones, can be used with the Arduino toolset. Chief among these are the ESP series of WiFi boards by Espressif (`www.espressif.com/en/products/devkits`).

While the modules vary in hardware capabilities and there is sure to be one to meet almost any need, most are compatible with Arduino and can be programmed in the same way. However, I urge caution when choosing these boards as your primary IoT device. I have encountered problems using them with several libraries, and the hardware can differ enough to make using them challenging for beginners.

However, some ESP boards may require you to modify your Arduino IDE by installing additional hardware support and libraries. Fortunately, Arduino has made this easy to do with their extensive hardware library manager.

That said, the most popular ESP boards include the ESP8266 and ESP32 chipsets. These can sometimes be found mounted as breakout boards, stand-alone, or part of another product. Figure 2-13 shows the ESP8266 WiFi module. They're tiny and really cheap and can be used in conjunction with other boards to provide WiFi capabilities.

Figure 2-13. *ESP8266 module (courtesy of Adafruit.com)*

The ESP32 chipset is often mounted on boards that have a more traditional header layout such as the Adafruit FeatherS2 board (Figure 2-14). This board is one of the more popular Adafruit boards and can be used for a variety of projects.

Figure 2-14. *Adafruit FeatherS2 (courtesy of Adafruit.com)*

Internet Shields

Recall the Arduino Uno layout supports add-on boards called shields. You will be using a shield later in this book, but of particular note are the shields that provide Internet capabilities for Arduino boards without WiFi.

While Arduino made several versions of shields that support WiFi and Ethernet, most have been discontinued due to the newer MKR boards and Uno supporting WiFi. However, you can still find Internet shields from vendors that still have the Arduino shields in stock, or you can buy one from SparkFun (they have an ESP8266 variant; `www.sparkfun.com/products/13287`) or Adafruit, which offers several varieties (`www.adafruit.com/category/828`). Figure 2-15 shows the ESP8266 WiFi shield from SparkFun. If you plan to use an Arduino that does not have WiFi, you may want to buy one of these shields for use later in this book when we discuss connecting your IoT device to the cloud.

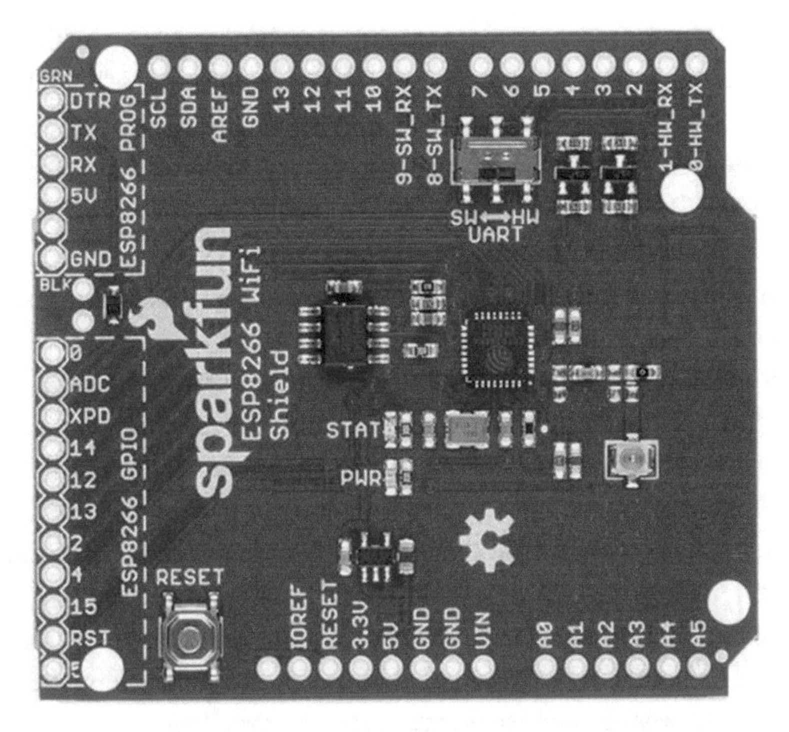

Figure 2-15. *ESP8266 WiFi shield (courtesy of SparkFun)*

Interestingly, the ESP8266 WiFi shield can be used independently because the ESP8266 pins are separated (called broken out) along one side. See `www.sparkfun.com/products/13287` for more details.

So Which Do I Buy?

If you're wondering which Arduino to buy, the answer depends on what you want to do. For most of the projects in this book, any Arduino Uno or similar clone that supports the standard shield headers is fine so long as you are able to use a shield for WiFi or Ethernet connections. You need not buy the larger Due or its predecessors, since the added memory and I/O pins aren't needed.

I use the Arduino Uno WiFi, Leonardo, or MKR boards for all the projects in this book. Although you can use an older board without issues, there are some issues with using the Leonardo board. I point these out as you encounter them. Most issues have to do with the relocated pins on the Leonardo board. For example, the SPI header pins (at upper left in Figure 2-3) have been moved on the Leonardo.

For future projects, there are some things you should consider before choosing the Arduino. For example, if your project is largely based on a breadboard or you want to keep the physical size of the project to a minimum and you aren't going to use any shields, the Arduino Mini may be the better choice. Conversely, if you plan to do a lot of programming to implement complex algorithms for manipulating or analyzing data, you may want to consider the Due for its added processing power and memory.

The bottom line is that most of the time your choice will be based on physical characteristics (size, shield support, etc.) and seldom on processing power or memory. SparkFun has an excellent buyer's guide in which you can see the pros and cons of each choice. See `www.sparkfun.com/pages/arduino_guide` for more details.

Where to Buy

Due to the popularity of the Arduino platform, many vendors sell Arduino and Arduino clone boards, shields, and accessories. The Arduino.cc website (`https://store.arduino.cc/usa`) also has a page devoted to approved distributors. If none of the resources listed here are available to you, you may want to check this page for a retailer near you.

Online Retailers

There are a growing number of online retailers where you can buy Arduino boards and accessories. The following lists a few of the more popular sites:

- *SparkFun*: From discrete components to the company's own branded Arduino clones and shields, SparkFun has just about anything you could possibly want for the Arduino platform (`www.sparkfun.com/`).
- *Adafruit*: Carries a growing array of components, gadgets, and more. It has a growing number of products for the electronics hobbyist, including a full line of Arduino products. Adafruit also has an outstanding documentation library and wiki to support all the products it sells (`www.adafruit.com/`).

You can also visit the manufacturers of some of the clone boards. Seeed Studio is the leading clone manufacturer (`www.seeedstudio.com/`):

- *Seeed Studio*: `www.seeedstudio.com/`

Retail Stores (USA)

There are also brick-and-mortar stores that carry Arduino products. Although there aren't as many as there are online retailers and their inventories are typically limited, if you need a new Arduino board quickly, you can find them at Micro Center[2] and some smaller electronics retailers. You may find additional retailers in your area. Look for popular hobby electronics stores:

- *Fry's*: An electronics superstore with a huge inventory of electronics, components, microcontrollers, computer parts, and more. If you have never had the chance to visit a Fry's store, you should be prepared to spend some time there. Fry's carries Arduino branded boards, shields, and accessories as well as products from Parallax, SparkFun, and many more (`http://frys.com/`).
- *Micro Center*: Micro Center is similar to Fry's, offering a huge inventory of products. However, most Micro Center stores have a smaller inventory of electronic components than Fry's (`www.microcenter.com/`).

Now that you have a better understanding of the hardware details and the variety of Arduino boards available, let's dive into how to use and program the Arduino. The next section provides a tutorial for installing the Arduino programming environment and programming the Arduino. Later sections present projects to build your skills for developing sensor networks.

[2] `www.microcenter.com/`

Arduino Tutorial

This section is a short tutorial on getting started using an Arduino. It covers obtaining and installing the IDE and writing a sample sketch. Rather than duplicate the excellent works that precede this book, I cover the highlights and refer readers who are less familiar with the Arduino to online resources and other books that offer a much deeper introduction. Also, the Arduino IDE has many sample sketches that you can use to explore the Arduino on your own. Most have corresponding tutorials on the Arduino.cc site.

Learning Resources

A lot of information is available about the Arduino platform. If you are just getting started with the Arduino, Apress offers an impressive array of books covering all manner of topics concerning the Arduino, ranging from getting started using the microcontroller to learning the details of its design and implementation. The following is a list of the more popular books. Some are a little older than you may expect but still quite useful:

- *Beginning Arduino* by Michael McRoberts (Apress, 2010)
- *Practical Arduino: Cool Projects for Open Source Hardware (Technology in Action)* by Jonathan Oxer and Hugh Blemings (Apress, 2009)
- *Arduino Software Internals: A Complete Guide to How Your Arduino Language and Hardware Work Together* by Norman Dunbar (Apress, 2020)
- *Arduino Internals* by Dale Wheat (Apress, 2011)

There are also some excellent online resources for learning more about the Arduino, the Arduino libraries, and sample projects. The following are some of the best:

- *Arduino.cc*: `http://arduino.cc/en/`
- *Adafruit*: `http://learn.adafruit.com/`
- *SparkFun*: `https://learn.sparkfun.com/`

The Arduino IDE

The Arduino IDE is available for download for the Mac, Linux (32- and 64-bit versions), and Windows platforms. You can download the IDE from `www.arduino.cc/en/software`. There are links for each platform as well as a link to the source code if you need to compile the IDE for a different platform. The current release is 1.8.13, but newer releases are produced periodically. So it's OK if you download a newer version than what is shown in this section.

Tip Interestingly, there is a web version of the IDE that you can use without installing it on your computer (`https://create.arduino.cc/editor`). This may be helpful if you want to use it on a PC where you don't want (or cannot) install the IDE. It is also an example of an IoT service.

Installing the IDE is straightforward. I omit the actual steps of installing the IDE for brevity, but if you require a walk-through of installing the IDE, you can see the Getting Started link on the download page or read more in *Beginning Arduino* by Michael McRoberts (Apress, 2010).

Once the IDE launches, you see a simple interface with a text editor area (a white background by default), a message area beneath the editor (a black background by default), and a simple button bar at the top. The buttons are (from left to right) *Compile, Upload, New, Open,* and *Save.* There is also a button to the right that opens the serial monitor. You use the serial monitor to view messages from the Arduino sent (or printed) via the Serial library. You see this in action in your first project. Figure 2-16 shows the Arduino IDE.

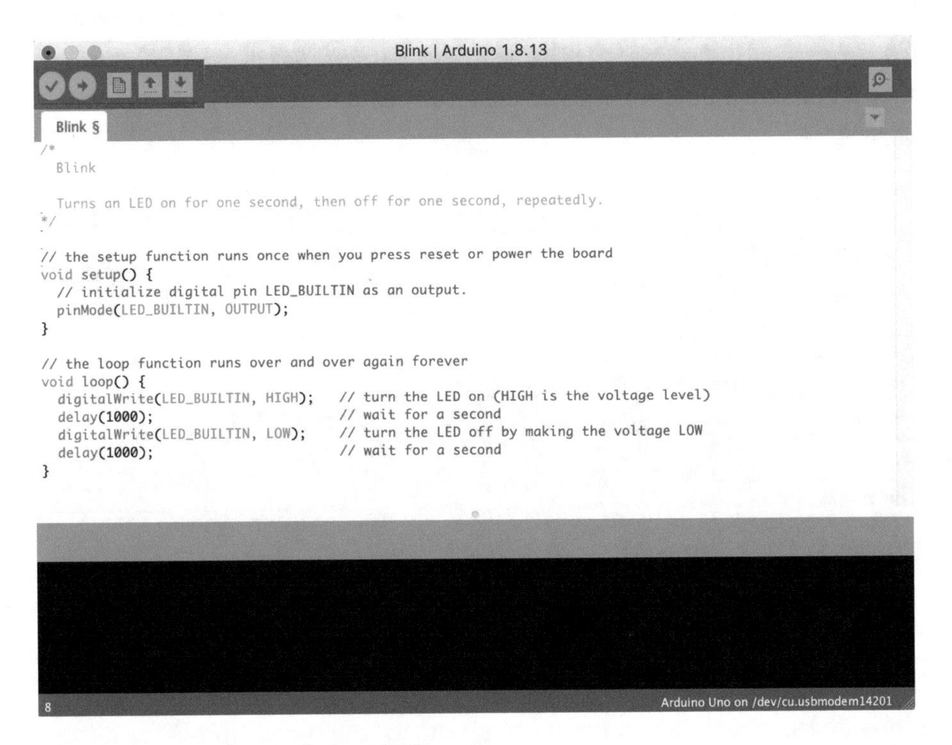

Figure 2-16. *The Arduino IDE*

Notice that in Figure 2-16 you see a sample sketch (called `blink`) and the result of a successful compile operation. I loaded this sketch by clicking *File ➤ Examples ➤ Basics ➤ Blink.* Notice also at the bottom that it tells you that you are programming an Arduino Uno board on a specific serial port.

Due to the differences in processor and supporting architecture, there are some differences in how the compiler builds the program (and how the IDE uploads it). Thus, one of the first things you should do when you start the IDE is choose your board from the *Tools* ➤ *Board* menu. Figure 2-17 shows a sample of selecting the board on the Mac. Here, we see some submenus that we use to locate the board we want. In this case, we want the *Arduino Uno WiFi Rev2* board, which is listed under the *Arduino megaAVR Boards* submenu. This is not uncommon, especially with the Arduino-compatible boards.

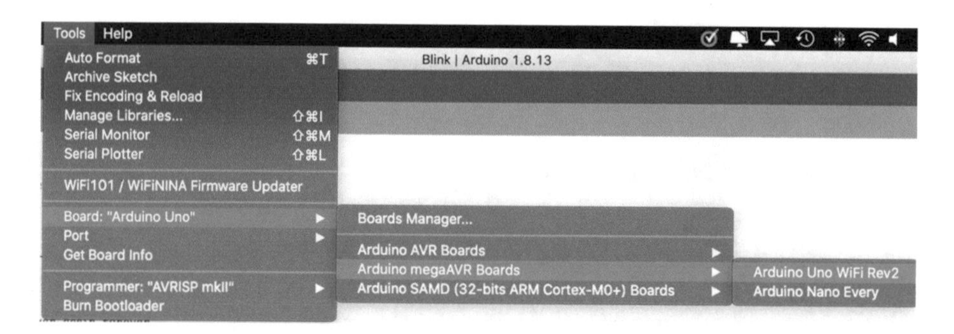

Figure 2-17. *Choosing the Arduino board*

Notice there are only a couple of boards available under the submenu. Be sure to choose the menus that match your board. If you are using a clone board, check the manufacturer's site for the recommended setting to use. If you choose the wrong board, you typically get an error during upload, but it may not be obvious that you've chosen the wrong board. Because I have so many different boards, I've made it a habit to choose the board each time I launch the IDE.

The next thing you need to do is choose the serial port to which the Arduino board is connected. To connect to the board, use the *Tools* ➤ *Port* menu option. Figure 2-18 shows an example on the Mac. If you do not see your board or you connected it after selecting the menu, you can fix it by simply unplugging the Arduino and plugging it back in and waiting until the computer recognizes the port.

Tip The list of available ports contains the name of the board connected. If you do not see the name of the board, you can still choose the port, and the IDE will attempt to use it.

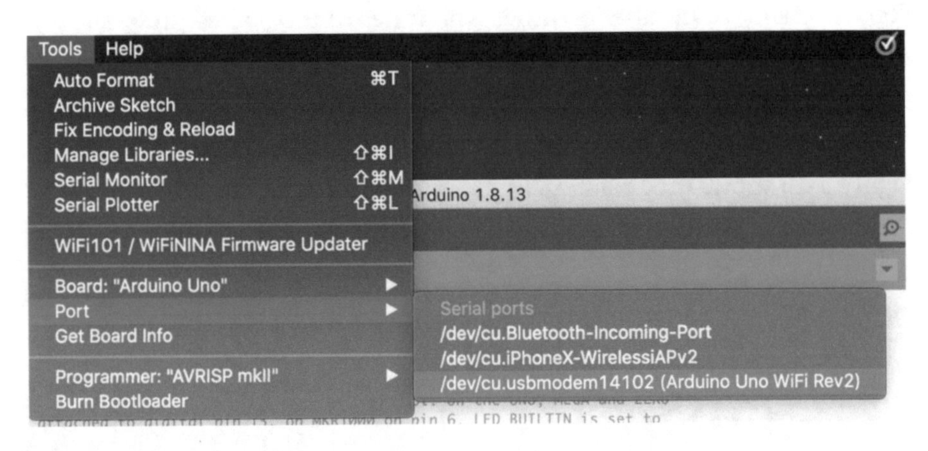

***Figure 2-18.** Choosing the serial port*

OK, now that you have your Arduino IDE installed, you can connect your Arduino and set the board and serial port. These are the first steps you should accomplish any time you want to work with an Arduino. Doing these steps first can help avoid problems like incorrect software libraries and strange compilation errors.

Once you connect your Arduino to your computer, you see the LEDs on the Arduino illuminate. This is because the Arduino is getting power from the USB. Thus, you do not need to provide an external power supply when the Arduino is connected to your computer.

Before we move on to some examples, let's explore how to modify the Arduino IDE.

Modifying the Arduino IDE

The Arduino IDE provides the ability to modify the environment as well as modify the hardware and software libraries. Most may not need to modify the environment, but you will likely need to modify the hardware and software libraries. Let's look at each.

Customize the Environment

You can change a number of things with the Arduino (called preferences) should you need to do so. There are two major categories: 1) settings such as language, font size, and editor and compiler behavior and 2) network connections. You should not need to change the network settings. Figure 2-19 shows the Preferences pane. You can open this under *File* ➤ *Preferences...* (Windows) or *Arduino* ➤ *Preferences...* (macOS).

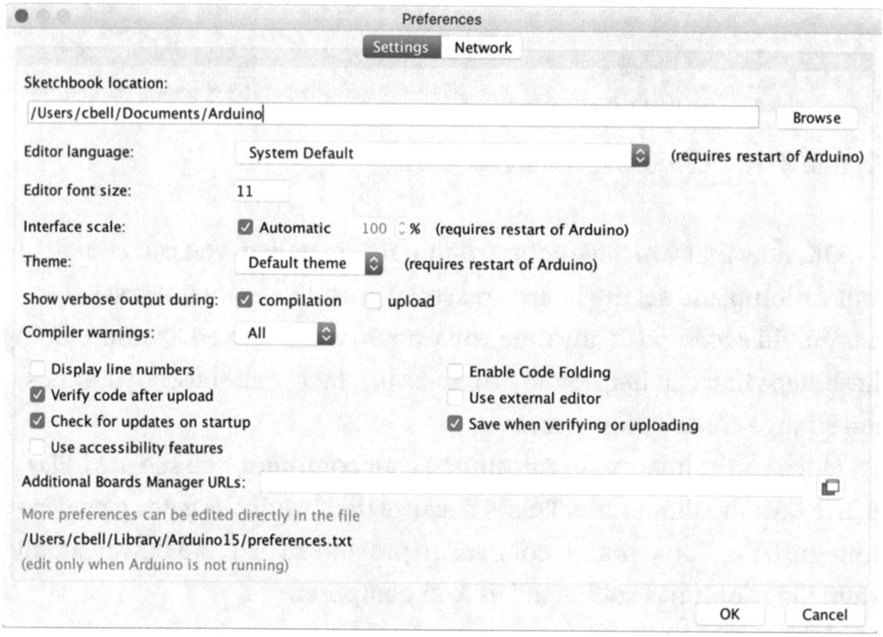

Figure 2-19. *Arduino IDE Preferences*

Notice the path to the *Sketchbook location*. This is where the Arduino IDE will look for your sketches and any software libraries you download. You should not need to change this, but it is possible on some platforms to alter this path if you want to store things in a different location.

You may want to change the font size and possibly the language for the editor and add line numbers, but most of these changes are for advanced users. Feel free to experiment with them if you are curious.

Manage the Hardware Libraries

The Arduino IDE comes with several hardware libraries to support a number of boards. However, as you begin using other boards such as newer boards from Arduino, those from SparkFun or Adafruit, etc., you may need to install new hardware libraries. Let's look at how to do this using an example.

Suppose you have a MKR board and you want to start using it. If you search the *Tools* ➤ *Board* menu, you may not find it. Fortunately, there is a way to add new hardware libraries. To do so, use the *Tools* ➤ *Board* ➤ *Boards Manager...* menu as shown in Figure 2-20.

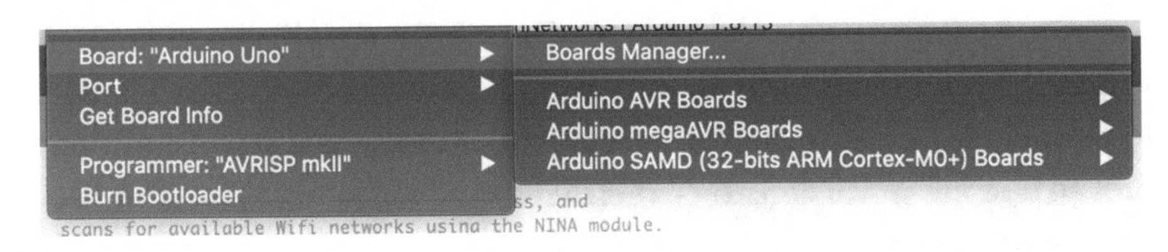

Figure 2-20. *Launching the Boards Manager*

This launches a dialog for the Boards Manager. If you want to add a new board, just type in the name (or part of the name) in the search box as shown in Figure 2-21.

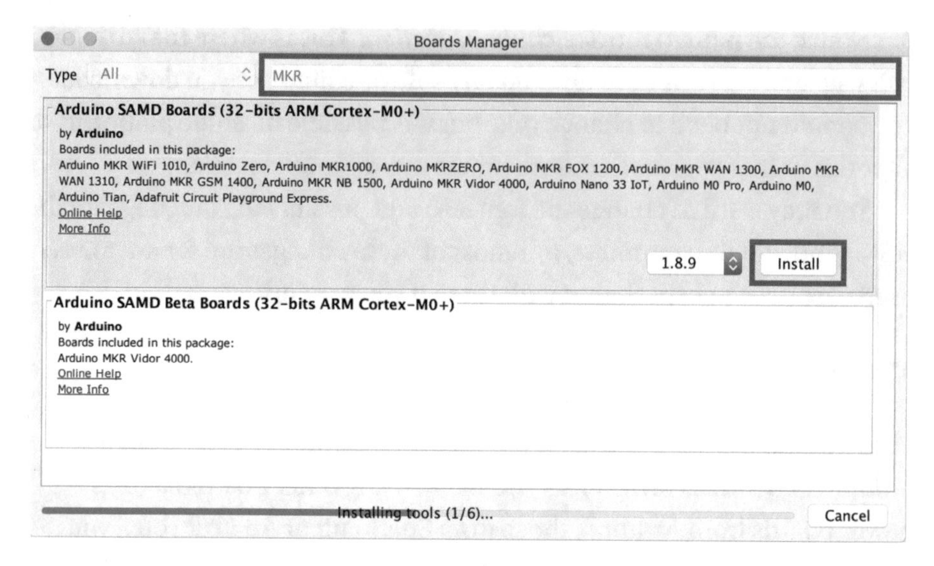

Figure 2-21. *Boards Manager (searching for MKR)*

You can then scroll through the entries until you the find the one that matches your board and then click the *Install* button to install it.

Once the install is compete, you will see the new board under the *Board* menu.

You may notice there is a version number beside each entry. This is so that you can install a different version of the library should you encounter a situation where it is required. For example, I found a case where I needed to downgrade one of my hardware libraries because a software library was broken by a newer version. Keep this in mind as you begin to work with many different boards.

Tip On some platforms, the Arduino IDE may detect that you need to install a new hardware library and will prompt you with a link to the Boards Manager.

Manage the Software Libraries

Software libraries are special programs (or additions, classes) designed to support additional add-on hardware like sensors, shields, and other modules. As you can imagine, the number of software libraries is quite large. Fortunately, like the Boards Manager, the Arduino IDE provides the Library Manager that works in the same way to install software libraries.

To see how to work with the Library Manager, let's look at an example.

Suppose you have a MKR1000 board that requires an older version of the WiFi library to work. In fact, it requires a retired library. Retired simply means it is no longer being maintained (except possibly for critical security issues), but is still available via the Library Manager.

In order to use WiFi in a sketch with the MKR1000, you need to install the WiFi101 library. To do so, click the *Sketch ➤ Include Library ➤ Manage Libraries...* menu as shown in Figure 2-22.

Figure 2-22. *Launching the Library Manager*

This launches the Library Manager. You can search for the library you want by typing in the name (or part of the name) of the library in the search box as shown in Figure 2-23. In this case, I want to see all of the WiFi libraries.

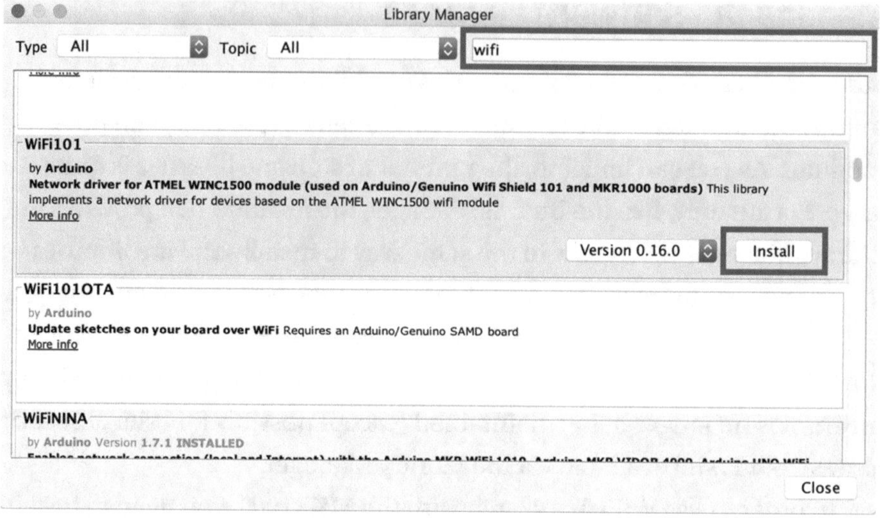

Figure 2-23. *Library Manager*

Since we used a rather short (and popular) term to search, there are a lot of entries. To find the one we need, you can scroll down to the WiFi101 library as shown and then click the *Install* button to install it.

Like the hardware libraries, you will see a version number beside each entry. This is so that you can install a different version of the library should you encounter a situation where it is required.

Once the install is compete, you will be able to use the library in your sketch (more on this in Chapter 3).

Now that we understand the basics of how to work with the Arduino IDE, let's try out a few of the built-in sketches to get started.

Example Sketch: Blink

The Arduino comes with a variety of example sketches, which we will explore in greater detail in Chapter 3. The example that is considered a simple "Hello, World" example[3] is the `Blink` sketch.

You can use almost any Arduino board for this example. It simply turns on and then off the onboard user LED (named `LED_BUILTIN` in the documentation) in an endless loop; hence, it "blinks" the LED. Most Arduino boards have built-in LEDs that are wired to pin 13. For example, Figure 2-24 shows the location of the LED on the Uno WiFi Rev2, and Figure 2-25 shows the location of the LED on the MKR1000.

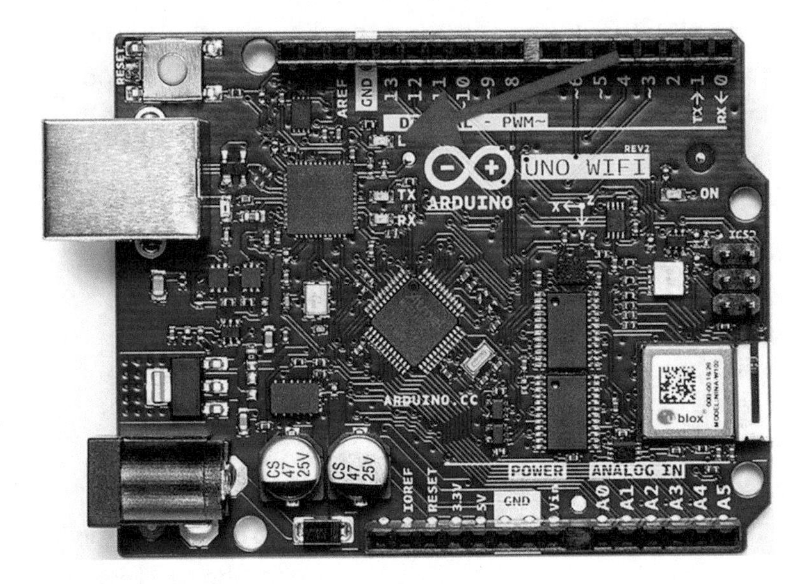

Figure 2-24. *User LED (Uno WiFi Rev2)*

[3] https://en.wikipedia.org/wiki/%22Hello,_World!%22_program

Figure 2-25. *User LED (MKR1000)*

Now that we know where the LED is, let's load the sketch. Recall we first plug our board into our PC and then set the board and port. Next, click *File* ➤ *Examples* ➤ *01.Basics* and select the *Blink* sketch. This will open a new window with the sketch loaded as shown in Listing 2-1 (comments removed for brevity).

Listing 2-1. Blink Sketch

```
void setup() {
  // initialize digital pin LED_BUILTIN as an output.
  pinMode(LED_BUILTIN, OUTPUT);
}

// the loop function runs over and over again forever
void loop() {
  digitalWrite(LED_BUILTIN, HIGH);   // turn the LED on (HIGH
                                        is the voltage level)
  delay(1000);                       // wait for a second
  digitalWrite(LED_BUILTIN, LOW);    // turn the LED off by
                                        making the voltage LOW
  delay(1000);                       // wait for a second
}
```

Don't worry if you have never seen code like this. Take a moment and read through it, and you will see it is not overly difficult. We see a `setup()` method that is responsible for getting things ready and then the `loop()` method, which means execute (run) repeatedly as its name suggests. Inside the `loop()` method, we see where the LED is turned on, the program waits, and the LED is turned off.

To run (execute) the example, we first must compile it and then upload it to our board. Fortunately, the Arduino provides a single button to do that. Just click the second from the left button (named Upload) as shown in Figure 2-26.

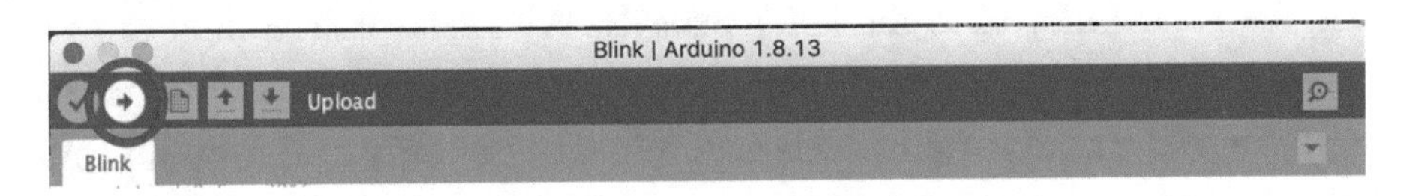

Figure 2-26. *Upload button*

Once you click that button, you will see a number of things happen including the compiler compiling the sketch and, after that succeeds, a message that the sketch is being uploaded to the board. Don't worry about the compiler messages at the bottom. We will see more about these in Chapter 3.

During the upload phase, you may see LEDs on the board flash briefly to indicate the board is receiving data. After the sketch is uploaded, it is executed, and you should see the user LED start to blink. It will continue to do so until you disconnect (power off) the board.

Congratulations! You've just programmed your first Arduino!

Note Once you upload a sketch to an Arduino, it will execute the sketch every time it is powered on until you upload a new sketch. You can see this by disconnecting your board and then connecting a power adapter. When you do that, the Blink sketch will execute. Try it yourself!

Caution Always double-check the power requirements for your board before applying external power. Most can run on 9V or 5V, but some require 3.3V (like the MKR series).

Let's try one more example that is a bit more complicated, but surprisingly helpful.

Example Sketch: Scan Networks

The Scan Networks example sketch uses the WiFi to scan for available networks in the area. It reports the name (SSID) of the network along with some helpful characteristics. This example requires either an Arduino Uno WiFi, MKR, or similar WiFi-enabled board. If you have another Arduino board without WiFi, you can still use it if you add a WiFi shield, but you may need to make modifications to the sketch to get it to work. More specifically, you may need to add a software library for the shield replacing the one in the sketch. We will see an example of this as we explore the MKR1000 board.

In fact, you may have to use a different example sketch depending on which board you use. We will see examples of two boards starting with the Uno WiFi Rev2.

Using the Uno WiFi Rev2

This board can use the example sketch under the WiFiNINA category. More specifically, you can click *File* ➤ *Examples* ➤ *WiFiNINA* ➤ *Scan Networks*. If you look through the sketch, you will see a lot more code that is a lot more complex than the previous example. Fortunately, we can use it without modification.

Once you've connected your board, you can click the Upload button to start the compilation and upload. It may take a bit longer to compile, but once it has uploaded, you won't see much other than some periodic, rapid flashes on some LEDs. Why? Because unlike the previous example, this one prints information to the serial monitor. To view the serial monitor, wait until the upload is complete and then click the button on the far right of the window as shown in Figure 2-27.

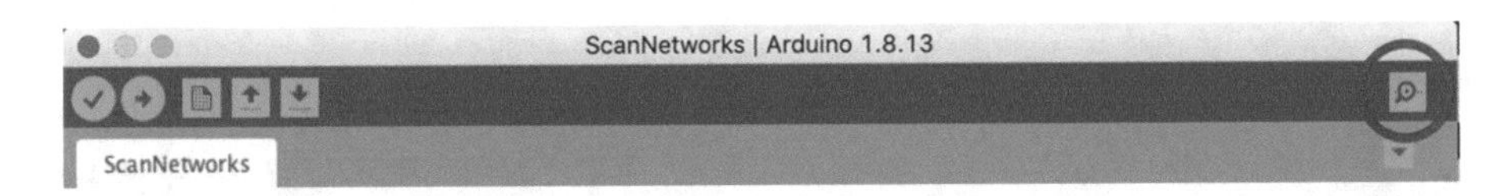

Figure 2-27. *Launching the serial monitor*

Once you click the button, you will see the serial monitor window open. If you don't see any output after a few seconds or you see some gibberish in the display, don't despair. This is because the serial monitor can operate at different speeds (baud rates).

So how do you know what baud rate to choose? It is set in the `setup()` method in the sketch. Look for this line of code. The number in the parentheses is the baud rate. You can change the baud rate using the drop-down box at the bottom right of the serial monitor as shown in Figure 2-28:

```
Serial.begin(9600);
```

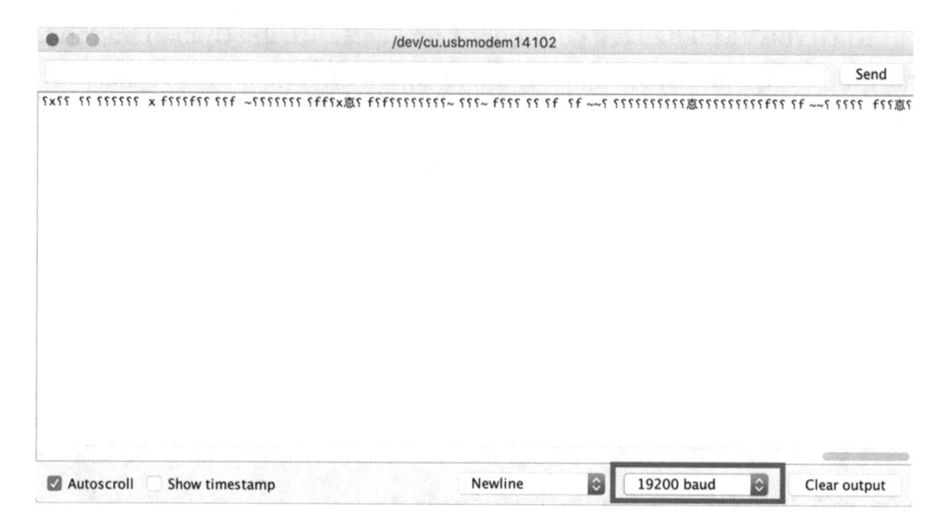

Figure 2-28. *Changing the baud rate in the serial monitor*

Notice you can also control the scrolling (stop autoscroll), add the timestamp, change the newline behavior, and clear the output.

Once you select the baud rate that matches the sketch, you will then see the correct output. Listing 2-2 shows an example of the output you will see in the serial monitor (obscured for security).

Listing 2-2. Example Scan Networks Output

```
Scanning available networks...
** Scan Networks **
number of available networks:5
0) RubberBiscuit1     Signal: -73 dBm     Encryption: WPA2
1) ATT-WIFI-EF2A    Signal: -76 dBm     Encryption: WPA2
2) DIRECT-FB-HP OfficeJet Pro 8730     Signal: -78 dBm
   Encryption: WPA2
3) RubberBiscuit2     Signal: -87 dBm    Encryption: WPA2
4) RubberBiscuit3     Signal: -96 dBm    Encryption: WPA2
```

If you got this to work, congratulations! You've just turned your Arduino into a simple WiFi network scanner. Cool!

Using the MKR1000

This board uses the example sketch under the WiFi101 category. However, recall that we must install the WiFi101 software library as shown in the previous section. If you're following along with your own MKR1000 board, do those steps now.

Let's use a different scan networking example. To load the example, click *File* ➤ *Examples* ➤ *WiFi101* ➤ *Scan Networks Advanced*. On some platforms, you may find this under the retired submenu (*File* ➤ *Examples* ➤ *Retired* ➤ *WiFi101* ➤ *Scan Networks Advanced*). Once again, if you look through the sketch, you will see a lot more code that is a lot more complex than the previous example. Fortunately, we can use it without modification.

When ready, connect your MKR1000 and upload the sketch. Wait until the upload is complete and then click the button on the far right of the window as shown in Figure 2-27. Once you click the button, you will see the serial monitor window open. Be sure to set the baud rate as needed (this example also uses 9600). You should start seeing a slightly different output with a bit more data like that shown in Listing 2-3.

Listing 2-3. Example Scan Networks Advanced Output

```
Scanning available networks...
** Scan Networks **
number of available networks: 3
1) Signal: -68 dBm    Channel: 11          BSSID: FE:32:75:
                                           FE:89:EE
   Encryption: WPA         SSID: RubberBiscuit1
2) Signal: -78 dBm    Channel: 6           BSSID: 42:FE:34:7E:
                                           98:FB
```

```
   Encryption: WPA             SSID: DIRECT-FB-HP OfficeJet Pro
                               8730
3) Signal: -64 dBm      Channel: 3              BSSID: FE:6A:FE:8F:
                                                75:FD
   Encryption: WPA             SSID: ATT-WIFI-EF2A
```

Once again, congratulations on experimenting with your Arduino for the first time. As you can see, there is still more to learn.

Summary

This chapter covered a lot of ground. You explored the Arduino platform, including the many forms available, and saw some example sketches (programs) to control the Arduino. This is the foundation of working with the Arduino. To become proficient with the Arduino requires a bit more than these basics.

In the next chapter, you will discover more about the Arduino platform including more about programming the Arduino along with more complex example sketches building up to writing your first sketch from scratch.

CHAPTER 3

Arduino Programming

If you're new to Arduino and programming in general, the source code for the Arduino is C++-like language[1] and the look of the code (the readability of the sketch) may be a bit overwhelming. It is OK if the thought of programming in a C++-like language sounds scary. Fortunately, it really isn't that difficult, and this chapter will get you there safely without inundating you with a plethora of computer science terms, practices, or theories. You really don't need all of that to succeed with the Arduino. Rather, you only need to learn the basic syntax and statements for the Arduino language. Once you learn that, you can learn to use libraries to expand the capabilities of your sketch and control devices.

In the last chapter, we saw a brief overview of the Arduino IDE including the location for the major functions you will need to use. In this chapter, we will take a closer look at the Arduino IDE including a tutorial of the language you will use to write your sketches as well as a look at how to begin writing Arduino sketches.

[1] You may also see it referenced as a "C-like" language, but C++ is more accurate since it supports objects. But it should in no way be confused as a general C++ language since it is written specifically for the Arduino platform.

C. Bell, *Beginning IoT Projects*, https://doi.org/10.1007/978-1-4842-7234-3_3

Getting Started

We've already seen how to install and use the basic functions of the Arduino IDE in the last chapter. Most works on the topic usually stop there and let you fend for yourself to learn the finer points of becoming productive on the Arduino platform.

That is unfortunate because most new to the platform do not have the background experience (or education) to pick things up and run with them. The Arduino was designed to be accessible for anyone no matter their experience. As such, it is important to take some time to learn some of the basics of developing Arduino sketches.

This includes learning more about the programming language, the libraries available, how to write sketches, and, more importantly, how to make everything work in the end. However, there are a few concepts we need to review and explore in more detail including how to work with sketches in the IDE and the basic layout (code design) of an Arduino sketch.

Working with Sketches in the Arduino IDE

While we saw a brief tour of the Arduino IDE in the last chapter, it is a good idea to review the basic operations that you will need to use to write sketches. Let's start with the basic controls in the Arduino IDE. Figure 3-1 shows the six menu buttons that you will use most often.

Figure 3-1. *Main menu buttons (Arduino IDE)*

These include, from left to right, the following operations, with a general description of when you would use each. The buttons are mirrors of menu items, which are also listed with each button:

- *Verify*: Compile your sketch to check for errors (does not upload). Menu: *Sketch ➤ Verify/Compile.*
- *Upload*: Compile and upload your sketch to your Arduino board. Menu: *Sketch ➤ Upload.*
- *New*: Open a new, blank sketch (it will contain the `setup()` and `loop()` functions). Menu: *File ➤ New.*
- *Open*: Open a sketch from your PC. This will cause a menu listing all of your sketches in your sketch library as well as the Examples submenu at the bottom. Menu: *File ➤ Open.*
- *Save*: Save the current sketch to your PC. If this is a new sketch, you will be permitted to name it. Menu: *File ➤ Save* or *File ➤ Save As* (to rename existing sketches).
- *Serial monitor*: Open the serial monitor to display the messages or interact with your sketch. Open this after your sketch has been uploaded. Menu: *Tools ➤ Serial Monitor.*

While these are the basic controls you will use most frequently, you may also encounter the need to add new libraries to your sketch (*Sketch ➤ Include Library*), which adds the #include statement into your sketch, or, if you want a new library, search for it and install it on your PC for use by your sketch.

Of course, if you add a new board, you may need to use the Boards Manager (*Tools ➤ Board [...] ➤ Boards Manager*) to download new hardware libraries as we discussed in Chapter 2.

Savvy readers will note that these menu items are a fraction of those available in the Arduino IDE. In fact, there is a lot more to the Arduino IDE that you can explore, and I encourage you to do so. However, for this book, these are the controls you will need to master.

Tip See `www.arduino.cc/en/Guide/Environment` for a complete list of all of the menu items and their operations in the Arduino IDE.

There is one more thing we must understand before we jump into writing sketches and learning the Arduino language: the layout of the editor window. Figure 3-2 shows the editor window with labels over the major parts.

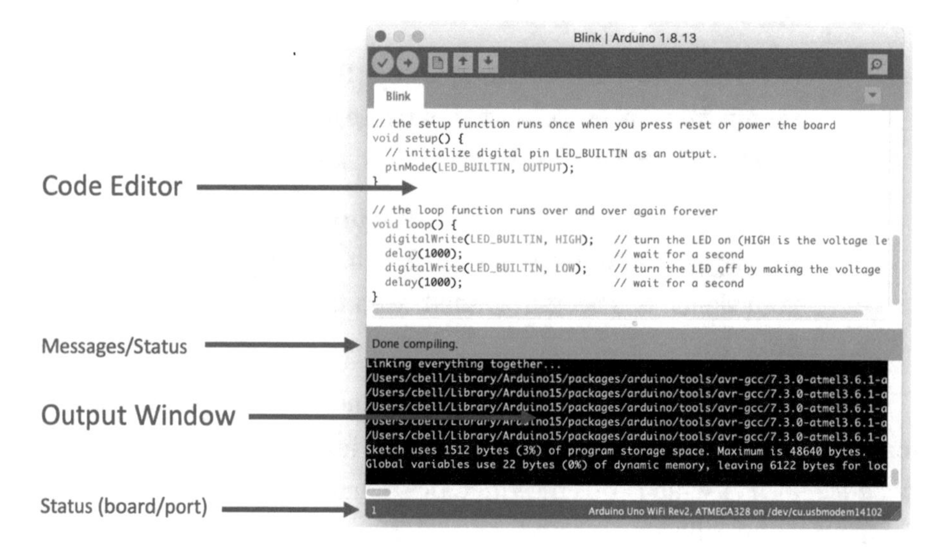

Figure 3-2. *Arduino IDE editor window*

Notice there are two major portions (windows). At the top is the code editor where you type in your sketch. At the bottom is the output window where messages from compilation are placed. Check this window whenever you compile your sketch. In fact, you should always click *Verify/ Compile* and check this window for errors before uploading your sketch to your board.

There are also two areas that are helpful. Between the code and output windows is an area where the Arduino IDE will display status messages. This is especially helpful to watch when you're uploading your sketch. For example, you should wait until you see *Done uploading* to open the serial monitor. This will ensure you will see all messages from your Arduino in the serial monitor (it is possible to start it late and miss messages, but you can click the reset button on the Arduino to force it to restart).

Notice the output window again. This is displaying the following statements:

```
Sketch uses 1512 bytes (3%) of program storage space. Maximum
is 48640 bytes.
Global variables use 22 bytes (0%) of dynamic memory, leaving
6122 bytes for local variables. Maximum is 6144 bytes.
```

Notice there is a lot of information there. What is important to note is the size of program storage space your sketch is using, not the actual size so much as the percentage of available space. Similarly, we want to watch the percentage of dynamic memory used. When either of these values reaches 80% or higher, you may encounter instability when your sketch runs. This is especially true for large, complex sketches that use a lot of memory. When the Arduino runs out of memory, it will either hang or reboot.

Fortunately, most sketches will remain well below that threshold, but if you find your sketches reaching these limits, you must either use a different Arduino board that has more memory or trim your sketch down to use less memory. Most often, reworking your sketch will solve the problem.

OK, now that we understand how to use the IDE, let's look at the layout of all sketches.

Basic Sketch Layout

The basic layout of every sketch includes the `setup()` and `load()` functions. When you create a new sketch from scratch (e.g., using *File ➤ New*), you will see the following in the code editor. Note that I've added some comments to help make the layout more informational:

```
// put all library headers (#include) and global variables here
void setup() {
  // put your setup code here, to run once:
}
void loop() {
  // put your main code here, to run repeatedly:
}
// include any additional functions you want to write here
   (optional)
```

Notice the comments (those lines that start with `//`) mention where certain things should go. For instance, at the top is where we list our `#include` statements and global variables (more about those later). At the bottom, we place any custom functions (functions) we want to write. Note that those can go anywhere, but I like to put them at the end. There is no requirement or rule there. It's only a matter of style.

Recall from Chapter 2 there are two built-in functions[2] that every sketch must contain even if there is nothing inside them (no code to execute). These include `setup()`, which is run once when the sketch is launched (started), and `loop()`, which is run repeatedly until the Arduino is rest or power is turned off. Let's examine more about each of these functions.

[2] Some use "method," while others insist on debating the terms "function" and "method" as having entirely separate meanings. For the purposes of anyone learning to write code, they are synonymous.

setup( )

This function is run once. It is where you should place all of the setup code for your sketch. Here is where you would initialize variables, configure libraries to work with hardware, and place any code that you want to execute once.

loop( )

This function is the heart of the sketch. You should place all of the code that you want to execute repeatedly here. This includes code to interact with devices, read sensors, save or display data, etc. You should try to design your code so that it forms a higher-level view of the operations of your sketch. If your sketch contains complex code, you can break the steps into separate functions so that the code in `loop()` function reads like an overview of the steps you want to execute (its algorithm). We'll discuss functions in a later section.

Now that we've reviewed the IDE controls and the basic layout of every sketch, let's learn more about the language we're using to write the code for our sketches.

Arduino Language Basics

Now let's learn some of the basic concepts of Arduino C++ programming. We will cover the building blocks of the language, such as comments, variables, and basic control structures, and then move on to the more complex concepts of data structures, libraries, and objects (classes). That last is a bit advanced, so we'll cover it at a high level. In fact, we'll be taking a high-level look at the language since it would require a tome of massive proportions to cover every nuance.[3]

[3] Drama aside, there is some truth in there.

While the material may seem to come at you in a rush, this crash course on Arduino C++ covers only the most fundamental knowledge of the language and how to use it for the Arduino platform. It is intended to get you started writing sketches and complete the projects in this book. Let's start with the basics.

The Basics

There are a number of basic concepts about the Arduino C++ programming language that you need to know in order to get started. In this section, I describe some of the fundamental concepts used, including basic syntax, how libraries are used, functions, and how to document your code.

Comments

One of the most fundamental concepts in any programming language is the ability to annotate your source code with text that not only allows you to make notes among the lines of code but also forms a way to document your source code.[4]

To add comments to your source code, use two slashes, // (no space between the slashes). Place them at the start of the line to create a comment for that line repeating the slashes for each subsequent line. This creates what is known as a block comment, as shown. Notice that I used a comment without any text to create whitespace. This helps with readability and is a common practice for block comments:

```
//
// Beginning IoT Projects
//
```

[4] If you ever hear someone claim, "My code is self-documenting," be cautious when using their code. Sure, plenty of good programmers can write code that is easy to understand (read), but all fall short of that lofty claim.

```
// Example Arduino C++ sketch to convert temperature to F or C.
//
// Created by Dr. Charles Bell
//
```

You can also use the double slash to add a comment at the end of a line of code. That is, the compiler ignores whatever is written after the double slash to the end of the line. You see an example of this next. Notice that I used the comment symbol (double slash) to comment out a section of code. This can be really handy when testing and debugging, but generally discouraged for final code. That is, don't leave any commented-out code in your deliverable (completed) source code. If it's commented out, it's not needed!

```
if (size < max_size) {
  size++;
} //else {
//  return -1;
//}
```

Writing good comments and indeed documenting your code well is a bit of an art form, one that I encourage you to practice regularly. Since it is an art rather than a science, keep in mind that your comments should be written to teach others what your code does or is intended to do. As such, you should use comments to describe any preconditions (or constraints) of using the code, limitations of use, errors handled, and how the parameters are used and what data is altered or returned from the code (should it be a function or class member).

You can also use the older /*...*/ notation for comments. This makes anything inside the /* and */ a comment and can span multiple lines. I've rewritten the block comment using this style. Some people prefer it over the //, which only applies to the line where it appears:

```
/*
  Beginning IoT Projects

  Example Arduino C++ sketch to convert temperature to F or C.

  Created by Dr. Charles Bell
*/
```

You may also see this appear like the following in one or more variations. Again, none is necessarily more correct, but the general style is to use the // marker:

```
/**
 * Beginning IoT Projects
 *
 * Example Arduino C++ sketch to convert temperature to F or C.
 *
 * Created by Dr. Charles Bell
 **/
```

Including Libraries

If you recall from the basic sketch layout, there is an area at the top of the sketch that indicates something is to be included. These are called preprocessor directives and often look like the following. They are called preprocessor directives because they signal the compilation process to perform some tasks before the code is compiled:

```
#include "SerialMonitorReader.h"
```

The directive does what it sounds like: it tells the compiler to include that file along with your source code. When the compiler encounters this directive, it "includes" that source file with your source code and compiles it. In other words, you are adding existing libraries (code) to your sketch that you can use.

The `#include` directive is one of the fundamental mechanisms that support modularity in Arduino C++. That is, you can create a library of source code that provides some functionality that resides in one or more separate source code files. Even if you do not create a new library, you can use modularity to split your source code into separate parts that form some high-level abstraction. More specifically, you would place like functionality together making the code easier to maintain or allowing more than yourself to work on it at the same time. However, you would not use modularity to separate random sections of code – that would gain you nothing except confusion as to where the bits of code reside.

The file that the preceding line of code includes is called a header file. A header file is named `.h` and contains only the declaration of the code. You can think of it as a blueprint or pattern for the code. A header file often contains only the primitives of the code that you will use, hence making it possible for the compiler to resolve any references to the features in the header file. A separate companion file called the source file is named `.cpp` (you can also use `.cc`, but most prefer `.cpp`) and contains the actual code for the features.

Curly Braces

Notice that the functions are implemented with a pair of curly braces that define the body of the function. Curly braces in Arduino C++ are used to define a block of code or simply to express grouping of code. Curly braces are used to define the body of functions, structures, classes, and more. Notice that you use them everywhere, even in the conditional statements (see the `if` statements).

Tip Some C++ programmers prefer to place the starting curly brace on the same line as the line of code to which it belongs. However, others prefer the open curly brace placed on the next line. Neither preference matters to the compiler; rather, this is an example of code style. You should choose the style you like best.

Functions

Notice that I've added a new function named `convertTemp()` that converts the temperature based on the scale chosen as shown on line 057 and repeated in the following. This effectively moves that logic out of the `loop()` function, thereby simplifying the code. This technique is a key technique you use when writing sketches. More specifically, in Arduino C++, sketches are built using functions:

```
057 // Convert temperature to Celsius or Fahrenheit
058 double convertTemp(char scale, double base_temp) {
059     if ((scale == 'c') || (scale == 'C')) {
060         return ((9.0 / 5.0) * base_temp) + 32.0;
061     }
062     else if ((scale == 'f') || (scale == 'F')) {
063         return (5.0 / 9.0) * (base_temp - 32.0);
064     }
065     return 0.0;
066 }
```

Notice the use of a comment at the start of the function along with the use of curly braces to define the body of the function as well as several blocks of code. We'll talk more about that in a moment, but draw your attention to the function declaration on line 058. Notice we have several parts. The general form of the function declaration is as follows:

```
<return type> <function_name>([<param_type> <param_name>,]) {
```

where we declare a return type or type of value returned when the function ends (you can use void if no value is to be returned), the name of the function, and zero or more parameters inside parentheses each consisting of a type and name separated by commas.

For example, suppose we have a function named doSomething(), which requires two parameters: an integer and a float. The function returns a float. The following shows how this function is defined:

```
float doSomething(int goal, float sourceValue)
```

To call this function, we would supply the values for the parameters and assign the result to a variable as shown in the following. Notice you can also use constants for the parameters when calling the function:

```
float val = 22.44;
float myFloat = doSomthing(12, val);
```

Tip Function parameters and values passed must match on type and order when called.

OK, we now have a good grasp of most basics, but what are these types and variables?

Variables and Types

No program would be very interesting if you did not use variables to store values for calculations. They are given a name that you make up. There are some names you cannot use such as reserved words (names for statements), and there are some rules to follow including that it cannot start with a number (must be a letter or underscore), and they are case sensitive. It should be noted that there are a number of style guides that introduce additional constraints such as the Google C++ Style Guide (`https://google.github.io/styleguide/cppguide.html`), but you don't have to follow those if you don't want to (but it will improve readability).

Variables are declared with a type and once defined with a specific type cannot be changed. Since Arduino C++ is strongly typed, the compiler ensures that anywhere you use the variable, it obeys its type. For example, the operation on the variable must be valid for the type. Thus, every variable must have a type assigned.

There are a number of simple types that the Arduino C++ language supports (often called *built-in types*). They are the basic building blocks for more complex types. Each type consumes a small segment of memory, which defines not only how much space you have to store a value but also the range of values possible.[5] Note that the values that can be stored are based on the board and its microcontroller as defined by the word size or the number of bits (e.g., 16- or 32-bit).

For example, on an Arduino Uno, an integer consumes 2 bytes (16 bits), and you can store values in the range -32,768 to 32,767. In this case, the integer variable is signed (the highest bit is used to indicate positive or negative values). An unsigned integer can store values in the range 0–65,535. Conversely, the Arduino Due uses 4 bytes (32 bits) for an integer. It has an integer range of -2,147,483,648 to 2,147,483,647 and an unsigned integer range of 0–4,294,967,295.

[5] For a complete list, see `www.arduino.cc/reference/en/#data-types`.

You can declare a variable by specifying its type first and then an identifier. The following shows a number of variables using a variety of types:

```
int num_fish = 0;           // number of fish caught
double max_length {0.0}; // length of the longest fish in feet
char fisherman[25];         // name of the fisherman
char rod_used[40];          // name or type of rod used
```

Notice also that I have demonstrated how to assign a value to the variable in the declaration. I demonstrate two widely used techniques: using a simple assignment and using the initialization mechanism available since C++11 (meaning it is the C++ standard adopted in 2011).

The assignment operator is the equals sign. All assignments must obey the type rules. That is, I cannot assign a floating-point number (e.g., `17.55`) to an integer value. The C++ initialization mechanism uses curly braces (called an initializer list) that contain the value you want to assign. The following shows an example:

```
int x {14};
```

Note that you can include the assignment operator with the curly braces (the compiler will not complain), but that is considered sloppy and discouraged. For example, the following code will compile, but it is considered a bad form:

```
int y = {15};
```

Table 3-1 shows a list of the built-in types that you commonly use in your applications.

Table 3-1. *Commonly Used Types in Arduino C++*

Symbol	Size in Bits	Range
bool/boolean	n/a	Has only two values: false and true
byte	8	0–255
char	8	–128 to 127 by default
unsigned char	8	0–255
double	32/64	–3.4028235E+38 to 3.4028235E+38 (32-bit)
Float	32	–3.4028235E+38 to 3.4028235E+38 (32-bit)
int	16/32	–32,768 to 32,767 (16-bit) –2,147,483,648 to 2,147,483,647 (32-bit)
unsigned int	16/32	0–65,535 (32-bit) 0–4,294,967,295 (32-bit)
long	32	–2,147,483,648 to 2,147,483,647
unsigned long	32	0–4,294,967,295
short	16	–32,768 to 32,767
size_t	n/a	Size of any object in bytes

Arithmetic

You can perform a number of mathematical operations in Arduino C++, including the usual primitives but also logical operations and operations used to compare values. Rather than discuss these in detail, I provide a quick reference in Table 3-2 that shows the operation and an example of how to use the operation.

***Table 3-2.** Arithmetic, Logical, and Comparison Operators in C++*

<table>
<tr><th>Type</th><th>Operator</th><th>Description</th><th>Example</th></tr>
<tr><td>Arithmetic</td><td>+</td><td>Addition</td><td>int_var + 1</td></tr>
<tr><td></td><td>-</td><td>Subtraction</td><td>int_var - 1</td></tr>
<tr><td></td><td>*</td><td>Multiplication</td><td>int_var * 2</td></tr>
<tr><td></td><td>/</td><td>Division</td><td>int_var / 3</td></tr>
<tr><td></td><td>%</td><td>Modulus</td><td>int_var % 4</td></tr>
<tr><td>Logical</td><td>&</td><td>Bitwise and</td><td>var1&var2</td></tr>
<tr><td></td><td>|</td><td>Bitwise or</td><td>var1|var2</td></tr>
<tr><td></td><td>^</td><td>Bitwise exclusive</td><td>var1^var2</td></tr>
<tr><td></td><td>~</td><td>Bitwise compliment</td><td>~var1</td></tr>
<tr><td></td><td>&&</td><td>Logical and</td><td>var1&&var2</td></tr>
<tr><td></td><td>||</td><td>Logical or</td><td>var1||var2</td></tr>
<tr><td>Comparison</td><td>==</td><td>Equal</td><td>expr1==expr2</td></tr>
<tr><td></td><td>!=</td><td>Not equal</td><td>expr1!=expr2</td></tr>
<tr><td></td><td><</td><td>Less than</td><td>expr1<expr2</td></tr>
<tr><td></td><td>></td><td>Greater than</td><td>expr1>expr2</td></tr>
<tr><td></td><td><=</td><td>Less than or equal</td><td>expr1<=expr2</td></tr>
<tr><td></td><td>>=</td><td>Greater than or equal</td><td>expr1>=expr2</td></tr>
</table>

Bitwise operations produce a result on the values performed on each bit. Logical operators (and, or) produce a value that is either true or false and are often used with expressions or conditions.

Finally, Arduino C++ has a concept called *constants*, where a value is set at compile time. There are two types of constants. One, signified by using the `const` keyword, creates a value (think variable) that will never be changed. The following are examples of constants:

```
const int fish_catch_limit {7};    // A constant whose value
                                      cannot change
const char slogan[] = "Teach a man to fish."  // A constant
                                               expression
```

Now that you understand variables and types, the operations permitted on them, and expressions, let's look at how you can use them in flow control statements.

Flow Control Statements

Flow control statements change the execution of the program. They can be conditionals that define gates using expressions that restrict execution to only those cases where the expression evaluates true (or negated), special constructs that allow you to repeat a block of code (loops), and functions to switch context to perform some special operations. You've already seen how functions work, so let's look at conditional and loop statements.

Conditionals

Conditional statements allow you to direct execution of your programs to sections (blocks) of code based on the evaluation of one or more expressions. There are two types of conditional statements in Arduino C++ - the `if` statement and the `switch` statement.

The following shows the general structure of the `if` statement:

```
if (expr1) {
  // execute only if expr1 is true
} else if ((expr2) || (expr3)) {
```

```
  // execute only if expr1 is false *and* either expr2 or expr3
     is true
} else {
  // execute if both sets of if conditions evaluate to false
}
```

Notice in the example that you can have one or more (optional) else phrases that you execute once the expressions for the conditions evaluate to false. You can chain `if/else` statements to encompass multiple conditions where the code executed depends on the evaluation of several conditions.

Although you can chain the statements as much as you want, use some care here because the more `else/if` sections you have, the harder it becomes to understand, maintain, and avoid logic errors in your expressions.

If you have a situation where you want to execute code based on one of several values for a variable or expression that returns a value (such as a function or calculation), you can use the `switch` statement. The following shows the structure of the `switch` statement:

```
switch (eval) {
  case <value1> :
     // do this if eval == value1
     break;
  case <value2> :
     // do this if eval == value2
     break;
  default :
     // do this if eval != any case value
     break;  // Not needed, but good form
 }
```

The case values must match the type of the thing you are evaluating. That is, case values must be the same type as eval. Notice the break statement. This is used to halt evaluation of the code once the case value is found. Otherwise, each successive case value will be compared. Finally, there is a default section for code that you want to execute should eval fail to match any of the values.

Tip Code style varies greatly in how to space/separate these statements. For example, some indent the case statements; some do not.

Loops

Loops are used to control the repetitive execution of a block of code. There are three forms of loops that have slightly different behavior. All loops use conditional statements to determine whether to repeat execution or not. That is, they repeat as long as the condition is true. The three types of loops are while, do, and for. I explain each with an example.

The while loop has its condition at the "top" or start of the block of code. Thus, while loops only execute the body if and only if the condition evaluates to true on the first pass. The following illustrates the syntax for a while loop. This form of loop is best used when you need to execute code only if some expression(s) evaluate to true, for example, iterating through a collection of things, where the number of elements is unknown (loop until you run out of things in the collection):

```
while (expression) {
   // do something here
 }
```

The do loop places the condition at the "bottom" of the statement, which permits the body of the loop to execute at least once. The following illustrates the do loop. This form of loop is handy for cases where you want to execute code that, depending on the results of that execution, may require repetition, for example, repeatedly asking the user for input that matches one or more known values, repeating the question if the answer doesn't match:

```
do {
  // do something here - always done once
} while (expression);
```

The for loop is sometimes called a counting loop because of its unique form. The for loop allows you to define a counting variable, a condition to evaluate, and an operation on the counting variable. More specifically, for loops allow you to define stepping code for a precise number of operations. The following illustrates the structure of the for loop. This form of loop is best used for a number of iterations for a known number (either at runtime or as a constant) and commonly used to step through memory, count things, and so forth:

```
for (<init> ; <expression> ; <increment>) {
// do something
}
```

The <init> section or counting variable declaration is executed once and only once. The <expression> is evaluated on every pass. The <increment> code is executed every pass except the last. The following is an example for loop:

```
for (int i; i < 10; i++) {
   // do something here
}
```

Now let's look at some commonly used data structures.

Basic Data Structures

What you have learned so far about Arduino C++ allows you to create sketches that do simple to moderately complex operations. However, when you start needing to operate on data (either from the user or from sensors and similar sources), you need a way to organize and store data and operations on the data in memory. The following introduces three data structures in order of complexity: arrays, structures, and classes.

Arrays allocate a contiguous area of memory for multiple storage of a specific type. That is, you can store several integers, characters, and so forth, set aside in memory. Arrays also provide an integer index that you can use to quickly access a specific element. The following illustrates how to create an array of integers and iterate through them with a for loop. Array indexes start at 0:

```
int num_array[10] {0,1,2,3,4,5,6,7,8,9};  // an array of 10
                                              integers
for (int i = 0; i < 10; i++) {
  // do something here
}
```

Notice the `i++` in the `for` loop. This is a shorthand for `i = i + 1` and is very common in Arduino C++. You can also define multiple-dimensional arrays (arrays of arrays). Arrays can be used with any type or data structure.

If you have a number of data items that you want to group together, you can use a special data structure called, amazingly, a `struct`. A `struct` is formed as follows:

```
struct <name> {
  // one or more declarations go here
};
```

You can add whatever declarations you want inside the `struct` body (defined by the curly braces). The following shows a crude example. Notice that you can use the structure in an array:

```
struct address {
  char first_name[30];
  char last_name[30];
  int street_num;
  char street_name[40];
  char city[40];
  char state[2];
  char zip_code[12];
};

address address_book[100];
```

Arrays and structures can increase the power of your programs by allowing you to work with more complex data types. However, there is one data structure that is even more powerful: the `class`.

A `class` is more than a simple data structure. You use classes to create abstract data types and to model concepts that include data and operations on data. Like structures, you can name the `class` and use that name to allocate (instantiate) a variable of that type. Indeed, structs and classes are closely related.

You use classes to break your programs down into modules. More specifically, you place the definition of a `class` in a header file and the implementation in a source file. The following shows the header file (`myclass.h`) for a simple and yet trivial class to store an integer and provide operations on the integer:

```
class MyClass {
  public:
    MyClass();
    int get_num();
```

```
    void inc();
    void dec();
  private:
    int num;
};
```

Notice several things here. First, the class has a name (`MyClass`) and a public section where anything in this area is visible (and usable) outside of the class. In this case, there are three functions. The function with the same name as the class is called a constructor, which is called whenever you instantiate a variable of the class (type). The private section is only usable from functions defined in the class (private or public).

The source code file (`myclass.cpp`) is where you implement the methods for the class as follows:

```
#include "myclass.h"

MyClass::MyClass() {
  num = 0;
}

int MyClass::get_num() {
  return num;
}

void MyClass::inc() {
  ++num;
}

void MyClass::dec() {
  --num;
}
```

Notice that you define the methods in this file prefixed with the name of the class and two colons (`MyClass::`). While missing in this example, you can also provide a destructor (noted as `~MyClass`) that is executed when the class instantiation is deallocated. Finally, notice at the top is the `#include` preprocessor directive to include the header file so that the compiler knows how to compile this code (using the class header or declaration). You can then use the class in the sketch, as follows:

```
#include "myclass.h"
...
void loop() {
    MyClass c = MyClass();
    c.inc();
    c.inc();
    Serial.print("contents of myclass: ");
    Serial.println(c.get_num());
}
```

Notice how you use the class. This is actually allocating memory for the class – both data and operations. Thus, you can use classes to operate on things or provide functionality when you need it saving you time and making your programs more sophisticated. Classes are used to form libraries of functionality that can be reused. Indeed, you have entire suites of libraries built using classes.

As you may have surmised, classes are the building block for object-oriented programming (OOP), and as you learn more about using classes, you can build complex libraries of your own.

Pointers

Pointers are one of the most difficult things for new programmers to understand. However, the following attempts to explain the basics of using pointers. There is a lot more that you can do with pointers, but this is the fundamental concept of simple pointers.

A pointer (also called a *pointer variable*) stores the memory address of a variable or data. Thus, a pointer "points to" a section of memory. You declare by type and the * symbol. All pointers must be typed, and any operation on what the pointer points to must obey the condition of that type. When you access the thing the pointer "points to," you call that dereferencing and use the * symbol to tell the compiler you want the value of the thing the pointer is "pointing to."[6] The following shows how you can declare a pointer and then dereference it. Note that you expect `int_ptr` to be assigned a value; otherwise, the code may not compile or exhibit side effects:

```
int *int_ptr; // pointer to an integer
int i = *int_ptr; // Store what int_ptr is pointing to
```

To store an address in a pointer variable, you use the & symbol (also called the *address of operator*). The following shows an example:

```
int *int_ptr = &i; // Store address of i in int_ptr
```

You can perform arithmetic and comparison on pointers. You can add or subtract an integer to change the address of the pointer (the actual value of the pointer variable, not the thing the pointer points to) by multiples of the size of the type. For example, adding 1 to an integer pointer advances (increases) the memory value by 4 bytes.

[6] I think you get the point.

You can also compare pointers to determine equality and subtract one pointer from another to find distance (in bytes) between the pointers. This could be handy for calculating distance for contiguous memory segments. When performing arithmetic on pointers, you should use parentheses to avoid nasty mistakes with operator precedence. For example, the following code is not equivalent. The second line increments the thing that the pointer points to, but the third line increments the value of the pointer variable (memory address). Be careful when performing math on pointers because you could unexpectedly end up dereferencing portions of memory:

```
*int_ptr = 10;        // set the thing that the pointer points
                         to = 10
i = *int_ptr + 1;     // add one to the thing that the pointer
                         points to
i = *(int_ptr + 1);   // add 4 bytes to the pointer variable
                         (size of integer) - ERROR? points to
                         nowhere!
```

Finally, always use `nullptr` to initialize a pointer variable when the address is not known, as follows:

```
int *int_ptr {nullptr};
```

Tip There are entire books written about pointers! If you want a more in-depth look at pointers or want to dive into the details of how to use pointers, see the book *Understanding and Using C Pointers* by Richard Reese (O'Reilly, 2013). The book was written for C programmers, and although some of the data is outdated, it is an excellent study on pointers.

Wow! That was a wild ride, wasn't it? I hope that this short crash course in Arduino C++ has explained enough about the language that you now know how it works. This crash course also forms the basis for understanding the other Arduino sketches in this book.

Now, let's walk through the temperature conversion example to apply what we've learned.

Practical Example

The temperature conversion sketch was designed to show you as many of the basics as possible without overwhelming. As such, it is missing some of the more advanced features of the language, but we've got to start somewhere. If you'd like to follow along, feel free to do so as we walk through this sketch, but this project is intended to demonstrate concepts rather than instruct how to build IoT projects. As such, we won't concentrate so much on the details of the logic in the sketch. We will spotlight a few points in the code in the process.

In this example, we will build a temperature conversion sketch that uses the Arduino to convert values in Fahrenheit to Celsius and Celsius to Fahrenheit using the serial monitor to input the data. That is, all you need is your Arduino and your PC.

The first thing to know is this sketch is built using three files: the main sketch file named `TemperatureConverter.ino` and two supporting code modules named `SerialMonitorReader.h` and `SerialMonitorReader.cpp`. These last two files are used to demonstrate how to use classes to break your code into smaller parts.

Let's examine each file and see how the code works.

TemperatureConverter.ino

This is the main sketch file and has the same name as the folder in which the sketch is saved (`TemperatureConverter`). Here is where our main logic is built along with the basic structure of the sketch (the `setup()` and `loop()` functions). Listing 3-1 shows the example sketch with line numbers. Take a few minutes to read through it. Most things should be clear, but there are some surprises.

Tip Notice the `\n` sequence in the print statements. This issues a newline character to the serial monitor, which is a special symbol that starts the next output on a new line.

Listing 3-1. TemperatureConverter Sketch Example

```
001 //
002 // Beginning IoT Projects
003 //
004 // Example Arduino C++ sketch to convert temperature to F
       or C.
005 //
006 // Created by Dr. Charles Bell
007 //
008
009 // Libraries
010 #include "SerialMonitorReader.h"
011
012 // Global variables
013 double temperature {0.0};
014 double converted_temp = 0.0;
015 char scale {'c'};
```

```
016 SerialMonitorReader *reader = new SerialMonitorReader();
017
018 // Initialize serial communications
019 void setup() {
020   Serial.begin(9600);
021   while (!Serial);
022   Serial.println("\nWelcome to the temperature conversion
      sketch.");
023   delay(500);
024 }
025
026 void loop() {
027     temperature = 0.0;
028     Serial.print("\nPlease choose a starting scale (F)
        or (C): ");
029     scale = reader->readChar();
030     Serial.println(scale);
031     // Throw error if scale is not one of the valid
           characters
032     if ((scale != 'c') && (scale != 'C') &&
033         (scale != 'f') && (scale != 'F')) {
034         Serial.print("\nERROR: I'm sorry, I don't
            understand '");
035         Serial.print(scale);
036         Serial.println("'.");
037     } else {
038         Serial.print("Please enter a temperature: ");
039         temperature = reader->readFloat();
040         Serial.println(temperature);
041         Serial.print(temperature);
042         converted_temp = convertTemp(scale, temperature);
```

```
043         if ((scale == 'c') || (scale == 'C')) {
044             Serial.print(" degrees Celsius = ");
045             Serial.print(converted_temp);
046             Serial.println(" Fahrenheit.");
047         }
048         else {
049             Serial.print(" degrees Fahrenheit = ");
050             Serial.print(converted_temp);
051             Serial.println(" Celsius.");
052         }
053     }
054     delay(1000);
055 }
056
057 // Convert temperature to Celsius or Fahrenheit
058 double convertTemp(char scale, double base_temp) {
059     if ((scale == 'c') || (scale == 'C')) {
060         return ((9.0 / 5.0) * base_temp) + 32.0;
061     }
062     else if ((scale == 'f') || (scale == 'F')) {
063         return (5.0 / 9.0) * (base_temp - 32.0);
064     }
065     return 0.0;
066 }
```

The following shows a sample of what the output would look like when you run the sketch. Values are entered in the input text box at the top of the serial monitor, and then you click *Send* to send the data to the sketch:

```
Welcome to the temperature conversion sketch.

Please choose a starting scale (F) or (C): f
Please enter a temperature: 67.35
```

```
Converting value from Fahrenheit to Celsius.
67.35 degrees Fahrenheit = 19.64 Celsius.

Please choose a starting scale (F) or (C): c
Please enter a temperature: 19.64
Converting value from Celsius to Fahrenheit.
19.64 degrees Celsius = 67.35 Fahrenheit.
```

Note If you've read through the code or tried to run it and determined something is missing, you're right! This example uses a custom class I created.

As you can see at the start of the code, we have a block of comments followed by the include statement to include the class and a list of global variables. We say they are "global" because every function in the sketch file can "see" the variables. Recall, if we had declared the variables inside a set of curly braces, they would only be visible to the code inside those curly braces.

Notice on line 016 we have this statement. That's our first surprise:

```
SerialMonitorReader *reader = new SerialMonitorReader();
```

This line of code declares a new pointer variable of type "points to `SerialMonitorReader`" and instantiates the object from the base class. Wow. So we now have a variable named `reader` that we can use to access all of the functions in the `SerialMonitorReader` class. Note that we could have split this statement into two as follows:

```
SerialMonitorReader *reader;
...
reader = new SerialMonitorReader(); // placed in setup()
```

Next, we see what is typically contained in the `setup()` function – initialization of the built-in `Serial` class communications on lines 019–024. We say "built-in" because we don't have to include the library or instantiate it as it is already available to the sketch by default. In this case, we call the `begin()` function specifying the speed we want to use on line 020. Next, we perform a `while` loop on line 021 that has no body (it just continues to do nothing while the condition is true). More specifically, we are waiting for the `Serial` object to be ready. We can then use the `Serial.print()` and `Serial.println()` functions to print (display) data to the serial monitor. The difference is the "`ln`" in the name implies that version prints a new line after the text.

Line 023 contains another surprise. The `delay()` function causes the sketch to pause or wait for the number of nanoseconds specified. This is helpful in tuning how fast you want the sketch to run so as not to overwhelm the serial communication (or the user). It can also be a nifty way to slow your sketch down to match your sensor sampling rate (how many times per unit of time you want to take a sample).

Next, we have the seemingly large `loop()` function on lines 026–054. This really isn't that long, and it does represent the basic logic of the example as follows. I included the line numbers for reference:

1. Get the scale to convert (028).
2. Get the value to convert (038).
3. Convert the value (042).
4. Display the value (043/048).
5. Repeat (1)-(4).

Reading from the serial monitor is not as clearly defined as some may expect. That's the reason we're using a class to hide that functionality so that the sketch isn't overly complicated (or long). A sketch with thousands of lines of code can be very difficult to diagnose (and test).

To read a character from the serial monitor, we use the readChar() function from the SerialMonitorReader class as follows. Here, we see the function returns a single character from the user and stores it in the variable named scale, which we will use a bit later:

```
029      scale = reader->readChar();
```

The code uses two conditionals. The first checks to see if the user entered one of the valid characters for the scale with the following statement:

```
032      if ((scale != 'c') && (scale != 'C') &&
033           (scale != 'f') && (scale != 'F')) {
```

Here, we see the use of the not equals (!=) or inequality sign as well as the logical and (&&). If you think about it, the only characters that cause the expression to evaluate to false are c, C, f, and F. Anything else evaluates to true, so we print an error and continue the loop.

Next, we read the temperature from the user using the readFloat() function from the SerialMonitorReader class and store it in a variable named temperature:

```
039           temperature = reader->readFloat();
```

Next, we make the conversion using a new function written to accomplish the goal. At the end of the file, we have a function named convertTemp(), which requires two parameters: a character type that specifies the scale and a double value that specifies the temperature to convert. When the function ends, it returns the converted temperature. The following shows how this function is called. I leave the explanation of the function as an exercise, but it is not difficult:

```
042           converted_temp = convertTemp(scale, temperature);
```

The second conditional comes next and is similar where we check if the scale entered signals a conversion from Celsius to Fahrenheit as shown in the following. We use the `else` clause for the Fahrenheit to Celsius conversion.

The `loop()` concludes by pausing for one second. Again, this is an optional interjection to help the flow of the sketch, but it is a practical choice more than a requirement.[7]

Now, let's look at the supporting files for the sketch.

SerialMonitorReader.h

This file contains the header or blueprint for the `SerialMonitorReader` class. As such, it defines the class and its methods. Both this file and the accompanying `.cpp` file are stored in the same folder as the sketch file.

Listing 3-2 shows the contents of the `SerialMonitorReader.h` file. There isn't much to see in the file other than how the functions are defined. There is one private function that is only used by the other two public functions for reading a character and reading a float value from the serial monitor.

Listing 3-2. The SerialMonitorReader Header

```
//
// Beginning IoT Projects
//
// Example Arduino C++ class header to read from the serial
   monitor.
//
// Created by Dr. Charles Bell
//
```

[7] However, delays may be important for communication logic, which sometimes needs to wait for data (send or receive).

```
class SerialMonitorReader {
private:
    void clearBuffer();
public:
    char readChar();
    float readFloat();
};
```

The class isn't a true object in the sense that it neither mimics a noun with verbs (something and operations on that something) nor encapsulates or hides data. Rather, it is simply an exercise in the syntax of writing classes for sketches.

Recall all header files have as a companion a source code file named the same with the `.cpp` extension.

SerialMonitorReader.cpp

The source code file for the class contains the code for all three methods. While the explanation of the code may be helpful to some, it is beyond the scope of demonstrating the basics. However, feel free to take a look to see if you can follow along.

Listing 3-3 shows the contents of the `SerialMonitorReader.cpp` file. The inspiration for this code was taken from the Arduino forums (`forum.arduino.cc`). If you'd like to understand how the code works, take a look at this tutorial series: `https://forum.arduino.cc/index.php?topic=396450`.

Listing 3-3. The SerialMonitorReader Source File

```
//
// Beginning IoT Projects
//
// Example Arduino C++ class to read from the serial monitor.
//
```

```
// Created by Dr. Charles Bell
//

#import <Arduino.h>
#import "SerialMonitorReader.h"

// Read character from serial monitor
char SerialMonitorReader::readChar() {
    char char_read;

    while (Serial.available() == 0);
    char_read = Serial.read();
    clearBuffer();
    return char_read;
}

// Read a floating point number from serial monitor
float SerialMonitorReader::readFloat() {
    char char_buffer[40];
    char char_read = ' ';
    byte index = 0;
    bool new_data = false;

    while (!new_data) {
        if (Serial.available() > 0) {
            char_read = Serial.read();
            if (char_read == '\n') {
                char_buffer[index] = '\0';
                new_data = true;
            } else {
                char_buffer[index] = char_read;
                index++;
            }
        }
    }
```

```
    clearBuffer();
    // Attempt to convert to float
    return atof(char_buffer);
}

// Clear the input buffer
void SerialMonitorReader::clearBuffer() {
    delay(50);
    while (Serial.available() > 0) {
      Serial.read();
    }
}
```

Let's discuss one more item that should be included in all basic knowledge tutorials, but often is overlooked: compiling your sketches.

Compiling Your Sketches

When you click Verify/Compile or Upload, the Arduino IDE proceeds to compile your sketch and all of its supporting code modules including any libraries you have referenced. If you look at the output window during this time, you will see a lot of messages passing by. Most of these you can safely ignore, but sometimes you may see either a warning or an error.

Rather than provide a step-by-step solution for all possible warnings and errors (an impossible feat), we will discuss strategies you can employ to help resolve them.

Warnings

A compiler warning is typically something the compiler identifies as either a violation of Arduino C++ coding or a flag that some logic error may exist. It is important to understand the compiler is not a style or code analyzer that can magically detect all possible issues in your code. So, when it does flag something as a warning, it is usually something important.

While some may say it is fine to ignore warnings, you should not do so out of hand. Rather, you should investigate the source of the warning and determine if it is something you can fix or not. There are three sources for compilation warnings: your code, software libraries, and hardware libraries. The most common compilation warnings are those from compiling software libraries and less likely hardware libraries.

When there are warnings in your code, you should fix them. Again, it could be an indication that there are potential logic or data errors in your code. So treat these as important especially if you've ignored them and discover later that your sketch fails at a really weird point (always an indication there is a logic error).

The following is an example of a compiler warning. Here we see the compiler has found a case where I have an unused variable. Perfectly harmless, yes? Well, maybe. Suppose I later added code that used that variable. Would that be a problem? Take a look at the statement again:

```
/Users/cbell/Documents/Writing/Books/Beginning IoT Projects/
source/Ch03/TemperatureConverter/TemperatureConverter.ino: In
function 'void loop()':
/Users/cbell/Documents/Writing/Books/Beginning IoT Projects/
source/Ch03/TemperatureConverter/TemperatureConverter.ino:27:9:
warning: unused variable 'i' [-Wunused-variable]
     int i = 0.1;
         ^
```

Do you see the issue? Look at the value being stored in the variable. It's a floating-point number, but the data type is integer. What do you think would happen here? Would the compiler flag this? Sadly, it doesn't, and you may never know there is a problem until your calculations using the variable fail. So pay attention to warnings in your code and fix them.

When warnings are generated from software libraries, it isn't always so easy to analyze what is wrong. Sometimes the problem is in how you're using the library. Maybe you're using the wrong data types for parameters or return values. It could also be the case that the library is out of date or isn't fully supported for the Arduino board (hardware). Whatever the case, it is best to contact the developer of the library to let them know (some have bug reporting facilities on their website) or conduct an Internet search for the warning to see if anyone else has found the problem. I've discovered most of these warnings can be explained or fixed. For the most part, warnings of this nature can be considered less harmful than warnings in your own code.

Warnings from hardware libraries are a bit more difficult to resolve. Most times, there isn't anything you can do about it except notify the vendor. So, in the general sense, these warnings may be ignored.

Errors

Compilation errors are things the compiler discovers that prevent it from building a binary executable of your sketch (it won't compile). There are many possible errors that can occur, but they can be generally classified as either syntactical or logical.

Syntax errors are the most common compilation errors. A typical syntax error is caused by a missing semicolon. The error message looks like the following. As you can see, the compiler helpfully identifies the file and even the line where the error occurred as well as the nature of the error:

```
In file included from /Users/cbell/Documents/Writing/Books/
Beginning IoT Projects/source/Ch03/TemperatureConverter/
TemperatureConverter.ino:10:0:
SerialMonitorReader.h:11:22: error: expected ';' at end of
member declaration
     void clearBuffer()
                      ^
exit status 1
expected ';' at end of member declaration
```

Most logical errors will be similar in that the compiler is able to identify what went wrong, but not always. It is likely you will encounter an error where either the explanation doesn't make sense or the code identified isn't the actual location of the error. Don't worry. This happens to everyone eventually.

It is also likely some errors will produce a long list of cascading errors or a list of seemingly unrelated errors. This happens sometimes when the compiler either attempts to continue compiling (in the case of warnings) or it is compiling multiple code modules together. It is sometimes possible the only error that is "correct" or valid is the first one. Fixing all of them at the same time can lead down a rabbit hole.

So what do you do? Simply, look at the first error and fix that one. Ignore the rest. Once you fix it, retry the compilation. Yes, you may still have other errors to fix, but at least this time you're negating any cascading effects the first error may have triggered. Follow this strategy, and your compilation woes can be mitigated.

If you cannot determine the exact cause of the error, try searching for the error message or a portion of the error message in the Arduino forums (`forum.arduino.cc`). There is a good chance someone else has encountered the same problem or something similar that will help you. In other words, don't be afraid to ask for help.

Finally, if you are working with newer or clone Arduino boards that require new hardware or software libraries, you may encounter one of two types of problems. Either the existing libraries have changed or the hardware library causes one or more aspects of your sketch to fail. Fortunately, errors generated from the hardware libraries are rare, but not so with software libraries because I've found they can change in unexpected ways.

When software libraries change, either the function names or their parameters change. Sometimes a function is removed and replaced with another, or a return type is changed. When any of these occur, your sketch will also need to change.[8] This can be frustrating, but normally the solution lies in reading the change notes for the new library.

When using new hardware libraries, errors may not be so easily fixed. I've seen cases where one hardware library doesn't support certain features altogether or they've been replaced with a different set of software libraries. For these types of issues, you will notice them most when trying to compile sketches from previous work on the new hardware. When this occurs, you should review the vendor's documentation on the hardware, and if you don't find an answer there, conduct an Internet search for help.

Example Sketches

There are many example sketches built into the Arduino IDE that you can explore. Most require some additional hardware or an Internet connection to work, but they each serve an excellent sample of what is possible. So long as you stick to the same basic flow of the example, you can use the examples as a starting point.

The Arduino development community encourages every developer to provide one or more example sketches with any library they want to publish and make available in the Arduino IDE. My own libraries include several examples you can use to learn how to work with the library.

[8] This is a big pet peeve for me. I was taught interfaces (class function definitions) should never change. If you must make a change, you should use polymorphism by adding a new function with the new parameters (or return type) so that existing code doesn't break and can continue to work with a newer version of the library. Sadly, it seems a few disagree, and they modify their libraries without a care to how they break things.

Thus, the example sketches should not be overlooked. Not only do they form a basis or pattern for how to write sketches that use the functionality, they are also helpful in testing that functionality.

Suppose you want to use a new sensor and discover someone (or the vendor) has written a software library for it. Rather than blindly plugging it in and trying to write a sketch from scratch to use it, you should open one of the example sketches and try it out. If everything works, you know two vital pieces of information: 1) the hardware works, and 2) the hardware and software libraries work with your board. I cannot tell you how many times I've encountered folks who claim a new sensor doesn't work when the problem lies in how they're trying to use it. Use the example sketches as a first stop when exploring things.

The bottom line is start with the examples to learn how to work with a device or software library and, once you verify everything is working, use those examples as guides along with the documentation to start writing your sketch.

Next, let's look at some helpful advice for writing your sketches.

Writing Your First Sketch

This section presents some much needed advice for those starting out on the Arduino. The following are some guiding principles and practices designed to help make your Arduino development easier. Take some time to read through these and apply them to your work going forward - you won't be sorry for taking the time. Some of these areas may seem a bit repetitive because many use the same techniques. In fact, I purposely cover some of the techniques from different viewpoints to give you a greater depth of understanding.

Note What follows is applicable to any development effort, not just Arduino. You will find the principles and practices, in general, apply to almost any project.

Keep It Simple[9]

One of the biggest mistakes beginners make is trying to write their entire sketch from scratch. When they attempt to execute it, they discover nothing works. This is a really bad idea and something you should train yourself to avoid at all costs. Why? If your sketch has more than one part (connecting to an electronic component, using additional libraries, complex coding, etc.), it is very likely one or more of these parts are either not coded correctly (syntax or logic errors), there are unforeseen incompatibilities, or you've assumed something works one way but doesn't work the way you expected.

All of this will lead you along the road to the circle of despair[10] where you will encounter one or more compilation errors, logic errors, or strange behavior. To avoid all of this, you must adopt a stepwise approach to writing your sketches.

Now, I do not expect you to drop everything and enroll in a computer science degree program so that you can flout about with terms like "bottom-up programming," "top-down design," "test-first development," or any of the myriad of programming disciplines one may encounter in computer science.[11] Rather, I want you to consider building your sketch one part at a time. It doesn't matter the order (well, sometimes it might if there are dependencies among the libraries); just build one part at a time, debug and test it until it works, and then add the next part. Or, better, make several sketches where each contains one or two parts. When everything is working, start by adding one part at a time to ensure everything continues to work and fix any errors or anomalies as you encounter them.

[9] Or more kindly spoken: keep it simple, silly!

[10] With apologies to my old dungeon master who said in her world, "All roads lead to the circle of despair."

[11] `https://en.wikipedia.org/wiki/Computer_science`

This will ensure that when an issue occurs, you can always point to the last thing you changed or added as the source. While that isn't strictly true for every issue (there could be cascading errors where the combination of libraries reveals an error in an early part that worked fine), it will help you know where to start debugging the problem.

Debugging is a skill you would do well to hone. It is both science (methodical approach) and, for the best developers, an art.

Debugging and Testing

Like the last section, debugging and testing is another area where novice Arduino developers make a critical mistake. They either expect the IDE to tell them exactly what is wrong with their code, or they expect to find the answer on the Internet or from the author of the library.[12]

Tip Compilation errors happen to everyone regardless of experience!

As you will see, debugging skills are general in nature and can be applied to almost any sort of development whether for the Arduino or another platform.

Debugging

Some explain debugging as simply finding the problem and fixing it. However, that definition neither does the skill justice nor does it explain how to go about it. We will discuss some basic techniques in this section to help you get started debugging your code.

[12] I've lost count of the number of requests I have received to fix issues that have nothing to do with the libraries I've published.

Note There is no "wrong" way to debug a problem. While sometimes circuitous with blind alleys, the path to resolution isn't as critical as the resolution itself.

Debugging, whether for resolving problems in your sketch, an electrical circuit, a leaking plumbing, or a weird noise from machinery, has at its core two critical goals: identification and resolution. We want to find the problem and fix it – that's debugging. Of course, debugging a strange noise in your car requires an entirely different set of tools and tests. However, the goals are the same.

The techniques you can employ to debug your code are numerous, but the following basic steps are a good starting point. Experience will likely be the deciding factor on the best approach for a given set of problems. Rather than pour out computer science terminologies, let's explore some of the key steps in debugging. The following are just one set of steps you can use. Before you do anything, make a copy of your sketch so that you can return to its original form. So often, I've encountered situations where multiple changes have altered the sketch behavior to the point where you can no longer reproduce the problem or you've introduced a different problem.

Caution When debugging, make sure you maintain a copy of the original sketch to ensure you don't inadvertently introduce changes.

---

1. *Inspect*: First, you should inspect your code to make sure that there aren't any mistakes in the logic and flow such as the incorrect function called or incorrect data types or maybe logic errors in your conditional statements. Sometimes you can find the odd error in logic simply by looking at the code. For example, if you mistakenly added one

instead of subtracting one in some equation, it can make a world of difference. Similarly, if you use the wrong inequality in a conditional, that make a big difference as well. Look for these and fix as many as you can find and run your sketch again to see if you have fixed the problem and haven't introduced a new problem.

2. *Isolate*: The next step is to remove any parts of the code that do not contribute to or are not executed in the area where the issue occurs. For example, if you have a complex sketch with many parts, try removing all the parts that are not needed to isolate the area where the problem occurs. This can be challenging, and we'll talk more about this in greater detail, but the point is to make sure that you remove any code that is not part of the problem. By doing this, you ensure that you can focus on the issue.

3. *Reproduce*: Once you have isolated the code to where the problem occurs, you can then set about trying to reproduce it. If your sketch is large with many parts and you've eliminated all the other parts except the one that has the issue, you may need to write additional code to supply missing values. For example, if the issue relies on another part of the code for input, you may need to write some code to supply dummy data (known good values) in order to reproduce the problem. More specifically, you may need to write code that initializes variables, supplies data from a sensor, etc. Once you are able to reproduce the problem, you can then work on trying to solve it.

4. *Experiment*: This step is where we spend our most time. Here, we want to try to figure out what went wrong and why. More specifically, we have the problem isolated and reproducible so that we can formulate possible fixes. The goal is to try to come up with one or more potential solutions for the problem. We then implement each one and see if the problem is resolved. This is another area where the novice can get stuck. Make sure you try one and only one change at a time, and if it doesn't fix the problem, return the code to the way it was in the previous step. Why? Because introducing multiple changes at the same time does not guarantee you are fixing the problem without introducing another! Experimenting with solutions can require several iterations to fix. However, recall at this point we are using the isolated code, not the complete sketch.

Caution Change one thing at a time! If it doesn't fix the problem, revert the code to its original form. Do not attempt to introduce multiple changes.

---

5. *Test*: Next, now that we have identified a solution, we want to attempt to solve the problem in the original sketch. Here is where we take the results of our experimentation and start adding back portions of the original sketch – one part at a time. We then test each iteration to ensure the problem has been fixed. While some may want to jump directly to making the change in the original sketch, it is important to rebuild your sketch with the proposed solution

to ensure the solution doesn't break another part of the code. If it does, go back two steps and try to figure out why. This is where the novice can become frustrated with the iterative nature of debugging complex sketches, but the methodical approach will help you reach a resolution.

6. *Resolve*: Now that we have corrected the problem and we're able to ensure the fix works, we can build it into our original sketch and test it again.

Now that you know what debugging is and how to go about it, let's talk about that testing thing that we mentioned.

Testing

Software testing is an area that has become more art than science. In fact, software testing and related fields (e.g., quality assurance) are subdisciplines of software development. Entire careers can be made in these areas. As you can imagine, there is a lot to learn about software test. However, for our purposes, testing is simply making sure your sketch is working correctly.

This may include simply running the sketch and observing its behavior, or it may require more careful instrumentation of the code. For your IoT projects, you're most likely going to simply run the sketch and ensure that it is reading sensors correctly and displaying the correct output. In case you're curious, this is known as black box testing.[13] In other words, you don't know how things work internally, but you can observe external behavior.

[13] https://en.wikipedia.org/wiki/Black-box_testing

When combined with debugging, testing may also require looking at certain values that occur in your code at key locations in order to resolve the problem. For example, we may want to introduce print statements to display the values of variables. This is known as white box texting[14] where you inspect the sketch as it runs internally. The following shows an example of print statements that are written to the serial monitor to inspect values. You would insert these in your code for debugging, but remove them once the problem is resolved:

```
ret_val = readSensor();
Serial.print("Value read from sensor: ");
Serial.println(ret_val);
```

How much testing you need to do to ensure that the sketch is working correctly depends on how your project is going to be used. For example, if you are building something to demonstrate to someone or maybe to experiment to find out if something is possible, you are likely to not do much testing other than to make sure it works – lights light up, sensors are read, displays show the correct data, etc.

However, if your project is something that's going to be used by other people or it is intended to run for a long period of time, then you would want to test it more thoroughly. For example, you may want to ensure that your project is going to work under all the conditions that it may encounter while running.

Testing of this nature is best accomplished by simply running the sketch and observing how it runs under real-world conditions (or as close as possible). For example, if you are building a project to measure freezing temperatures, testing it in your living room is hardly adequate (but maybe a few hours in your freezer might be closer).

---

[14] https://en.wikipedia.org/wiki/White-box_testing

Accordingly, you may not be able to test all conditions; more specifically, you may not be able to simulate rain, temperature ranges, or any kind of similar environmental condition. In those cases, you may want to create additional copies of your sketch where you use additional code to inject certain data to simulate those conditions.

For example, you can add lines of code that return extreme values in place of your sensors so that it appears that the sensor is operating in whatever condition that you require. That way, you can test your sketch to ensure that any calculations or resulting output is correct for the state. This is also known as fault injection where we purposely introduce errors into our code to see how it performs, but in this case, we are introducing environmental changes in order to simulate conditions without having to wait until those conditions occur naturally. This technique is great for the most complex of IoT projects.

Whether you are simply running your sketch or have introduced simulated data, if it does not do what you expect, you can take what you have learned and begin debugging the problem.

For example, sometimes a sketch will work perfectly for expected results, but may suddenly fail for unexpected results. For example, if a sensor returns a value outside of an expected range, it may cause your sketch to fail. Similarly, your sketch may run fine for a few hours or days but suddenly stop. In this case, you're most likely encountering problems with the hardware or possibly running out of memory for your variables.

Whatever the case, the testing is used to ensure your sketch is working by running your sketch as often as you can under as many real-work conditions as you can simulate to ensure that it works for its intended purpose and conditions. If it fails, go back and try to figure out why and fix it and then test the sketch again until it works.

Now you see why software testing is an art. While there are a lot of methods, practices, and principles one can learn and employ, there is somewhat of an art to ensuring a project works the way it should.

We briefly discussed debugging and testing, and now we understand how to go about isolating our code, reducing it to the nearest essentials to ensure that it's working. However, there are times when we may not know what is wrong or you know what is wrong but simply can't figure out how to fix it. In those cases, we can turn to the community and ask for help. Fortunately, there are a number of Arduino supporting communities out there that you can use to get your answers.

Getting Help

Despite being careful to build, debug, and test your sketch in a stepwise fashion, you may still encounter an issue you cannot easily solve. Don't worry if this happens (it happens to us all). Rather than blame someone else like the author of the library you're using (or the nice folks at `arduino.cc`), take some time to fully understand the problem as best as you can.

This is where we employ what we have learned from debugging and testing. More specifically, there is a process you can follow to help you get your answers. While the process is very similar to debugging and testing, the process usually occurs after you've attempted to fix the problem.

The goal here is to get help fixing a problem, but you must first be able to successfully communicate the problem. There are three steps: isolate and reproduce, research potential solutions, and ask the right question.

If you follow this process, you should be able to solve your problem quickly. It all begins with pruning your code down to the essential issue. If this reads a lot like debugging, that isn't a coincidence. Recall the second and third steps in debugging are isolate and reproduce.

Isolate and Reproduce

First, isolate your code to the most essential lines of code to reproduce the problem. Don't submit thousands of lines of code on a forum and expect people to read and debug it for you. They won't take the time. However, if you have isolated the problem to a few dozen lines, more people would be inclined to help.

This is the hardest part, but the most crucial. Why? Because sketches that include one or more devices or software libraries or complex logic can fail in most unexpected ways. Even if you've read all of the documentation, there is still the possibility you will encounter something that doesn't work as expected.

When this happens, you need to dissect your code. The first effort should be to remove any nonessential code, that is, code that is not needed for the device or software library to work. Isolate this code to a new sketch. If you've done it correctly, your resulting sketch should resemble one of the example sketches (but don't rely on the example sketches - use your own code as you could have a logic error or something different in your code).

However, don't lose sight of the goal - to reproduce the problem so that we can identify potential solutions.

Research Potential Solutions

Second, do your research in advance of asking a question. While you may end up with more questions than answers or sometimes conflicting answers, you should at least be able to understand what may work vs. what should not work. For example, if you are encountering a problem with an Arduino class or library, knowing how others have worked with and fixed similar issues will either provide you an answer or at least educate you into what won't fix the problem. It is important to note this is where you query the very forums where you want to seek help to ensure you're not repeating a question.

This is one area where novice developers fail the most. They simply don't do their homework. There is a plethora of information on the Internet that you can use to your advantage, but your job is to find it.

It is important that you first review as many solutions as you can find and try them before reaching out for help. You will avoid a lot of annoying "see <url>" responses that way.

Ask the Right Question

Third, once you've isolated the code to a reproducible snippet and have done your homework, you can now ask intelligent questions of the community. Visit `forums.arduino.cc` or a similar Arduino forum and state your problem in as few sentences as possible to completely describe the problem, list what you have tried to do to fix the problem, and paste in as small a snippet as possible that reproduces the problem.

For example, don't be this guy: "I wrote this sketch (containing 1200+ lines of code pasted into the channel), and it doesn't work. The library is broken, and I need it to pass my XYZ class or I won't graduate."

OK. What does a statement like that tell us? Nothing, sadly. The individual should have isolated the problem to as few lines as possible preferably in a new sketch, researched the potential solutions, and reported only those items. Things like why you wrote the sketch, what it means to your class grade, etc. are not helpful and should be avoided.

When you get a response (don't expect one to occur immediately or even within a couple of days), be sure to read the responders' comments carefully and take their suggestions as just that. Don't assume everyone who responds has the correct answer to your issue. It may be the case that the responder's suggestion worked for them, but it may not work for you.

You may also contact the vendor for your board or hardware device or the developer of the software library if you have a mechanism for doing so, but the same preceding rules apply. A vendor or developer has a vested interest in helping you, but only if you first try to help yourself.

Above all, be nice and play well with others.

Summary

Learning how to program the Arduino is not as difficult as some may contend. As you have seen, the programming language is not overly complex. It retains enough "C"-like to be familiar to those who've programmed in C, C++, and similar languages, but not so much as to deter someone from learning without any such experience.

In fact, the Arduino platform with its programming language, its built-in libraries, and a growing list of third-party libraries has become very powerful and accessible for anyone who wants to work with microcontrollers to control hardware. That, I think, is what makes the Arduino the perfect choice for artists, students, and hobbyists.

In this chapter, we learned more about the Arduino IDE, its programming language, and how to write sketches. We explored several examples along the way and even took a short detour into some tips for writing good sketches.

In the next chapter, we will discover another development platform that you may want to use: the Raspberry Pi.

CHAPTER 4

Introducing the Raspberry Pi

The Raspberry Pi is one of the latest disruptive devices in recent years that has changed the way that we think about and design embedded solutions and the IoT. In fact, the Raspberry Pi has had tremendous success among hobbyists and enthusiasts. This is partly due to its low cost but also because it is a full-fledged computer running an open source operating system that has a wide audience: Linux.

In fact, given the popularity of the Raspberry Pi, it is likely that you will encounter example projects and resources that are written for or only work with the Raspberry Pi. Thus, learning more about the Raspberry Pi and its native environment allows you to leverage the plethora of data for the Raspberry Pi.

WHAT ABOUT THE PICO?

By now, you may have heard of a new Raspberry Pi board called the Pico. This is a small microcontroller board designed for IoT and other microcontroller applications. It runs on MicroPython and is rapidly becoming an excellent choice for learning microcontrollers and electronics. Look for my upcoming book *Beginning MicroPython* (Apress, late 2021) where I cover how to get started with this exciting new offering from Raspberry Pi.

C. Bell, *Beginning IoT Projects*, https://doi.org/10.1007/978-1-4842-7234-3_4

This chapter introduces the Raspberry Pi and explains how to set up and configure the Raspberry Pi using the Linux operating system. You'll also discover a few key concepts of how to work with Linux. Let us begin with an in-depth look at the Raspberry Pi.

What Is a Raspberry Pi?

The Raspberry Pi is a small, inexpensive personal computer, also called a low-cost computing board or simply low-cost computer. Although it lacks the capacity for memory expansion and can't accommodate onboard devices such as CD, DVD, and hard drives, it has everything a simple personal computer requires. That is, it has USB ports, an Ethernet port, HDMI, and even an audio connector for sound. There are various models, and some also include Bluetooth and WiFi!

The Raspberry Pi has a micro-SD drive that you can use to boot the computer into any of several Linux operating systems (the default operating system is called Raspberry Pi OS – formerly named Raspbian). All you need is an HDMI monitor (or DVI with an HDMI-to-DVI adapter), a USB keyboard and mouse, and a 5V power supply – and you're off and running.

The Raspberry Pi Model B boards cost as little as $35. They can be purchased online from electronics vendors such as SparkFun and Adafruit. Most vendors have a host of accessories that have been tested and verified to work with the Raspberry Pi. These include small monitors, miniature keyboards, and cases for protecting the board.

This section explores the origins of the Raspberry Pi, tours the hardware connections, and covers the accessories needed to get started using the Raspberry Pi. Fortunately, you can use the same keyboard, mouse, and monitor as you would for a typical PC.

Raspberry Pi Origins

The Raspberry Pi was designed to be a platform to explore topics in computer science. The designers saw the need to provide inexpensive, accessible computers that could be programmed to interact with hardware such as servomotors, display devices, and sensors. They also wanted to break the mold of having to spend hundreds of dollars on a personal computer and thus make computers available to a much wider audience.

The designers observed a decline in the experience of students entering computer science curriculums. Instead of having some experience in programming or hardware, students are entering their academic years having little or no experience working with computer systems, hardware, or programming. Rather, students are well versed in Internet technologies and applications. One of the contributing factors cited is the higher cost and greater sophistication of the personal computer, which means parents are reluctant to let their children experiment on the family PC.

This poses a challenge to academic institutions, which have to adjust their curriculums to make computer science palatable to students. They have had to abandon lower-level hardware and software topics due to students' lack of interest or ability. Students no longer wish to study the fundamentals of computer science such as assembly language, operating systems, theory of computation, hardware, and concurrent programming. Rather, they want to learn higher-level languages to develop applications and web services. Thus, some academic institutions are no longer offering courses in fundamental computer science.[1] This could lead to a loss of knowledge and skill sets in future generations of computer professionals.

To combat this trend, the designers of the Raspberry Pi felt that, equipped with the right platform, today's youth could return to experimenting with personal computers and electronics as in the days

[1] Sadly, my alma mater is a fine example of this decline.

when PCs required a much greater commitment to learning the hardware, system components, and programming it in order to meet your needs. For example, the venerable Commodore 64, Amiga, and early Apple and IBM PC computers had very limited software offerings. Having owned a number of these machines, I was exposed to the wonder and discovery of hardware and programming at an early age. Perhaps that is why I find low-cost computing boards so fascinating – they pack a lot of features into a tiny board.

WHY IS IT CALLED RASPBERRY PI?

The name was partly derived from design committee contributions and partly chosen to continue a tradition of naming new computing platforms after fruits (think about it). The Pi portion comes from Python, because the designers intended Python to be the language of choice for programming the computer. However, other programming language choices are available.

The Raspberry Pi is an attempt to provide an inexpensive platform that encourages experimentation. The following sections further explore the Raspberry Pi, discussing topics such as the required accessories and where to buy the boards.

Raspberry Pi Boards

There are currently several versions of Raspberry Pi boards with two model classifications: Model A and Model B. The early Model A boards were the first mass-produced boards with 256MB of RAM, one USB port, and no Ethernet port. This was followed closely by the first Model B board, which had 512MB of RAM, two USB ports, and an Ethernet port. Figure 4-1 shows the version 3 variant of the Model A board designated as Raspberry Pi 3A+.

***Figure 4-1.** Raspberry Pi 3A+ (courtesy of the Raspberry Pi Foundation)*

WHAT DOES THE "+" MEAN?

The "+" symbol in the model designation indicates it is a newer release of the same version only with some improvements. For example, the 3B+ included a slightly faster processor and a host of minor refinements. Typically, the boards are effectively the same, and you may not notice a difference, but if you want the "latest" or "better" board, you'll want the one with the "+" designation.

The latest boards include the Raspberry Pi 3 Model B and Raspberry Pi 4 Model B. The Raspberry Pi 3B and 4B are very similar. In fact, they are very hard to tell apart without reading the label on the top of the board. This is because they share the same layout (Model B) with the same connectors. While the boards appear nearly identical, the Raspberry Pi 4B has a faster 64-bit quad-core processor, uses USB-C for the power source, and is available with 2, 4, or 8GB of RAM (vs. 1GB for the 3B). There are a number of smaller changes, but these are by far the most significant differences. Let's look at these boards in more detail.

Figure 4-2 shows the version 3 Model B board designated as Raspberry Pi 3B+. Notice the board is a bit larger and has more connections.

Figure 4-2. *Raspberry Pi 3B+ (courtesy of the Raspberry Pi Foundation)*

Figure 4-3 shows the latest Model B board designated as the Raspberry Pi 4B. The figure depicts some of the improvements from the 3B+ model including more RAM, USB-C power, two HDMI ports, and USB-3 support. Plus, it is the fastest Raspberry Pi computer to date!

Figure 4-3. *Raspberry Pi 4B*

You can often find the Raspberry Pi 3A+ at online retailers and auction sites for a bit less than the Raspberry Pi 3B+ board, which can be found for less than the Raspberry Pi 4B. The newest Raspberry Pi 4B boards are still in demand, so you may pay more for those boards, but shop around to find retailers that offer the board at suggested retail prices of $35 (2GB), $55 (4GB), and $75 (8GB). If you plan to use the Raspberry Pi for experimentation and do not need the extra memory to run memory-intensive applications, you can use the Raspberry Pi 3A+.

There is a new offering that may be of interest to some. It's called the Raspberry Pi 400, which is an official Raspberry Pi keyboard with the equivalent of a Raspberry Pi 4B 4GB board built in. This lists for around $70 and can be found anywhere Raspberry Pi is sold. Figure 4-4 shows the Raspberry Pi 400.

Figure 4-4. *Raspberry Pi 400*

Online retailers often offer Raspberry Pi bundles that include everything you need to get started. For example, you can get a Raspberry Pi 400 starter kit for about $125 where all you need to add is an HDMI monitor (`https://thepihut.com/collections/raspberry-pi-kits-and-bundles/products/raspberry-pi-400-personal-computer-kit`).

Tip It is recommended to use the Raspberry Pi 3B+ or the newest Raspberry Pi 4B for the projects in this book. The examples in the remaining chapters use the Model B variant – either the Raspberry Pi 3B+ or 4B.

A Tour of the Board

Not much larger than a deck of playing cards, the Raspberry Pi board contains a number of ports for connecting devices. This section presents a tour of the board. If you want to follow along with your board, hold it with the Raspberry Pi logo faceup. I work around the board clockwise. Figure 4-5 depicts the board with all the major connectors labeled.

***Figure 4-5.** Raspberry Pi 3 Model B (courtesy of raspberrypi.org)*

Let's begin by looking at the bottom edge of the board (looking from above). In the center of the bottom side, you see two HDMI connectors (for dual monitors). To the left of the HDMI connector is the USB-C power connector. The power connector is known to be a bit fragile on some boards, so take care plugging and unplugging it. Be sure to avoid putting extra strain on this cable while using your Raspberry Pi. To the right of the HDMI connector is the camera ribbon cable connector, and next to that is the audio connector.

On the left side of the board is the LCD ribbon cable connector. You can use this connector with the Raspberry Pi 7-inch touch LCD and similar devices. On the underside of the board is the micro-SD card drive. When installed, the SD card protrudes a few millimeters out of the board. If you plan to use a case for your Raspberry Pi, be sure the case provides access to the SD card drive (some do not).

On the top edge of the board is the general-purpose input/output (GPIO) header (a double row of 20 pins each), which can be used to attach to sensors and other electronic components and devices. You will work with this connector later in this chapter. On the right side of the board are two USB connectors with two USB 2 and two USB 3 ports and the Gigabit Ethernet connector.

Take a moment to examine the top and bottom faces of the board. As you can see, components are mounted on both sides. This is a departure from most printed circuit boards (PCBs) that have components on only one side. The primary reason the Raspberry Pi has components on both sides is that it uses multiple layers for trace runs (the connecting wires on the board). Stacking the trace runs on multiple levels means that you don't have to worry about crossing paths. It also permits the board to be much smaller and enables the use of both surfaces. This is probably the most compelling reason to consider using a case – to protect the components on the bottom of the board and thus avoid shorts (accidental connection of contacts or pins) that can lead to board failure.

Caution Because the board is small, it is tempting to use it in precarious places, like in a moving vehicle or on a messy desk. Ensure that your Raspberry Pi is in a secure location. The USB power, HDMI, and SD card slots seem to be the most vulnerable connectors.

Required Accessories

The Raspberry Pi is sold as a bare system board with no case, power supply, or peripherals. Depending on how you plan to use the Raspberry Pi, you need a few commonly available accessories. If you have been accumulating computer and electronic spares like me, a quick rummage through your stores may locate most of what you need.

If you want to use the Raspberry Pi in console mode (no graphical user interface), you need a USB power supply, a keyboard, and an HDMI monitor. The power supply should have a minimal rating of 2.5A or greater. If you want to use the Raspberry Pi with a graphical user interface, you also need a pointing device (such as a mouse).

If you have to purchase these items, stick to the commonly available brands and models without extra features. For example, avoid the latest multifunction keyboard and mouse because they may require drivers that are not available for the various operating system choices for the Raspberry Pi.

You also must have a micro-SD card. I recommend a 16GB or higher version. Recall that the micro-SD is the only on-board storage medium available. You need to put the operating system on the card, and any files you create are stored on the card.

If you want to use sound in your applications, you also need a set of powered speakers that accept a standard 3.5mm audio jack. Finally, if you want to connect your Raspberry Pi to the Internet, you need an Ethernet cable, or if you are using a Raspberry Pi 3B or 4B, you need a WiFi network.

Recommended Accessories

I highly recommend, at a minimum, adding small 5–10mm rubber or silicone self-adhesive bumpers to the bottom side of the board over the mounting holes to keep the board off your desk. On the bottom of the board are many sharp prongs that can come into contact with conductive materials, which can lead to shorts or, worse, a blown Raspberry Pi. They can also damage your desktop, skin, and clothing. Small self-adhesive bumpers are available at most home improvement and hardware stores.

If you plan to move the board from room to room or you want to ensure that your Raspberry Pi is well protected against accidental damage, you should consider purchasing a case to house the board. Many cases are available, ranging from simple snap-together models to models made from laser-cut acrylic or even milled aluminum.

Tip If you plan to experiment with the GPIO pins or require access to the power test pins or the other ports located on the interior of the board, you may want to consider either using the self-adhesive bumper option or ordering a case that has an open top to make access easier. Some cases are prone to breakage if opened and closed frequently.

Aside from a case, you should also consider purchasing (or pulling from your spares) a powered USB hub. The USB hub power module should be 2–2.5A or more. Even though the Raspberry Pi 2 and 3 have four USB ports, a powered hub is required if you plan to use USB devices that draw a lot of power, such as a USB hard drive or a USB toy missile launcher.

Where to Buy

The Raspberry Pi boards are plentiful and can be found on many websites serving many continents. Chances are there is an online store available near you. To find out, go to `www.raspberrypi.org/products/raspberry-pi-4-model-b/` and click *Buy now* and use the drop-down list to look for a country or city near you. Fortunately, those online retailers who stock it offer a host of accessories that are known to work with the Raspberry Pi. The following are some of the more popular online retailers with links to their Raspberry Pi catalog entry:[2]

- *Adafruit*: `www.adafruit.com/category/105`
- *SparkFun*: `www.sparkfun.com/categories/233`

---

[2] You can often find the 2 and 3B listed, but quantities may be limited. Online auction sites sometimes have excellent deals on the older boards.

- *The Pi Hut*: `thepihut.com`
- *Pi Shop US*: `pishop.us`

Recall you can also find a host of Raspberry Pi offerings from Micro Center. If you have one of those stores near you or within driving distance, I highly recommend the trip.

The next section presents a short tutorial on getting started using the Raspberry Pi. If you have already learned how to use the Raspberry Pi, you can skip to the following section to begin learning how to use your board.

Setting Up the Raspberry Pi

The Raspberry Pi is a personal computer with a surprising amount of power and versatility. You may be tempted to consider it a toy or a severely limited platform, but that is far from the truth.[3] With the addition of onboard peripherals like USB, Ethernet, and HDMI video (as well as Bluetooth and WiFi for the Raspberry Pi 3 and later), the Raspberry Pi has everything you need for a lightweight desktop computer. If you consider the addition of the GPIO header, the Raspberry Pi becomes more than a simple desktop computer and fulfills its role as a computing system designed to promote hardware experimentation.

The following sections present a short tutorial on getting started with your new Raspberry Pi, from a bare board to a fully operational platform. A number of excellent works cover this topic in much greater detail. If you find yourself stuck or wanting to know more about beginning to use the Raspberry Pi and more about Raspberry Pi OS, read *Learn Raspberry Pi with Linux* by Peter Membrey and David Hows (Apress, 2012). If you want to know more about using the Raspberry Pi in hardware projects, excellent resources include *Practical Raspberry Pi* by Brendan Horan (Apress, 2013) and *Computing with the Raspberry Pi* by Brian Schell (Apress, 2019).

[3] Especially considering the Raspberry Pi 4B 4GB and 8GB versions.

As mentioned in the "Required Accessories" section, you need a micro-SD card of 16GB or larger, a USB power supply rated at 2.5A or better with a male micro-USB connector (for the Raspberry Pi 3B) or USB-C connector (for the Raspberry Pi 4B and variants), a keyboard, a mouse, and an HDMI monitor. However, before you can boot your Raspberry Pi and bask in its brilliance, you need to create a boot image for your micro-SD card.

Choosing a Boot Image (Operating System)

The first thing you need to do is decide which operating system variant you want to use. There are several excellent choices, including the standard Raspberry Pi OS. Each is available as a compressed file called an image or card image. You can find a list of recommended images along with links to download each on the Raspberry Pi Foundation download page: `www.raspberrypi.org/software/operating-systems/`. The following images are available at the site:

- *Raspberry Pi OS*: A Debian-based official operating system and contains a graphical user interface, development tools, and rudimentary multimedia features.
- *Ubuntu Desktop*: Features the Ubuntu desktop and a scaled-down version of the Ubuntu operating system. If you are familiar with Ubuntu, you will feel at home with this version.
- *Ubuntu Core*: The developer's edition of core Ubuntu system. It is the same as Mate with addition of the developer core utilities.
- *Ubuntu Server*: The developer's edition of core Ubuntu server system.

- *LibreELEC (Libre Embedded Linux Entertainment Center)*: A Linux-based platform for turning your Raspberry Pi into a media center.
- *RetroPie*: A Unix-like operating system that allows you to turn your Raspberry Pi intro a retro arcade gaming rig.

Tip If you are just starting with the Raspberry Pi and haven't used a Linux operating system, you should use the Raspberry Pi OS image as it is the most popular choice and more widely documented in examples.

There are a few other image choices, including a special variant of the Raspberry Pi OS image from Adafruit. Adafruit calls their image "Occidentalis" that includes a number of applications and utilities preinstalled, including WiFi support and several utilities. Some Raspberry Pi examples - especially those from Adafruit - require the Occidentalis image. You can find out more about the image and download it at `http://learn.adafruit.com/adafruit-raspberry-pi-educational-linux-distro/overview`. Before choosing an alternative image, check to see if the vendor is actively supporting it or is planning future releases. Some images have been abandoned or are not updated as regularly as Raspberry Pi OS.

Wow! That's a lot of choices, isn't it? As you can see, the popularity of the Raspberry Pi is very wide and diverse. While you may not use these operating systems, it is good to know what choices are available should you need to explore them.

Now let's see how to install the base operating system. As you will see, it is very easy.

Creating the Boot Image

There are several methods to get your Raspberry Pi boot image created, but for this book, we will use the recommended method via the Raspberry Pi Imager software from raspberrypi.org. This special software will allow you to choose the operating system as well as the SD card, which it will format, and download your choice onto the SD card. Figure 4-6 shows the Raspberry Pi Imager software.

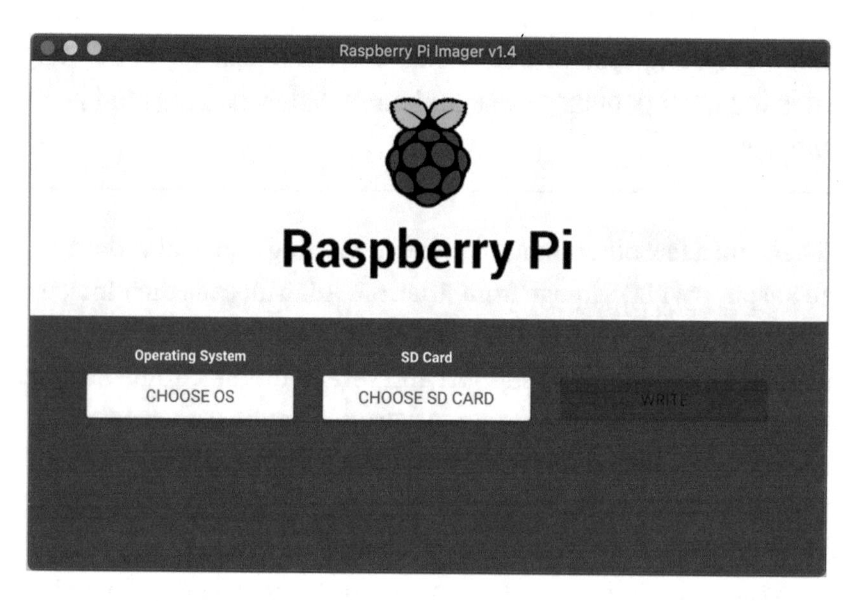

Figure 4-6. *Raspberry Pi Imager (macOS)*

The Imager software is available for most platforms including macOS, Windows, and Linux (Ubuntu). Simply visit `www.raspberrypi.org/software/` and download and then install the version for your PC.

Once you've installed Imager, open it and click the *CHOOSE OS* button. You will be presented with a number of choices as outlined previously. Select the top entry for Raspberry Pi OS (32-bit) as shown in Figure 4-7.

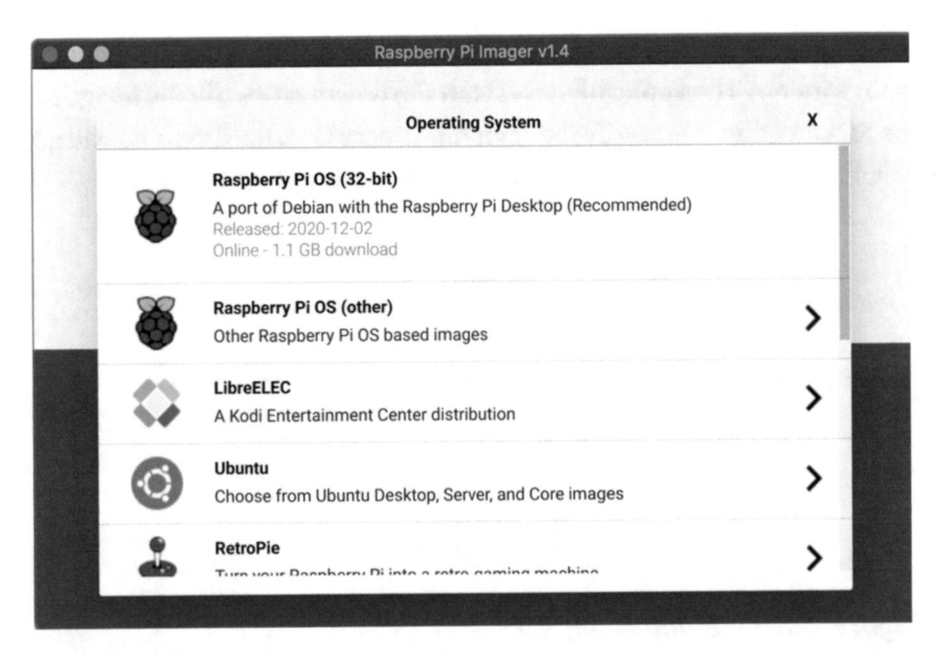

Figure 4-7. *Selecting the OS in Raspberry Pi Imager*

Next, insert your SD card and select it using the *CHOOSE SD CARD* button. If you do not see your SD card in the list, try reinserting it and try again. Once you've selected the SD card, you will see a warning that the card will be erased as shown in Figure 4-8. Click *YES* to acknowledge the box and continue.

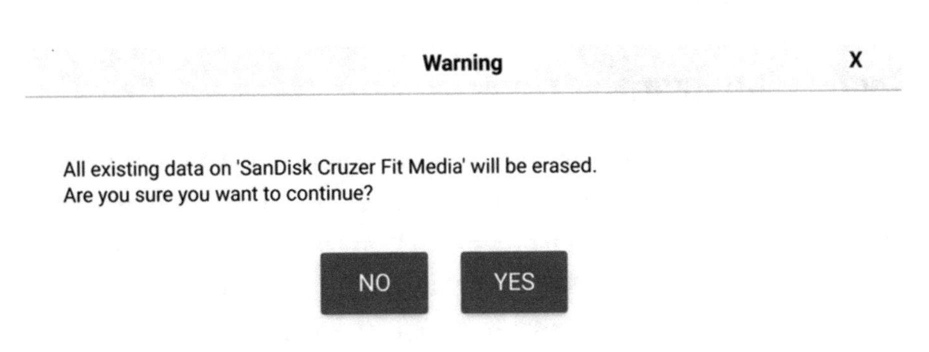

Figure 4-8. *Choosing the SD Card in Raspberry Pi Imager*

To download and write the image to the SD card, click *WRITE* in the main window. The application will start the download, write the image to the SD card, and then verify it. When this process has finished, you will get a dialog similar to Figure 4-9 stating it is safe to remove the SD card.

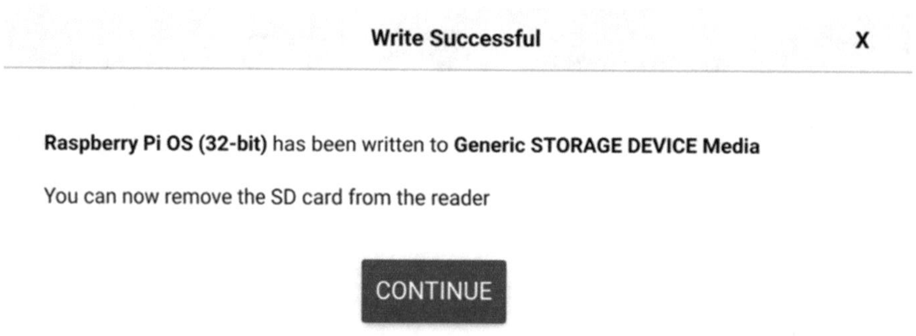

Write Successful X

Raspberry Pi OS (32-bit) has been written to **Generic STORAGE DEVICE Media**

You can now remove the SD card from the reader

CONTINUE

Figure 4-9. *Operation complete*

Tip For those that want to explore the other OS options for the Raspberry Pi, visit `www.raspberrypi.org/software/operating-systems/` for additional information. You will also find three options for installing Raspberry Pi OS on the site: a full version, the version used previously, and a lite version for smaller boards.

Booting the Board

You are now ready to hook up all of your peripherals. I like to keep things simple and only connect a monitor, keyboard, and mouse. If you want to download an operating system other than Raspberry Pi OS, you also need to connect your Raspberry Pi to your network. If you are planning to use the WiFi option on the Raspberry Pi 3B or later, you'll need to set up your WiFi configuration after you boot up. I'll show you how to do that in a moment.

When you power on the Raspberry Pi, the system bootstraps and then starts loading the OS. You may also see a message about resizing the boot device, and your Raspberry Pi may reboot. This is automatic and nothing to be concerned about. In fact, it is ensuring the boot volume is expanded to the maximum size your micro-SD supports.

Once it reboots, you may see a dialog with a list of statements that communicate the status of each subsystem as it is loaded followed by a welcome banner. You don't have to try to read or even understand all the rows presented,[4] but you should pay attention to any errors or warnings. When the boot sequence is complete, you will see the Raspberry Pi OS desktop as shown in Figure 4-10.

Figure 4-10. *Raspberry Pi OS desktop*

[4] They go by so fast it is unlikely you can read them anyway. Basically, they're noise unless there is an error, and those usually appear in the last few lines displayed.

Notice there is a dialog open in the center. Once again, these steps will execute only once on first boot. The steps include the following.

- *Welcome*: Click *Next* to start the setup. You can cancel and run the setup later.
- *Set Country*: Choose your country, language, and time zone. Click *Next* to continue.
- *Set/Change Password*: Choose the password for the default user. Click *Next* to continue.
- *Set Up Screen*: If your screen shows a black rectangle around the edge, you can tick the checkbox to have the video adapter synchronize properly on next boot. Click *Next* to continue.
- *Select WiFi Network*: Choose your WiFi access point to connect to the Internet. You can click *Skip* to skip the step or click *Next* to continue.
- *Update Software*: If you have connected to the Internet, you can optionally download and install updates for the system. This is highly recommend, and when you choose this option, you will go through several more informational dialogs that show you the progress of the updates. You can click *Skip* to skip the step. Click *Next* to continue when done.
- *Setup Complete*: The setup is done. Click *Next* to continue, and if you selected any options that require a reboot, the system will reboot now.

Figure 4-11 shows each step starting from the upper left working left to right.

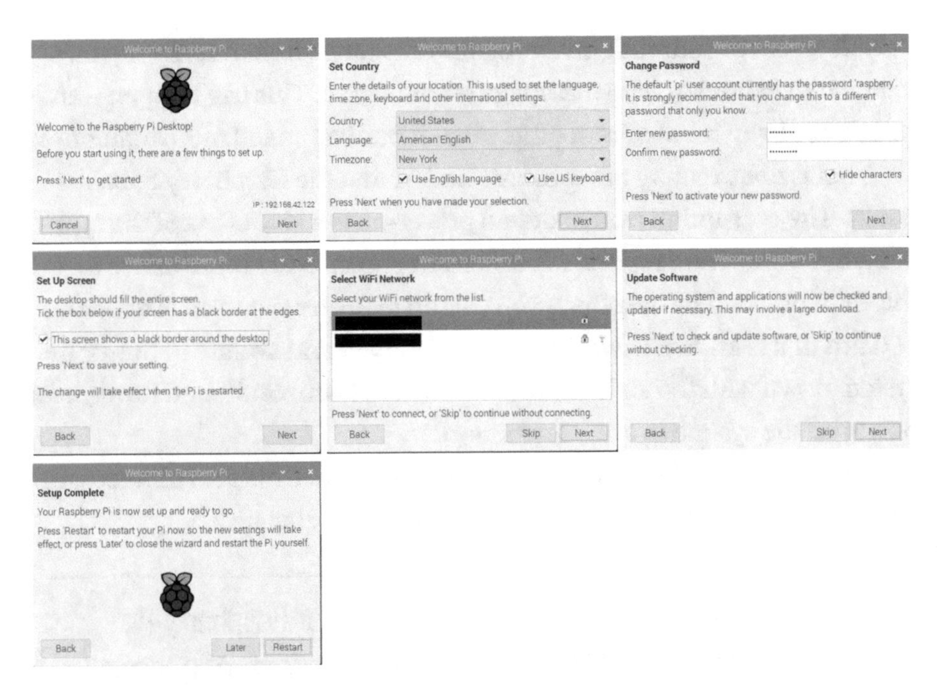

Figure 4-11. *First boot setup sequence*

When the system next boots, you will see the Raspberry Pi OS desktop with your settings configured. If you set up a WiFi connection, it will automatically reconnect. Nice.

Care and Feeding of the SD Card

Imagine this scenario. You're working away on creating files, downloading documents, and so on. Your productivity is high, and you're enjoying your new low-cost, supercool Raspberry Pi. Now imagine the power cable accidentally gets kicked out of the wall, and your Raspberry Pi loses power. No big deal, yes? Well, most of the time.

The SD card is not as robust as your hard drive. You may already know that it is unwise to power off a Linux system abruptly, because doing so can cause file corruption. Well, on the Raspberry Pi, it can cause a complete loss of your disk image. Symptoms range from minor read errors to inability to boot or load the image on bootstrap. This can happen - and there have been reports from others that it has happened more than once.

That is not to say all SD cards are bad or that the Raspberry Pi has issues. The corruption on accidental power-off is a side effect of the type of media. Some have reported that certain SD cards are more prone to this than others. The best thing you can do to protect yourself is to use an SD card that is known to work with Raspberry Pi and be sure to power the system down with the `sudo shutdown -h now` command - and never, ever power off the system in any other manner.

You can also make a backup of your SD card. See `http://elinux.org/RPi_Beginners#Backup_your_SD_card` for more details.

Tip If you need any help at all when using your Raspberry Pi, there are very helpful articles at `www.raspberrypi.org/help/`, and the official documentation is at `www.raspberrypi.org/documentation/`.

To shut down or reboot Raspberry Pi OS, click the Pi, and then choose *Logout* and then *Shutdown*. If you want to shut down from a terminal (command line), use the command `shutdown -h now` to shut down the system.

Getting Started with Raspberry Pi OS

OK, now you have a Raspberry Pi booting Linux (Raspberry Pi OS) into the desktop environment. And although it looks cool, it can be a bit confusing and intimidating. The best way to learn the GUI is to simply spend some time clicking your way through the menus. You'll find the most basic of features, including productivity tools.

However, working with hardware typically requires knowledge of basic commands used in a terminal (also called the command line). This section describes a number of the more basic commands you need to use. This is by no means meant to be a complete or thorough coverage of all of the commands. Rather, it gives you the basics that you need to get started.

Thus, this primer is more like a 10-minute guided tour of an automobile. You cannot possibly learn all of the maintenance requirements and internal components in 10 minutes. You would need to have an automotive technician's training or years of experience before you could begin to understand everything. What you get in a lightning tour is more of a bird's-eye view with enough information to permit you to know where the basic maintenance items are located, not necessarily how they work.

I recommend you read through the following sections to familiarize yourself with the commands that you may need. You can refer back to these sections should you need to recall the command name. Oftentimes it is simply a matter of learning a different name for the same commands (conceptually) that you're familiar with from Windows. As you will see, many of these commands are familiar in concept, as they also exist on Windows albeit with a different name and parameters.

Tip If you want to master the Linux command-line commands, tools, and utilities, read the book *Beginning the Linux Command Line* by Sander van Vugt (Apress, 2015).

Let's begin with how to get help about commands.

Getting Help

Linux provides help for all commands by default. While it can be a bit terse, you can always get more information about a command by using the manual command as shown in Listing 4-1. Here you want more help with the list directories command (ls).

Listing 4-1. Getting Help from the man (Manual) Command

```
pi@raspberrypi:~ $ man ls

LS(1) User Commands LS(1)

NAME
  ls - list directory contents

SYNOPSIS
  ls [OPTION]... [FILE]...

DESCRIPTION
  List  information  about  the FILEs (the current directory by
  default).
  Sort entries alphabetically if none of -cftuvSUX nor
  --sort  is  speci fied.

  Mandatory  arguments  to  long  options are mandatory for
  short options too.

  -a, --all
   do not ignore entries starting with .

  -A, --almost-all
   do not list implied . and ..

...
```

File and Directory Commands

Like any operating system, some of the most basic commands are those that allow you to manipulate files and directories. These include operations such as copying, moving, and creating files and directories. I list a few of the most common commands in the following sections and provide an example of each. If you want to know more about each, try using the manual command (`man`). Just use the name of the command you want to know more about as the option. For example, to learn more about ls, enter man ls.

List Directories and Files

The first command you will likely need is the ability to list files and directories. In Linux, we use the `ls` (list files and directories) command. Without any options, the command lists all of the files and directories in the current location. There are many options available, but the ones I find most helpful are show long listing format (`-l`), sort the output (`-s`), and show all files (`-a`). You can combine these options in a single string, such as `-lsa`.

The command uses color and highlighting to help distinguish directories from files, executable files, and more. The long listing format also shows you the permissions for the file (the series of `rwx` values). The first character in the directory list refers to the file type (`d` means a directory and `-` is a regular file), the next three characters refer to file owner permissions, the next three are group permissions, and the final three are for other users' permissions. Figure 4-12 shows an example of the `ls -lsa` command output.

```
pi@raspberrypi:~/source $ ls -lsa
total 24
4 drwxr-xr-x  3 pi pi 4096 Mar 13 17:43 .
4 drwxr-xr-x 20 pi pi 4096 Mar 12 00:34 ..
4 -rw-r--r--  1 pi pi    8 Mar 13 17:39 me.txt
4 -rw-r--r--  1 pi pi    8 Mar 13 17:40 my.txt
4 drwxr-xr-x  2 pi pi 4096 Mar 12 00:37 python
4 -rw-r--r--  1 pi pi   16 Mar 13 17:40 that.txt
pi@raspberrypi:~/source $ 
```

Figure 4-12. *Output of list directories (ls) command*

Change Directory

You can change from one directory to another by using the cd command, which is quite familiar:

```
pi@raspberrypi:~ $ cd source
pi@raspberrypi:~/source $
```

Tip The Linux path separator is a /, which can take some getting used to.

Copy

You can copy files with the cp command with the usual expected parameters of <from_file> <to_file>, as shown next. You can also use full paths to copy files from one directory to another:

```
pi@raspberrypi:~/source $ ls
me.txt  python
pi@raspberrypi:~/source $ cp me.txt my.txt
pi@raspberrypi:~/source $ ls
me.txt  my.txt  python
```

Tip Use the * symbol as a wildcard to specify all files (synonymous with *.* in Windows). For example, to copy all of the files from one folder to another, use the cp ./old/* ./new command.

Move

If you want to move files from one folder to another, you can use the mv command with the usual expected parameters of <from> <to>, as shown next. You can also use full paths to move files from one directory to another:

```
pi@raspberrypi:~/source $ ls
me.txt  my.txt  python  this.txt
pi@raspberrypi:~/source $ mv this.txt that.txt
pi@raspberrypi:~/source $ ls
me.txt  my.txt  python  that.txt
```

Create Directories

Creating directories can be accomplished with the mkdir command. If you do not specify a path, the command executes in the current directory:

```
pi@raspberrypi:~/source $ ls
me.txt  my.txt  python  that.txt
pi@raspberrypi:~/source $ mkdir test
pi@raspberrypi:~/source $ ls
me.txt  my.txt  python  test  that.txt
```

Delete Directories

If you want to delete a directory, use the rmdir command. This command requires that the directory be empty. You will get an error if the directory contains any files or other directories:

```
pi@raspberrypi:~/source $ ls
me.txt  my.txt  python  test  that.txt
pi@raspberrypi:~/source $ rmdir test
pi@raspberrypi:~/source $ ls
me.txt  my.txt  python  that.txt
```

Create (Empty) Files

Sometimes you may want to create an empty file for use in logging output or just to create a placeholder for editing later. The touch command allows you to create an empty file:

```
pi@raspberrypi:~/source $ ls ./test
pi@raspberrypi:~/source $ touch ./test/new_file.txt
pi@raspberrypi:~/source $ ls ./test
new_file.txt
pi@raspberrypi:~/source $ rmdir test
rmdir: failed to remove 'test': Directory not empty
```

Delete Files

If you want to delete a file, use the rm command. There are a number of options for this command, including recursively deleting files in subfolders (-r) and options for more powerful (thorough) cleaning:

```
pi@raspberrypi:~/source $ rm ./test/new_file.txt
pi@raspberrypi:~/source $ ls ./test
```

Caution You can use the `rm` command with the force option to remove directories, but you should use such options with extreme caution. Executing `sudo rm * -rf` in a directory will permanently delete all files!

System Commands

The Linux operating system provides a huge list of system commands to do all manner of operations on the system. Mastering all of the system commands can take quite a while. Fortunately, there are only a few that you may want to learn to use Linux with a minimal of effort.

Show (Print) Working Directory

The system command I use most frequently is the print working directory (`pwd`) command. This shows you the full path to the current working directory:

```
pi@raspberrypi:~/new_source $ pwd
/home/pi/new_source
```

Command History

The one system command that you may find most interesting and helpful is the `history` command. This command lists the commands that you have entered over time. So, if you find that you need to issue some command you used a month ago, use the history command to show all of the commands executed until you find the one you need. This is especially helpful if you cannot remember the options and parameters! However, this list is only for the current user. The following is an excerpt of the history for my Raspberry Pi 3:

```
pi@raspberrypi:~/source $ history
  1  sudo apt-get update
  2  sudo apt-get upgrade
  3  sudo shutdown -r now
  4  rpi-update
  5  sudo
  6  sudo rpi-update
  7  sudo apt-get dist-upgrade
  8  sudo shutdown -r now
  9  startx
 10  ls /lib/firmware/brcm
...
```

Tip Use the *Up* and *Down* keys on the keyboard to call back the last command issued and scroll forward and backward through the history one command at a time.

Archive Files

Occasionally, you may need the ability to archive or unarchive files, which you can do with a system command (utility). The tape archive (`tar`) command shows the longevity of the Linux (and its cousin/predecessor, Unix) operating system from the days when offline storage included tape drives[5] (no disk drives existed at the time). The following shows how to create an archive and extract it. The first tar command creates the archive and the second extracts it:

```
pi@raspberrypi:~ $ tar -cvf archive.tar ./source/
./source/
./source/test/
./source/my.txt
./source/that.txt
./source/python/
./source/python/blink_me.py
./source/me.txt
pi@raspberrypi:~ $ mkdir new_source
pi@raspberrypi:~ $ cd new_source
pi@raspberrypi:~/new_source $ tar -xvf ../archive.tar
./source/
./source/test/
./source/my.txt
./source/that.txt
./source/python/
./source/python/blink_me.py
./source/me.txt
pi@raspberrypi:~/new_source $ ls
source
```

There are a host of options for the tape archive command. The most basic are the create (-cvf) and extract (-xvf) option strings, as shown in the preceding code. See the manual for the tape archive command if you want to perform more complicated operations.

[5] Anyone remember punch cards?

Administrative Commands

Like the system commands, there is a long list of administrative operations that you may need to perform. I list those commands that you may need to perform more advanced operations, starting with the run as administrator equivalent command.

Run as Super User

To run a command with elevated privileges, use the sudo command. Some commands and utilities require sudo. For example, to ping another computer, install software, change permissions, and so forth, you need elevated privileges:

```
pi@raspberrypi:~/new_source $ sudo ping localhost
PING localhost (127.0.0.1) 56(84) bytes of data.
64 bytes from localhost (127.0.0.1): icmp_seq=1 ttl=64
time=0.083 ms
64 bytes from localhost (127.0.0.1): icmp_seq=2 ttl=64
time=0.068 ms
64 bytes from localhost (127.0.0.1): icmp_seq=3 ttl=64
time=0.047 ms
^C
--- localhost ping statistics ---
3 packets transmitted, 3 received, 0% packet loss, time 1998ms
rtt min/avg/max/mdev = 0.047/0.066/0.083/0.014 ms
```

Change File/Directory Permissions

In Linux, files and directories have permissions, as described in the previous section. You can see the permissions with the list directories command. To change the permissions, use the chmod command as shown in the following code. Here we use a series of numbers to determine the bits of the permissions. That is, 7 means rwx, 6 means rw, and so forth. For a complete list of these numbers and an alternative form of notation, see the manual for chmod:[6]

```
pi@raspberrypi:~/source $ ls -lsa
total 12
4 drwxr-xr-x  3 pi pi 4096 Mar 13 18:07 .
4 drwxr-xr-x 21 pi pi 4096 Mar 13 18:02 ..
0 -rw-r--r--  1 pi pi 0 Mar 13 18:07 cmd
4 drwxr-xr-x  2 pi pi 4096 Mar 12 00:37 python
pi@raspberrypi:~/source $ chmod 0777 cmd
pi@raspberrypi:~/source $ ls -lsa
total 12
4 drwxr-xr-x  3 pi pi 4096 Mar 13 18:07 .
4 drwxr-xr-x 21 pi pi 4096 Mar 13 18:02 ..
0 -rwxrwxrwx  1 pi pi 0 Mar 13 18:07 cmd
4 drwxr-xr-x  2 pi pi 4096 Mar 12 00:37 python
```

Change Owner

Similarly, you can change ownership of a file with the chown command if someone else created the file (or took ownership). You may not need to do this if you never create user accounts on your Raspberry Pi, but you should be aware of how to do this in order to install some software such as MySQL:

```
pi@raspberrypi:~/source $ ls -lsa
total 12
4 drwxr-xr-x  3 pi pi 4096 Mar 13 18:07 .
4 drwxr-xr-x 21 pi pi 4096 Mar 13 18:02 ..
```

[6] For more information, see the "Numerical permissions" section at https://en.wikipedia.org/wiki/Chmod

```
0 -rwxrwxrwx  1 pi pi 0 Mar 13 18:07 cmd
4 drwxr-xr-x  2 pi pi 4096 Mar 12 00:37 python
pi@raspberrypi:~/source $ sudo chown chuck cmd
pi@raspberrypi:~/source $ ls -lsa
total 12
4 drwxr-xr-x  3 pi pi 4096 Mar 13 18:13 .
4 drwxr-xr-x 21 pi pi 4096 Mar 13 18:02 ..
0 -rw-r--r--  1 chuck pi 0 Mar 13 18:13 cmd
4 drwxr-xr-x  2 pi pi 4096 Mar 12 00:37 python
pi@raspberrypi:~/source $ sudo chgrp chuck cmd
pi@raspberrypi:~/source $ ls -lsa
total 12
4 drwxr-xr-x  3 pi pi 4096 Mar 13 18:13 .
4 drwxr-xr-x 21 pi pi 4096 Mar 13 18:02 ..
0 -rw-r--r--  1 chuck chuck 0 Mar 13 18:13 cmd
4 drwxr-xr-x  2 pi pi 4096 Mar 12 00:37 python
```

Tip You can change the group with the `chgrp` command.

Install/Remove Software

The second most used administrative operation is installing or removing software. To do this on Raspberry Pi OS (and similar Linux distributions), you use the apt-get command, which requires elevated privileges.

Linux maintains a list of header files that contain the latest versions and locations of the source code repositories for all components installed on your system. Occasionally, you need to update these references, and you can do so with the following options. Do this before you install any software. In fact, most documentation for software requires you to run this command. You must be connected to the Internet before running the command, and it could take a few moments to run:

```
sudo apt-get update
```

To install software on Linux, you use the install option (conversely, you can remove software with the remove option). However, you must know the name of the software you want to install, which can be a challenge. Fortunately, most software providers tell you the name to use. Interestingly, this name can be the name of a group of software. For example, the following command initiates the installation of MySQL 5.5 (the latest version is 8.0), which involves a number of packages:

```
pi@raspberrypi:~/source $ sudo apt-get install mysql-server
Reading package lists... Done
Building dependency tree
Reading state information... Done
The following extra packages will be installed:
  libaio1 libdbd-mysql-perl libdbi-perl libhtml-template-perl
  libmysqlclient18
  libterm-readkey-perl mysql-client-5.5 mysql-common mysql-
  server-5.5
  mysql-server-core-5.5
Suggested packages:
  libclone-perl libmldbm-perl libnet-daemon-perl libsql-
  statement-perl
  libipc-sharedcache-perl mailx tinyca
The following NEW packages will be installed:
  libaio1 libdbd-mysql-perl libdbi-perl libhtml-template-perl
  libmysqlclient18
  libterm-readkey-perl mysql-client-5.5 mysql-common mysql-
  server
  mysql-server-5.5 mysql-server-core-5.5
```

```
0 upgraded, 11 newly installed, 0 to remove and 0 not upgraded.
Need to get 8,121 kB of archives.
After this operation, 88.8 MB of additional disk space will be
used.
Do you want to continue? [Y/n]
```

Shutdown

Finally, you want to shut down your system when you are finished using it or perhaps reboot it for a variety of operations. For either operation, you need to run with elevated privileges (`sudo`) and use the shutdown command. This command takes several options: use `-r` for reboot and `-h` for halt (`shutdown`). You can also specify a time to perform the operation, but I always use the now option to initiate the command immediately.

To reboot the system, use this command:

```
sudo shutdown -r now
```

To shut down the system, use this command:

```
sudo shutdown -h now
```

Useful Utilities

There are a number of useful utilities that you need at some point during your exploration of Linux. Those that I use most often are described in the following list, which includes editors. There are, of course, many more examples, but these will get you started for more advanced work:

- *Text editor* (`nano`): A simple, easy-to-use text editor. It has a help menu at the bottom of the screen. Some operations may seem odd after using Windows text editors, but it is much easier to use than some other Linux text editors.

- *File search* (`find`): Locates files by name in a directory or path.
- *File/text search* (`grep`): Locates a text string in a set of files or directory.
- *Archive tools* (`gzip`, `gunzip`): A zip file archive tool (an alternative to tar).
- *Text display tools* (`less`, `more`): Less shows the last portion of a file; more shows the file contents of a page (console page) at a time.

As you can see, working with Raspberry Pi OS and the Raspberry Pi isn't complicated and in many ways similar to other PC platforms making it easy to use. Plus, it's a real computer.

Summary

There can be little argument that the Raspberry Pi has contributed greatly to the world of embedded hardware and the IoT. With its low cost, GPIO headers, and robust peripheral support, the Raspberry Pi is an excellent choice for building your IoT solutions. Due to its increasing popularity, there are tons of information available for those who want to learn how to work with hardware.

In this chapter, you explored the origins of the Raspberry Pi, including a tour of the hardware and a short primer on how to use its native operating system. You learned these things about the Raspberry Pi so that you can leverage the many examples and libraries that support IoT projects. Plus, the Raspberry Pi is a fun computer to use.

The next chapter introduces the most popular software development language for the Raspberry Pi – the Python programming language.

CHAPTER 5

Python Programming for the Raspberry Pi

Now that we know a little more about the Raspberry Pi, we can learn more about one of the programming languages on the Raspberry Pi when developing your IoT solutions. One of those languages is Python – a very robust and powerful language that you can use to write very powerful applications. Mastering is very easy, and some may suggest it doesn't require any formal training to use. This is largely true, and thus you should be able to write IoT applications with only a little bit of knowledge about Python.

Thus, this chapter presents a crash course on the basics of Python programming including an explanation about some of the most commonly used language features. As such, this chapter will provide you with the skills you need to understand the growing number of IoT project examples available on the Internet.

The chapter concludes with a walk-through of an example project that will show you how to work with Python on the Raspberry Pi.

Tip If you already know the basics of Python programming, feel free to skim through this chapter.

C. Bell, *Beginning IoT Projects*, https://doi.org/10.1007/978-1-4842-7234-3_5

Getting Started

Python is a high-level, interpreted, object-oriented scripting language. One of the biggest tenets of Python is to have a clear, easy-to-understand syntax that reads as close to English as possible. That is, you should be able to read a Python script and understand it even if you haven't learned Python. Python also has less punctuation (special symbols) and fewer syntactical machinations than other languages.

Tip While you can terminate a statement with the semicolon (`;`), it is not needed and considered a bad form.

Here are a few of the key features of Python.

An interpreter processes Python at runtime. No compiler is used:

- Python supports object-oriented programming constructs by way of a class.
- Python is a great language for the beginner-level programmers and supports the development of a wide range of applications.
- Python is a scripting language but can be used for a wide range of applications.
- Python is very popular and used throughout the world giving it a huge support base.
- Python has few keywords, a simple structure, and a clearly defined syntax. This allows the student to pick up the language quickly.
- Python code is more clearly defined and visible to the eyes.

Python was developed by Guido van Rossum from the late 1980s to the early 1990s at the National Research Institute for Mathematics and Computer Science in the Netherlands and maintained by a core development team at the institute. It was derived from and influenced by many languages including Modula-3, C, C++, and even Unix shell scripting languages.

A fascinating fact about Python is it was named after the BBC show *Monty Python's Flying Circus* and has nothing to do with the reptile by the same name.[1] Quoting Monty Python in source code documentation (and even a humorous diversion for error messages) is very common, and while some professional developers may cringe at the insinuation, it's considered by Pythonistas as showing your Python street cred.[2]

Python is available for download (`python.org/downloads`) for just about every platform that you may encounter or use – even Windows! Fortunately, you don't need to install anything to use Python on the Raspberry Pi (it's already installed). We will be using the Raspberry Pi as the platform for the examples in this chapter.

Should you require more in-depth knowledge of Python, there are a number of excellent books on the topic. I list a few of my favorites in the following. A great resource is the documentation on the Python site – python.org/doc/:

- *Pro Python* Second Edition (Apress, 2014) by J. Burton Browning and Marty Alchin
- *Learning Python* Fifth Edition (O'Reilly Media, 2013) by Mark Lutz

---

[1] Monty Python refers to a group of comedians and not a single individual. However, the comedy is undeniably brilliant (`https://en.wikipedia.org/wiki/Monty_Python`).

[2] Pythonistas are expert Python developers and advocates for all things Python.

- *Automate the Boring Stuff with Python: Practical Programming for Total Beginners* (No Starch Press, 2015) by Al Sweigart

Now, let's jump into a tutorial on how to program in Python.

Python Primer

Now let's learn some of the basic concepts of Python programming. We will begin with the building blocks of the language such as variables, modules, and basic statements and then move on to the more complex concepts of flow control and data structures.

While the material may seem to come at you in a rush (hence the crash part), this crash course on Python covers only the most fundamental knowledge of the language and how to use it for writing Python for IoT projects on the Raspberry Pi. If you find you want to write more complex applications than the examples in this book, I encourage you to acquire one or more of the preceding resources to learn more about Python programming.

The following sections present many of the basic features of Python programming that you will need to know in order to understand example projects and are vital to successfully implementing the Python projects in this book.

The Basics

Python is a very easy language to learn with very few constructs that are even mildly difficult to learn. Rather than toss out a sample application, let's approach learning the basics of Python in a Python-like way: one step at a time.

The first thing you should learn is that Python does not use a code block demarcated with symbols like other languages. More specifically, code that is local to a construct such as a function or conditional or loop is designated using indentation. Thus, the lines below that are indented (by spaces or, *gag*, tabs) so that the starting characters align for the code body of the construct.

In my experience and travels through geekdom, few programmers prefer tabs over spaces. Indeed, many coding guidelines prohibit the use of tabs because they can interfere with certain offline code analysis tools. Plus, they can really mess up if your editor is set differently than others or if tabs are used inconsistently

The following shows this concept in action:

```
if (expr1):
    print("inside expr1")
    print("still inside expr1")
else:
    print("inside else")
    print("still inside else")

print("in outer level")
```

Here we see a conditional or if statement. Notice the function call `print()` – a common way to display output to the console – is indented. This signals the interpreter that the line belongs to the construct above it. For example, the two print statements that mention expr1 form the code block for the if condition (expression evaluates to true). Similarly, the next two print statements form the code block for the else condition. Finally, the non-indented line is not part of the conditional and thus is executed after either the if or else depending on the expression evaluation.

As you can see, indentation is a key concept to learn when writing Python. Even though it is very simple, making mistakes in indentation can result in code executing that you did not expect or worse errors from the interpreter.

Note I use "program" and "application" interchangeably with "script" when discussing Python. While technically Python code is a script, we often use it in contexts where "program" or "application" is more appropriate.

There is one special symbol that you will encounter frequently. Notice the use of the colon (`:`) in the preceding code. This symbol is used to terminate a construct and signals the interpreter that the declaration is complete and the body of the code block follows.

Now let's look at how we can use variables in our programs (scripts).

Variables

Python does not have a formal-type specification mechanism like other languages. However, you can still define variables to store anything you want. In fact, Python permits you to create and use variables based on context. However, you can use initialization to "set" the data type for the variable. The following shows several examples:

```
# Numbers
float_value = 9.75
integer_value = 5

# Strings
my_string = "He says, he's already got one."
```

```
print("Floating number: {0}".format(float_value))
print("Integer number: {0}".format(integer_value))
print(my_string)
```

For situations where you need to convert types or want to be sure values are typed a certain way, there are many functions for converting data. Table 5-1 shows a few of the more commonly used type conversion functions. I discuss some of the data structures in a later section.

Table 5-1. *Type Conversion in Python*

Function	Description
`int(x [,base])`	Converts x to an integer. Base is optional (e.g., 16 for hex).
`long(x [,base] )`	Converts x to a long integer.
`float(x)`	Converts x to a floating point.
`str(x)`	Converts object x to a string.
`tuple(t)`	Converts t to a tuple.
`list(l)`	Converts l to a list.
`set(s)`	Converts s to a set.
`dict(d)`	Creates a dictionary.
`chr(x)`	Converts an integer to a character.
`hex(x)`	Converts an integer to a hexadecimal string.
`oct(x)`	Converts an integer to an octal string.

However, you should use these conversion functions with care to avoid data loss or rounding. For example, converting a float to an integer can result in truncation.

Including Modules

Python applications can be built from reusable libraries that are provided by the Python environment. They can also be built from custom modules or libraries that you create yourself or download from a third party. When we want to use a library (function, class, etc.) that is part of a module, we use the import keyword and list the name of the module. The following shows some examples:

```
import os
import sys
import mysql.client.tools
```

The first two lines demonstrate how to import a base or common module provided by Python. In this case, we are using or importing the os and sys modules (operating system and Python system functions).

Next, we see a special dotted notation in use. The dotted notation is realized by using folders. In this case, we're using a module written for MySQL. Specifically, we're using the module located in the `mysql/client` folder named `tools`. Indeed, if you were to locate that module, you would find a file named `tools.py`. The starting point of the module dotted notation reference is any path in the PYTHONPATH environment variable.

Tip It is customary (but not required) to list your imports in alphabetical order with built-in modules first, then third-party modules, and then your own modules.

Comments

One of the most fundamental concepts in any programming language is the ability to annotate your source code with text that not only allows you to make notes among the lines of code but also forms a way to document your source code.

To add comments to your source code, use the pound sign (#). Place the sign at the start of the line to create a comment for that line repeating the # symbol for each subsequent line. This creates what is known as a block comment as shown. Notice I used a comment without any text to create whitespace. This helps with readability and is a common practice for block comments:

```
#
# Beginning IoT Projects
#
# Example Python application.
#
# Created by Dr. Charles Bell
#
```

You can also place comments on the same line as the source code. The compiler will ignore anything from the pound sign to the end of the line. For example, the following shows a common method for documenting variables:

```
zip = 35012                 # Zip or postal code
address1= "123 Main St."  # Store the street address
```

Arithmetic

You can perform a number of mathematical operations in Python including the usual primitives but also logical operations and operations used to compare values. Rather than discuss these in detail, I provide a quick reference in Table 5-2 that shows the operation and an example of how to use the operation.

Table 5-2. *Arithmetic, Logical, and Comparison Operators in Python*

<table>
<tr><th>Type</th><th>Operator</th><th>Description</th><th>Example</th></tr>
<tr><td>Arithmetic</td><td>+</td><td>Addition</td><td>int_var + 1</td></tr>
<tr><td></td><td>-</td><td>Subtraction</td><td>int_var - 1</td></tr>
<tr><td></td><td>*</td><td>Multiplication</td><td>int_var * 2</td></tr>
<tr><td></td><td>/</td><td>Division</td><td>int_var / 3</td></tr>
<tr><td></td><td>%</td><td>Modulus</td><td>int_var % 4</td></tr>
<tr><td></td><td>-</td><td>Unary subtraction</td><td>-int_var</td></tr>
<tr><td></td><td>+</td><td>Unary addition</td><td>+int_var</td></tr>
<tr><td>Logical</td><td>&</td><td>Bitwise and</td><td>var1&var2</td></tr>
<tr><td></td><td>|</td><td>Bitwise or</td><td>var1|var2</td></tr>
<tr><td></td><td>^</td><td>Bitwise exclusive</td><td>var1^var2</td></tr>
<tr><td></td><td>~</td><td>Bitwise compliment</td><td>~var1</td></tr>
<tr><td></td><td>&&</td><td>Logical and</td><td>var1&&var2</td></tr>
<tr><td></td><td>||</td><td>Logical or</td><td>var1||var2</td></tr>
<tr><td>Comparison</td><td>==</td><td>Equal</td><td>expr1==expr2</td></tr>
<tr><td></td><td>!=</td><td>Not equal</td><td>expr1!=expr2</td></tr>
<tr><td></td><td><</td><td>Less than</td><td>expr1<expr2</td></tr>
<tr><td></td><td>></td><td>Greater than</td><td>expr1>expr2</td></tr>
<tr><td></td><td><=</td><td>Less than or equal</td><td>expr1<=expr2</td></tr>
<tr><td></td><td>>=</td><td>Greater than or equal</td><td>expr1>=expr2</td></tr>
</table>

Bitwise operations produce a result on the values performed on each bit. Logical operators (and, or) produce a value that is either true or false and are often used with expressions or conditions.

Now that we understand variables and types and operations permitted on them and expressions, let's see how we can use them in flow control statements.

Flow Control Statements

Flow control statements change the execution of the program. They can be conditionals that use expressions that restrict execution to only those cases where the expression evaluates true (or negated), special constructs that allow you to repeat a block of code (loops), and functions to switch context to perform some special operations. We've already seen how functions work, so let's look at conditional and loop statements.

Conditionals

Conditional statements allow us to direct execution of our programs to sections (blocks) of code based on the evaluation of one or more expressions. There are two types of conditional statements in Python – the `if` statement and the `switch` statement.

We have seen the `if` statement in action in our example code. Notice in the example we can have one or more (optional) `else` phrases that we execute once the expressions for the `if` conditions evaluate to false. We can chain `if`/`else` statements to encompass multiple conditions where the code executed depends on the evaluation of several conditions. The following shows the general structure of the `if` statement:

```
if (expr1):
    # execute only if expr1 is true
elif ((expr2) || (expr3)):
    # execute only if expr1 is false *and* either expr2 or
      expr3 is true
else:
    # execute if both sets of if conditions evaluate to false
```

While you can chain the statements as much as you want, use some care here because the more `elif` sections you have, the harder it will become to understand, maintain, and avoid logic errors in your expressions.

Loops

Loops are used to control the repetitive execution of a block of code. There are three forms of loops that have slightly different behavior. All loops use conditional statements to determine whether to repeat execution or not. That is, they repeat as long as the condition is true. The three types of loops are `while`, `do`, and `for`. I explain each with an example.

The `while` loop has its condition at the "top" or start of the block of code. Thus, `while` loops only execute the body if and only if the condition evaluates to true on the first pass. The following illustrates the syntax for a `while` loop. This form of loop is best used when you need to execute code only if some expression(s) evaluate to true, for example, iterating through a collection of things, where the number of elements is unknown (loop until we run out of things in the collection):

```
while (expression):
   # do something here
```

The `for` loop variants are sometimes called counting loops because of their unique form. The `for` loop allows you to define a counting variable and a range or list to iterate over. The following illustrates the structure of the `for` loop. This form of loop is best used for performing an operation in a collection. In this case, Python will automatically place each item in the collection in the variable for each pass of the loop until no more items are available:

```
for variable_name in list:
  # do something here
```

You can also use range loops or counting loops. This uses a special function called range() that takes up to three parameters, range([start], stop[, step]), where start is the starting number (an integer), stop is the last number in the series, and step is the increment. So you can count by 1, 2, 3, etc. through a range of numbers. The following shows a simple example:

```
for i in range(2,9):
   # do something here
```

There are other uses for range() that you may encounter. See the documentation at https://docs.python.org/3/library/functions.html for more information on this function and other built-in functions.

Python also provides a mechanism for controlling the flow of the loop (e.g., duration or termination) using a few special keywords as follows:

- break: Exit the loop body immediately.
- continue: Skip to next iteration of the loop.
- else: Execute code when the loop ends.

There are some uses for these keywords, particularly break, but it is not the preferred method of terminating and controlling loops. That is, professionals believe the conditional expression or error handling code should behave well enough to not need these options.

Functions

Python allows you to use modularization in your code. While it supports object-oriented programming by way of classes (a more advanced feature that you are unlikely to encounter for most Python GPIO examples), on a more fundamental level, you can break your code into smaller chunks using functions.

A special keyword construct (rare in Python) is used to define a function. We simply use `def` followed by a name and a parameter list in parentheses. The colon is used to terminate the declaration. The following shows an example:

```
def print_dictionary(the_dictionary):
    for key, value in the_dictionary.items():
      print("'{0}': {1}".format(key, value))
```

You may be wondering what this strange code does. Notice the loop is assigning two values from the result of the `items()` function. This is a special function available from the dictionary object. (Yes, dictionaries are objects! So are tuples and lists and many other data structures.) The `items()` function returns the key/value pair, hence the names of the variables.

The next line prints out the values. The use of formatting strings where the curly braces define the parameter number starting at 0 is common for Python 3.x applications. See the Python documentation for more information about formatting strings.

Here the body of the function is indented. All statements indented under this function declaration belong to the function and are executed when the function is called. We can call functions by name providing any parameters as follows:

```
print_dictionary(my_dictionary)
```

This example, when executed, generates the following output as shown in Listing 5-1. I generated this by writing the function in the Python interpreter on my Raspberry Pi and executing it. To run the interpreter, search for Python and select the Python command window. Thus, this demonstrates how to use the Python interpreter to execute Python code on the fly (as you type each line and press Enter). Notice I used the `python3` command to launch the Python command window. This tells the computer to use the Python version 3.x installation. This is because the Raspberry Pi (and many PCs) has multiple versions installed and version 2.x is soon to be obsolete. Thus, it is always best to use the latest version of Python.

Listing 5-1. Using the Python Interpreter

```
pi@raspberrypi:~ $ python3
Python 3.7.3 (default, Jul 25 2020, 13:03:44)
[GCC 8.3.0] on linux
Type "help", "copyright", "credits" or "license" for more
information.
>>> def print_dictionary(the_dictionary):
...   for key, value in the_dictionary.items():
...     print("'{0}':{1}".format(key, value))
...
>>> my_dictionary = {
...   'name': "Chuck",
...   'age': 38,
... }
>>> print_dictionary(my_dictionary)
'name':Chuck
'age':38
>>> quit()
```

Tip Functions (methods) provided by objects (classes) can be called using the syntax `object_name.method_name()`.

Now let's look at some commonly used data structures including this strange thing called a dictionary.

Basic Data Structures

What you have learned so far about Python is enough to write the most basic programs and indeed more than enough to tackle the example projects in later chapters. However, when you start needing to operate on data – either from the user or from sensors and similar sources – you will need a way to organize and store data and operations on the data in memory. The following introduces data structures in order of complexity: lists, tuples, etc. I will demonstrate the three you will encounter most in the following.

Lists

Lists are a way to organize data in Python. It is a free-form way to build a collection. That is, the items (or elements) need not be the same data type. Lists also allow you to do some interesting operations such as adding things at the end, at the beginning, or at a special index. The following demonstrates how to create a list:

```
# List
my_list = ["abacab", 575, "rex, the wonder dog", 24, 5, 6]
my_list.append("end")
my_list.insert(0,"begin")
for item in my_list:
  print("{0}".format(item))
```

Here we see I created the list using square brackets ([ ]). The items in the list definition are separated by commas. Note that you can create an empty list simply by setting a variable equal to [ ]. Since lists, like other data structures, are objects, there are a number of operations available for lists such as the following:

- `append(x)`: Add x to the end of the list.
- `extend(l)`: Add all items to the end of the list.

- `insert(pos,item)`: Insert an item at a position pos.
- `remove(value)`: Remove the first item that matches (==) the value.
- `pop([i])`: Remove the item at position `i` or end of list.
- `index(value)`: Return the index of the first item that matches.
- `count(value)`: Count occurrences of value.
- `sort()`: Sort the list (ascending).
- `reverse()`: Reverse-sort the list.

Lists are like arrays in other languages and very useful for building dynamic collections of data.

Tuples

Tuples, on the other hand, are a more restrictive type of collection. That is, they are built from a specific set of data and do not allow manipulation like a list. The following shows an example of a tuple and how to use it:

```
# Tuple
my_tuple = (0,1,2,3,4,5,6,7,8,"nine")
for item in my_tuple:
  print("{0}".format(item))
if 7 in my_tuple:
  print("7 is in the list")
```

Here we see I created the tuple using parentheses `()`. The items in the tuple definition are separated by commas. Note that you can create an empty tuple simply by setting a variable equal to `()`. Since tuples, like other data structures, are objects, there are a number of operations available such as the following including operations for sequences such as inclusion, location, etc.:

- `x in t`: Determine if `t` contains `x`.
- `x not in t`: Determine if `t` does not contain `x`.
- `s + t`: Concatenate tuples.
- `s[i]`: Get element `i`.
- `len(t)`: Length of `t` (number of elements).
- `min(t)`: Minimal (smallest value).
- `max(t)`: Maximal (largest value).

If you want even more structure with storing data in memory, you can use a special construct (object) called a dictionary.

Dictionaries

A dictionary is a data structure that allows you to store key/value pairs where the data is assessed via the keys. Dictionaries are a very structured way of working with data and the most logical form we will want to use when collecting complex data. Listing 5-2 shows an example of a dictionary.

Listing 5-2. Dictionary Example

```
# Dictionary
my_dictionary = {
  'first_name': "Chuck",
  'last_name': "Bell",
  'age': 36,
  'my_ip': (192,168,1,225),
  42: “What is the meaning of life?”,
```

```
}
# Access the keys:
print(my_dictionary.keys())
# Access the items (key, value) pairs
print(my_dictionary.items())
# Access the values
print(my_dictionary.values())
# Create a list of dictionaries
my_addresses = [my_dictionary]
```

There is a lot going on here! We see a basic dictionary declaration that uses curly braces to create a dictionary. Inside that, we can create as many key/value pairs we want separated by commas. Keys are defined using strings (I use single quotes by convention, but double quotes will work) or integers; values can be any data type we want. For the `my_ip` attribute, we are also storing a tuple.

Following the dictionary, we see a number of operations performed on the dictionary from printing the keys, printing all of the values, and printing only the values. The following shows the output of executing this code snippet from the Python interpreter:

```
[42, 'first_name', 'last_name', 'age', 'my_ip']
[(42, 'what is the meaning of life?'), ('first_name', 'Chuck'),
('last_name', 'Bell'), ('age', 36), ('my_ip', (192, 168, 1,
225))]
['what is the meaning of life?', 'Chuck', 'Bell', 36, (192,
168, 1, 225)]
'42': what is the meaning of life?
'first_name': Chuck
'last_name': Bell
'age': 36
'my_ip': (192, 168, 1, 225)
```

As we have seen in this example, there are a number of operations (methods) available for dictionaries including the following. Together, this list of operations makes dictionaries a very powerful programming tool:

- `len(d)`: Number of items in `d`.
- `d[k]`: Item of `d` with key `k`.
- `d[k] = x`: Assign key `k` with value `x`.
- `del d[k]`: Delete item with key `k`.
- `k in d`: Determine if `d` has an item with key `k`.
- `d.items()`: Return a list of the (key, value) pairs in `d`.
- `d.keys()`: Return a list of the keys in `d`.
- `d.values()`: Return a list of the values in `d`.

Best of all, objects can be placed inside other objects. For example, you can create a list of dictionaries like I did previously, a dictionary that contains lists and tuples, and any combination you need. Thus, using lists, tuples, and dictionaries is a powerful way to manage data for your program.

Classes and Objects

You may have heard that Python is an object-oriented programming language. But what does that mean? Simply, Python is a programming language that provides facilities for describing objects (things) and what you can do with the object (operations). Objects are an advanced form of data abstraction where the data is hidden from the caller and only manipulated by the operations (methods) the object provides.

The syntax we use in Python is the `class` statement, which you can use to help make your projects modular. By modular, we mean the source code is arranged to make it easier to develop and maintain. Typically, we place classes in separate modules (code files), which helps organize the code better. While it is not required, I recommend using this technique of placing a class in its own source file. This makes modifying the class or fixing problems (bugs) easier.

So what are Python classes? Let's begin by considering the construct as an organization technique. We can use the class to group data and methods together. The name of the class immediately follows the keyword class followed by a colon. You declare other class methods like any other method except the first argument must be self, which ties the method to the class instance when executed.

METHOD OR FUNCTION?

I prefer to use terms that have been adopted by the language designers or community of developers. For example, some use "function," but others may use "method." Still others may use subroutine, routine, procedure, etc. It doesn't matter which term you use, but you should strive to use terms consistently. One example, which can be confusing to some, is I use the term method when discussing object-oriented examples. That is, a class has methods, not functions. However, you can use function in place of method, and you'd still be correct (mostly).

Accessing the data is done using one or more methods by using the class (creating an instance) and using dot notation to reference the data member or function. Let's look at an example. Listing 5-3 shows a complete class that describes (models) the most basic characteristics of a vehicle used for transportation. I created a file named `vehicle.py` to contain this code.

Listing 5-3. Vehicle Class

```
#
# Beginning IOT Projects
#
# Class Example: A generic vehicle
#
# Dr. Charles Bell
#
class Vehicle:
    """Base class for defining vehicles"""
    axles = 0
    doors = 0
    occupants = 0

    def __init__(self, num_axles, num_doors):
        self.axles = num_axles
        self.doors = num_doors

    def get_axles(self):
        return self.axles

    def get_doors(self):
        return self.doors

    def add_occupant(self):
        self.occupants += 1

    def num_occupants(self):
        return self.occupants
```

Notice a couple of things here. First, there is a method with the name `__init__()`. This is the constructor and is called when the class instance is created. You place all your initialization code like setting variables in this method. We also have methods for returning the number of axles, doors, and occupants. We have one method in this class to add occupants.

Also notice we address each of the class attributes (data) using `self.<name>`. This is how we can ensure we always access the data that is associated with the instance created.

Let's see how this class can be used to define a family sedan. Listing 5-4 shows code that uses this class. We can place this code in a file named sedan.py.

Listing 5-4. Using the Vehicle Class

```
#
# Beginning IoT Projects
#
# Class Example: Using the generic Vehicle class
#
# Dr. Charles Bell
#
from vehicle import Vehicle

sedan = Vehicle(2, 4)
sedan.add_occupant()
sedan.add_occupant()
sedan.add_occupant()
print("The car has {0} occupants.".format(sedan.num_
occupants()))
```

Notice the first line imports the `Vehicle` class from the vehicle module. Notice I capitalized the class name but not the file name. This is a very common naming scheme. Next in the code, we create an instance of the class.

Notice I passed in 2, 4 to the class name. This will cause the `__init__()` method to be called when the class is instantiated. The variable, sedan, becomes the class instance variable (object) that we can manipulate, and I do so by adding three occupants and then printing out the number of occupants using the method in the `Vehicle` class.

We can run the code on our Raspberry Pi (or our PC) using the following command. As we can see, it tells us there are three occupants in the vehicle when the code is run. Nice:

```
$ python3 ./sedan.py
The car has 3 occupants.
```

OBJECT-ORIENTED PROGRAMMING (OOP) TERMINOLOGY

Like any technology or concept, there are a certain number of terms that you must learn to be able to understand and communicate with others about the technology. The following briefly describes some of the terms you will need to know to learn more about object-oriented programming:

- *Attribute*: A data element in a class.
- *Class*: A code construct used to define an object in the form of attributes (data) and methods (functions) that operate on the data. Methods and attributes in Python can be accessed using dot notation.
- *Class instance variable*: A variable that is used to store an instance of an object. They are used like any other variable and, combined with dot notation, allow us to manipulate objects.
- *Instance*: An executable form of a class created by assigning a class to a variable initializing the code as an object.
- *Inheritance*: The inclusion of attributes and methods from one class in another.

- *Instantiation*: The creation of an instance of a class.
- *Method overloading*: The creation of two or more methods with the same name but with a different set of parameters. This allows us to create methods that have the same name but may operate differently depending on the parameters passed.
- *Polymorphism*: Inheriting attributes and methods from a base class adding additional methods or overriding (changing) methods.

There are many more OOP terms, but these are the ones you will encounter most often.

Now, let's see how we can use the Vehicle class to demonstrate inheritance. In this case, we will create a new class named `PickupTruck` that uses the Vehicle class but adds specialization to the resulting class. Listing 5-5 shows the new class. I placed this code in a file named `pickup_truck.py`. As you will see, a pickup truck is a type of vehicle.

Listing 5-5. PickupTruck Class

```
#
# Beginning IoT Projects
#
# Class Example: Inheriting the Vehicle class to form a
# model of a pickup truck with maximum occupants and maximum
# payload.
#
# Dr. Charles Bell
#
from vehicle import Vehicle
```

```
class PickupTruck(Vehicle):
    """This is a pickup truck that has:
    axles = 2,
    doors = 2,
    __max occupants = 3
    The maximum payload is set on instantiation.
    """
    occupants = 0
    payload = 0
    max_payload = 0

    def __init__(self, max_weight):
        super().__init__(2,2)
        self.max_payload = max_weight
        self.__max_occupants = 3

    def add_occupant(self):
        if (self.occupants < self.__max_occupants):
            super().add_occupant()
        else:
            print("Sorry, only 3 occupants are permitted in the
            truck.")

    def add_payload(self, num_pounds):
        if ((self.payload + num_pounds) < self.max_payload):
            self.payload += num_pounds
        else:
            print("Overloaded!")

    def remove_payload(self, num_pounds):
        if ((self.payload - num_pounds) >= 0):
            self.payload -= num_pounds
```

```
        else:
            print("Nothing in the truck.")

    def get_payload(self):
        return self.payload
```

Notice a few things here. First, notice the class statement: `class PickupTruck(Vehicle):`. When we want to inherit from another class, we add the parentheses with the name of the base class. This ensures Python will use the base class allowing the derived class to use all its accessible data and memory. If you want to inherit from more than one class (called multiple inheritance), you can just list the classes with a comma-separated list.

Next, notice the `__max_occupants` variable. Using two underscores in a class for an attribute or a method, through convention, makes the item private to the class. That is, it should only be accessed from within the class. No caller of the class (via a class variable/instance) can access the private items nor can any class derived from the class. It is always a good practice to hide the attributes (data).

You may be wondering what happened to the occupant methods. Why aren't they in the new class? They aren't there because our new class inherited all that behavior from the base class. Not only that, but the code has been modified to limit occupants to exactly three occupants.

I also want to point out the documentation I added to the class. We use documentation strings (strings that use a set of three double quotes before and after) to document the class. You can put documentation here to explain the class and its methods. We'll see a good use of this a bit later.

Finally, notice the code in the constructor. This demonstrates how to call the base class method, which I do to set the number of axles and doors. We could do the same in other methods if we wanted to call the base class method's version.

Now, let's write some code to use this class. Listing 5-6 shows the code we use to test this class. Here, we create a file named pickup.py that creates an instance of the pickup truck, adds occupants and payload, and then prints out the contents of the truck.

Listing 5-6. Using the PickupTruck Class

```
#
# Beginning IoT Projects
#
# Class Example: Exercising the PickupTruck class.
#
# Dr. Charles Bell
#
from pickup_truck import PickupTruck

pickup = PickupTruck(500)
pickup.add_occupant()
pickup.add_occupant()
pickup.add_occupant()
pickup.add_occupant()
pickup.add_payload(100)
pickup.add_payload(300)
print("Number of occupants in truck = {0}.".format(pickup.num_
occupants()))
print("Weight in truck = {0}.".format(pickup.get_payload()))
pickup.add_payload(200)
pickup.remove_payload(400)
pickup.remove_payload(10)
```

Notice I add a couple of calls to the add_occupant() method, which the new class inherits and overrides. I also add calls so that we can test the code in the methods that check for excessive occupants and maximum payload capacity. When we run this code, we will see the results as shown in the following:

```
$ python3 ./pickup.py
Sorry, only 3 occupants are permitted in the truck.
Number of occupants in truck = 3.
Weight in truck = 400.
Overloaded!
Nothing in the truck.
```

Once again, I ran this code on my Raspberry Pi, but you can run all this code on your PC and will see the same results.

There is one more thing we should learn about classes: built-in attributes. Recall the __init__() method. Python automatically provides several built-in attributes each starting with __ that you can use to learn more about objects. The following lists a few of the operators available for classes:

- __dict__: Dictionary containing the class namespace
- __doc__: Class documentation string
- __name__: Class name
- __module__: Module name where the class is defined
- __bases__: The base class(es) in order of inheritance

The following shows what each of these attributes returns for the preceding PickupTruck class. I added this code to the pickup.py file:

```
print("PickupTruck.__doc__:", PickupTruck.__doc__)
print("PickupTruck.__name__:", PickupTruck.__name__)
```

```
print("PickupTruck.__module__:", PickupTruck.__module__)
print("PickupTruck.__bases__:", PickupTruck.__bases__)
print("PickupTruck.__dict__:", PickupTruck.__dict__)
```

When this code is run, we see the following output:

```
PickupTruck.__doc__: This is a pickup truck that has:
    axles = 2,
    doors = 2,
    max occupants = 3
    The maximum payload is set on instantiation.

PickupTruck.__name__: PickupTruck
PickupTruck.__module__: pickup_truck
PickupTruck.__bases__: (<class 'vehicle.Vehicle'>,)
PickupTruck.__dict__: {'__module__': 'pickup_truck', '__doc__':
'This is a pickup truck that has:\n    axles = 2,\n    doors
= 2,\n    max occupants = 3\n    The maximum payload is set
on instantiation.\n    ', 'occupants': 0, 'payload': 0, 'max_
payload': 0, ' _PickupTruck__max_occupants': 3, '__init__':
<function PickupTruck.__init__ at 0x1018a1488>, 'add_occupant':
<function PickupTruck.add_occupant at 0x1018a17b8>, 'add_
payload': <function PickupTruck.add_payload at 0x1018a1840>,
'remove_payload': <function PickupTruck.remove_payload at
0x1018a18c8>, 'get_payload': <function PickupTruck.get_payload
at 0x1018a1950>}
```

You can use the built-in attributes whenever you need more information about a class. Notice the `_PickupTruck__max_occupants` entry in the dictionary. Recall that we made a pseudo-private variable, `__max_occupants`. Here, we see how Python refers to the variable by prepending the class name to the variable. Remember variables that start with two underscores (not one) should be considered private to the class and only usable from within the class.

Tip See `https://docs.python.org/3/tutorial/classes.html` for more information about classes in Python.

Wow! That was a wild ride, wasn't it? I hope that this short crash course in Python has explained enough about the sample programs shown so far that you now know how they work. This crash course also forms the basis for understanding the other Python examples in this book.

OK, now it's time to see some of these fundamental elements of Python in action.

Example Scripts

Now, let's work on a couple of example Python scripts that you can use to experiment with writing Python on the Raspberry Pi. None require additional hardware, and you can execute them on any PC if you don't have a Raspberry Pi setup. We'll start with a few simple examples, then move on to the temperature conversion example from Chapter 3, and close with an example using classes.

I explain the code in detail for each example and show example output when you execute the code. I encourage you to implement these examples and figure out the challenge yourself as practice for the projects later in this book.

Example 1: Using Loops

This example demonstrates how to write loops in Python using the `for` loop. The problem we are trying to solve is converting integers from decimal to binary, hexadecimal, and octal. Often with IoT projects, we need to see values in one or more of these formats, and in some cases the sensors we use (and the associated documentation) use hexadecimal rather than decimal. Thus, this example can be helpful in the future not only for how to use the for loop but also how to convert integers into different formats.

Write the Code

The example begins with a tuple of integers to convert. Tuples and lists can be iterated through (values read in order) using a `for` loop. Recall a tuple is read-only, so in this case since it is input, it is fine. However, in other cases where you may need to change values, you will want to use a list. Recall the syntactical difference between a tuple and a list is the tuple uses parentheses and a list uses square brackets.

The `for` loop demonstrated here is called a "for each" loop. Notice we used the syntax "for value in values," which tells Python to iterate over the tuple named values fetching (storing) each item into the value variable each iteration through the tuple.

Finally, we use the `print()` and `format()` functions to replace two placeholders `{0}` and `{1}` and to print out a different format of the integer using the methods `bin()` for binary, `oct()` for octal, and `hex()` for hexadecimal that do the conversion for us.

Listing 5-7 shows the completed code for this example.

Listing 5-7. Converting Integers

```
#
# Beginning IoT Projects
#
# Example: Convert integer to binary, hex, and octal
#
# Dr. Charles Bell
#
```

```
# Create a tuple of integer values
values = (12, 450, 1, 89, 2017, 90125)

# Loop through the values and convert each to binary, hex, and
octal
for value in values:
    print("{0} in binary is {1}".format(value, bin(value)))
    print("{0} in octal is {1}".format(value, oct(value)))
    print("{0} in hexadecimal is {1}".format(value,
    hex(value)))
```

Execute the Code

You can save this code in a file named conversions.py, then open a terminal (console window), and run the code with the command python3 ./conversions.py. Listing 5-8 shows the output.

Listing 5-8. Conversions Example Output

```
$ python3 ./conversions.py
12 in binary is 0b1100
12 in octal is 0o14
12 in hexadecimal is 0xc
450 in binary is 0b111000010
450 in octal is 0o702
450 in hexadecimal is 0x1c2
1 in binary is 0b1
1 in octal is 0o1
1 in hexadecimal is 0x1
89 in binary is 0b1011001
89 in octal is 0o131
89 in hexadecimal is 0x59
2017 in binary is 0b11111100001
```

```
2017 in octal is 0o3741
2017 in hexadecimal is 0x7e1
90125 in binary is 0b10110000000001101
90125 in octal is 0o260015
90125 in hexadecimal is 0x1600d
```

Notice all the values in the tuple were converted.

Example 2: Using Complex Data and Files

This example demonstrates how to work with the JavaScript Object Notation[3] (JSON) in Python. In short, JSON is a markup language used to exchange data. Not only is it human readable, it can be used directly in your applications to store and retrieve data to and from other applications, servers, and even MySQL. In fact, JSON looks familiar to programmers because it resembles other markup schemes. JSON is also very simple in that it supports only two types of structures: 1) a collection containing (name, value) pairs and 2) an ordered list (or array). Of course, you can also mix and match the structures in an object. When we create a JSON object, we call it a JSON document.

The problem we are trying to solve is writing and reading data to/from files. In this case, we will use a special JSON encoder and decoder module named json that allows us to easily convert data in files (or other streams) to and from JSON. As you will see, accessing JSON data is easy by simply using the key (sometimes called fields) names to access the data. Thus, this example can be helpful in the future not only for how to read and write files but also how to work with JSON documents.

[3] www.json.org/json-en.html

Write the Code

This example stores and retrieves data in files. The data is basic information about pets including the name, age, breed, and type. The type is used to determine broad categories like fish, dog, or cat.

We begin by importing the JSON module (named `json`), which is built into Python. Next, we prepare some initial data by building JSON documents and storing them in a Python list. We use the `json.loads()` method to pass in a JSON-formatted string. The result is a JSON document that we can add to our list. The examples use a very simple form of JSON documents – a collection of (name, value) pairs. The following shows one of the JSON-formatted strings used:

```
{"name":"Violet", "age": 6, "breed":"dachshund", "type":"dog"}
```

Notice we enclose the string inside curly braces and use a series of key names and values and a colon separated by commas. If this looks familiar, it's because it is the same format as a Python dictionary. This demonstrates my comment that JSON syntax looks familiar to programmers.

The JSON method, `json.loads()`, takes the JSON-formatted string, then parses the string checking for validity, and returns a JSON document. We then store that document in a variable and add it to the list as shown in the following:

```
parsed_json = json.loads('{"name":"Violet", "age": 6,
                          "breed":"dachshund", "type":"dog"}')
pets.append(parsed_json)
```

Once the data is added to the list, we then write the data to a file named `my_data.json`. To work with files, we first open the file with the `open()` function, which takes a file name (including a path if you want to put the file in a directory) and an access mode. We use "`r`" for read and "`w`" for write. You can also use "a" for append if you want to open a file and add to the end. Note that the "w" access will overwrite the file when you write to it. If the open() function succeeds, you get a file object that permits you to call additional functions to read or write data. The open() will fail if the file is not present (and you have requested read access) or you do not have permissions to write to the file.

Once the file is open, we can write the JSON documents to the file by iterating over the list. Iteration means to start at the first element and access the elements in the list one at a time in order (the order they appear in the list). Recall iteration in Python is very easy. We simply say "for each item in the list" with the for loop as follows:

```
for pet in pets:
  # do something with the pet data
```

To write the JSON document to the file, we use the json.dumps() method, which will produce a JSON-formatted string writing that to the file using the file variable and the write() method. Thus, we now see how to build JSON documents from strings and then decode (dump) them to a string.

Once we've written data to the file, we then close the file with the close() function and then reopen it and read data from the file. In this case, we use another special implementation of the for loop. We use the file variable to read all of the lines in the file with the readlines() method and then iterate over them with the following code:

```
json_file = open("my_data.json", "r")
for pet in json_file.readlines():
  # do something with the pet string
```

We use the json.loads() method again to read the JSON-formatted string as read from the file to convert it to a JSON document, which we add to another list. We then close the file. Now the data has been read back into our program, and we can use it. Finally, we iterate over the new list and print out data from the JSON documents using the key names to retrieve the data we want. Listing 5-9 shows the completed code for this example.

Listing 5-9. Writing and Reading JSON Objects to/from Files

```
#
# Beginning IoT Projects
#
# Example: Storing and retrieving JSON objects in files
#
# Dr. Charles Bell
#

import json

# Prepare a list of JSON documents for pets by converting JSON
to a dictionary
pets = []
parsed_json = json.loads('{"name":"Violet", "age": 6,
"breed":"dachshund", "type":"dog"}')
pets.append(parsed_json)
parsed_json = json.loads('{"name": "JonJon", "age": 15,
"breed":"poodle", "type":"dog"}')
pets.append(parsed_json)
parsed_json = json.loads('{"name": "Mister", "age": 4,
"breed":"siberian khatru", "type":"cat"}')
pets.append(parsed_json)
parsed_json = json.loads('{"name": "Spot", "age": 7,
"breed":"koi", "type":"fish"}')
pets.append(parsed_json)
parsed_json = json.loads('{"name": "Charlie", "age": 6,
"breed":"dachshund", "type":"dog"}')
pets.append(parsed_json)
```

```
# Now, write these entries to a file. Note: overwrites the file
json_file = open("my_data.json", "w")
for pet in pets:
    json_file.write(json.dumps(pet))
    json_file.write("\n")
json_file.close()

# Now, let's read the JSON documents then print the name and
age for all of the dogs in the list
my_pets = []
json_file = open("my_data.json", "r")
for pet in json_file.readlines():
    parsed_json = json.loads(pet)
    my_pets.append(parsed_json)
json_file.close()

print("Name, Age")
for pet in my_pets:
    if pet['type'] == 'dog':
        print("{0}, {1}".format(pet['name'], pet['age']))
```

Notice the loop for writing data. We added a second `write()` method passing in a strange string (it is actually an escaped character). The `\n` is a special character called the newline character. This forces the JSON-formatted strings to be on separate lines in the file and helps with readability.

Tip For a more in-depth look at how to work with files in Python, see `https://docs.python.org/3/tutorial/inputoutput.html#reading-and-writing-files`.

So what does the file look like? The following is a dump of the file using the more utility, which shows the contents of the file. Notice the file contains the JSON-formatted strings just like we had in our code:

```
$ more my_data.json
{"age": 6, "breed": "dachshund", "type": "dog", "name":
"Violet"}
{"age": 15, "breed": "poodle", "type": "dog", "name": "JonJon"}
{"age": 4, "breed": "siberian khatru", "type": "cat", "name":
"Mister"}
{"age": 7, "breed": "koi", "type": "fish", "name": "Spot"}
{"age": 6, "breed": "dachshund", "type": "dog", "name": "Charlie"}
```

Now, let's see what happens when we run this script.

Execute the Code

You can save this code in a file named `rw_json.py`, then open a terminal (console window), and run the code with the command `python3 ./rw_json.py`. The following shows the output:

```
$ python3 ./rw_json.py
Name, Age
Violet, 6
JonJon, 15
Charlie, 6
```

While the output may not be very impressive, by completing the example, you've learned a great deal about working with files and structured data using JSON documents.

Example 3: Temperature Conversion

Recall from Chapter 3 an example where we wrote a sketch to convert temperature on the Arduino. Let's see how to do that same operation in Python. In this example, we will build a temperature conversion script that converts values in Fahrenheit to Celsius and Celsius to Fahrenheit using the terminal for input and output.

Write the Code

This example is very similar to the Arduino example in Chapter 3. In fact, it is a rewrite using nearly the same logic. What differs most showcases some of the versatility of Python over the Arduino language.

For example, reading input from the user is much easier with the Python `input()` function. Similarly, the `print()` method is also easier to use and can operate on one or more variables, thereby allowing to use fewer statements.

Aside from that, the code also includes a check for repeating the conversion instead of an endless loop. This means we can ask the user if they want to continue and, if not, to exit the script. The rest of the code is merely a conversion from the Arduino language syntax to Python. Take some time and read through the code comparing it to the sketch from Chapter 3. You should see it is very similar and you may even agree easier to read. Listing 5-10 shows the completed code for the example.

Listing 5-10. Temperature Conversion

```
#
# Beginning IoT Projects
#
# Example Python sketch to convert temperature to F or C.
#
# Created by Dr. Charles Bell
#
```

```
# Imports
import time

# Convert temperature to Celsius or Fahrenheit
def convert_temp(scale_read, base_temp):
    """convert_temp"""
    if scale_read in {'c', 'C'}:
        return ((9.0 / 5.0) * base_temp) + 32.0
    return (5.0 / 9.0) * (base_temp - 32.0)

converted_temp = 0.0
done = False
while not done:
    temperature = 0.0
    scale = input("\nPlease choose a starting scale (F) or (C): ")
    # Throw error if scale is not one of the valid characters
    if scale not in {'c', 'C', 'f', 'F'}:
        print("\nERROR: I'm sorry, I don't understand
        '{0}'.".format(scale))
    else:
        temperature = float(input("Please enter a temperature: "))
        converted_temp = convert_temp(scale, temperature)
        from_scale = ''
        to_scale = ''
        if scale in {'c', 'C'}:
            from_scale = "Celsius"
            to_scale = "Fahrenheit."
        else:
            from_scale = "Fahrenheit."
            to_scale = "Celsius"
```

```
        print("{0} degrees {1} = {2:4.2f} {3}".format
        (temperature, from_scale, converted_temp, to_scale))
    time.sleep(1)
    done = input("\nConvert another value (Y)? ") not in {'y', 'Y'}
```

Now, let's see what happens when we run this script.

Execute the Code

You can save this code in a file named temperature_converter.py, then open a terminal (console window), and run the code with the command python3 ./temperature_converter.py. The following shows the output:

```
$ python3 ./temperature_converter.py

Please choose a starting scale (F) or (C): F
Please enter a temperature: 55.44
55.44 degrees Fahrenheit. = 13.02 Celsius

Convert another value (Y)? n
```

Example 4: Using Classes

This example ramps up the complexity considerably by introducing an object-oriented programming concept: classes. Recall from earlier that classes are another way to modularize our code. Classes are used to model data and behavior on that data. Further, classes are typically placed in their own code module (file) that further modularizes the code. If you need to modify a class, you may need to only change the code in the class module.

The problem we're exploring in this example is how to develop solutions using classes and code modules. We will be creating two files: one for the class and another for the main code.

Write the Code

This example is designed to convert Roman numerals to integers. That is, we will enter a value like VIII, which is eight, and expect to see the integer 8. To make things more interesting, we will also take the integer we derive and convert it back to Roman numerals.

Roman numerals are formed as a string using the characters I for 1, V for 5, X for 10, L for 50, C for 100, D for 500, and M for 1000. Other numbers are formed by adding the character values together (e.g., 3 = III) or putting a single, lower character before a higher character to indicate the representative minus that character (e.g., 4 = IV). The following shows some examples of how this works:

```
3 = III
15 = XV
12 = XII
24 = XXIV
96 = LXLVI
107 = CVII
```

This may sound like a lot of extra work, but consider this: if we can convert from one format to another, we should be able to convert back without errors. More specifically, we can use the code for one conversion to validate the other. If we get a different value when converting it back, we know we have a problem that needs to be fixed.

To solve the problem, we will place the code for converting Roman numerals into a separate file (code module) and build a class called `RomanNumerals` to contain the methods. In this case, the data is a mapping of integers to Roman numerals:

```
# Private dictionary of roman numerals
__roman_dict = {
    'I': 1,
    'IV': 4,
    'V': 5,
    'IX': 9,
    'X': 10,
    'XL': 40,
    'L': 50,
    'XC': 90,
    'C': 100,
    'CD': 400,
    'D': 500,
    'CM': 900,
    'M': 1000,
}
```

Notice the two underscores before the name of the dictionary. This is a special notation that marks the dictionary as a private variable in the class. This is a Python aspect for information hiding, which is a recommended technique to use when designing objects; always strive to hide data that is used inside the class.

Notice also that instead of using the basic characters and their values, I used several other values too. I did this to help make the conversion easier (and cheat a bit). In this case, I added the entries that represent the previous one-value conversions such as 4 (IV), 9 (IX), etc. This makes the conversion a bit easier (and more accurate).

We will also add two methods: `convert_to_int()`, which takes a Roman numeral string and converts it to an integer, and `convert_to_roman()`, which takes an integer and converts it to a Roman numeral. Rather than explain every line of code in the methods, I leave it to you to read the code to see how it works.

Simply, the convert to integer method takes each character and gets its value from the dictionary summing the values. There is a trick there that requires special handling for the lower value characters appearing before higher values (e.g., IX). The convert to Roman method is a bit easier since we simply divide the value by the highest value in the dictionary until we reach zero. Listing 5-11 shows the code for the class module, which is saved in a file named `roman_numerals.py`.

Listing 5-11. RomanNumerals Class

```
#
# Beginning IoT Projects
#
# Example: Roman numerals class
#
# Convert integers to roman numerals
# Convert roman numerals to integers
#
# Dr. Charles Bell
#

class RomanNumerals:

    # Private dictionary of roman numerals
    __roman_dict = {
        'I': 1,
        'IV': 4,
        'V': 5,
        'IX': 9,
```

```
        'X': 10,
        'XL': 40,
        'L': 50,
        'XC': 90,
        'C': 100,
        'CD': 400,
        'D': 500,
        'CM': 900,
        'M': 1000,
    }

    def convert_to_int(self, roman_num):
        value = 0
        for i in range(len(roman_num)):
            if i > 0 and \
                    self.__roman_dict[roman_num[i]] > \
                    self.__roman_dict[roman_num[i - 1]]:
                value += self.__roman_dict[roman_num[i]] - 2 \
                         * self.__roman_dict[roman_num[i - 1]]
            else:
                value += self.__roman_dict[roman_num[i]]
        return value

    def convert_to_roman(self, int_value):
        # First, get the values of all of entries in the
          dictionary
        roman_values = list(self.__roman_dict.values())
        roman_keys = list(self.__roman_dict.keys())
        # Prepare the string
        roman_str = ""
        remainder = int_value
        # Loop through the values in reverse
        for i in range(len(roman_values)-1, -1, -1):
```

```
            count = int(remainder / roman_values[i])
            if count > 0:
                for j in range(0,count):
                    roman_str += roman_keys[i]
                remainder -= count * roman_values[i]
        return roman_str
```

Now let's look at the main code. For this, we simply need to import the new class from the code module as follows. This is a slightly different form of the import directive. In this case, we're telling Python to include the `roman_numerals` class from the file named `RomanNumerals`.

```
from roman_numerals import RomanNumerals
```

Note If the code module were in a subfolder, say `roman`, we would have written the import statement as `from roman import Roman_Numerals` where we list the folders using dot notation instead of slashes.

The rest of the code is straightforward. We first ask the user for a valid Roman numeral string, then convert it to integer, and use that value to convert back to a Roman numeral string printing the result. So, you see, having the class in a separate module has simplified our code making it shorter and easier to maintain. Listing 5-12 shows the complete main code.

Listing 5-12. Converting Roman Numerals

```
#
# Beginning IoT Projects
#
# Example: Convert roman numerals using a class
#
```

```
# Convert integers to roman numerals
# Convert roman numerals to integers
#
# Dr. Charles Bell
#

from roman_numerals import RomanNumerals

roman_str = input("Enter a valid roman numeral: ")
roman_num = RomanNumerals()

# Convert to roman numerals
value = roman_num.convert_to_int(roman_str)
print("Convert to integer:        {0} = {1}".format
(roman_str, value))

# Convert to integer
new_str = roman_num.convert_to_roman(value)
print("Convert to Roman Numerals: {0} = {1}".format(value,
new_str))

print("bye!")
```

Now, let's see what happens when we run this script.

Execute the Code

You can save this code in a file named `roman.py`, then open a terminal (console window), and run the code with the command `python3 ./roman.py`. The following shows the output:

```
$ python3 ./roman.py
Enter a valid roman numeral: MVXIII
Convert to integer:        MVXIII = 1008
Convert to Roman Numerals: 1008 = MVIII
bye!
```

Go ahead and try this example a few more times. See what happens when you enter various Roman numerals. Hint: Try entries that are not quite valid like XIIIIV and see what happens. Does it work, and if so, can you figure out why (or why not)?

Summary

If you are learning how to work with IoT projects and don't know how to program with Python, learning Python can be fun given its easy-to-understand syntax. While there are many examples on the Internet you can use, very few are documented in such a way as to provide enough information for someone new to Python to understand or much less get started or even compile and deploy the sample!

This chapter has provided a crash course in Python that covered the basics of the things you will encounter when examining most of the smaller example projects. We discovered the basic syntax and constructs of a Python application including a few simple examples to illustrate the concepts.

In the next chapter, we'll take a closer look at one exciting new component system designed to make building IoT and similar projects easier – the Qwiic Component System.

PART II

The Qwiic and STEMMA QT Component Systems

This part introduces the Qwiic and STEMMA QT component systems including a series of chapters containing example projects that detail the steps needed to implement them with the Arduino and Raspberry Pi. While the example projects are not complete IoT solutions in that they are not integrated with the cloud, they are a good starting point to learn how to program IoT projects for the Arduino and Raspberry Pi.

CHAPTER 6

Introducing Qwiic and STEMMA QT

Building IoT projects can be quite challenging and even more so for those without any electronics experience. While almost everyone can, given time, learn how to work with electronic components by following a detailed example, few may have the time to devote, or some may simply not be interested in learning more than is minimally necessary.

For those readers, you can rejoice! There are several component systems that have been developed to accomplish these goals including the following:

- *Qwiic* system from SparkFun (`www.sparkfun.com/qwiic`)
- *STEMMA/STEMMA QT* from Adafruit (`www.adafruit.com/category/1005`)
- *Grove* from Seeed Studio (`https://wiki.seeedstudio.com/Grove/`)

All three support a vast array of components and each an excellent choice. Since they are compatible, we will cover Qwiic and STEMMA QT in this chapter and Grove in Chapter 12.

C. Bell, *Beginning IoT Projects*, https://doi.org/10.1007/978-1-4842-7234-3_6

Overview

In this section, we will discover the Qwiic and STEMMA QT component systems. We will learn about the capabilities and limitations of the systems as well as examples of the components available. The chapter also includes details on how to start using the components in projects.

Both systems are designed to make building projects faster using pluggable modules containing sensors, input, output, and other functions. They both implement a modularized Inter-Integrated Circuit (I2C) (sometimes written I^2C) communication protocol.

What Is I2C?

I2C is a fast digital protocol that uses two wires (plus power and ground) to read data from circuits (or devices). The I2C protocol is perhaps the most common protocol that you will find on breakout boards[1] (a separate printed circuit board (PCB) with pins mounted).

Most host boards such as the Arduino and Raspberry Pi support the I2C protocol, and the pins can be found in the GPIO header. In fact, this is how you would connect I2C devices if you were not using a component system. For example, Figure 6-1 shows the I2C pins including power and ground on the Arduino.

[1] https://theorycircuit.com/breakout-boards-electronics/

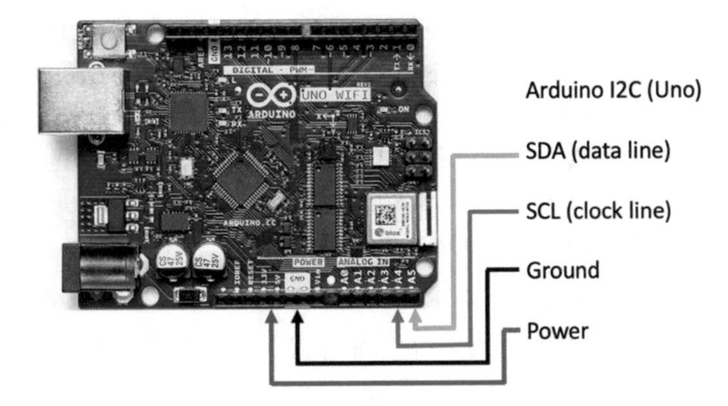

Figure 6-1. *I2C hardware interface location (Arduino)*

Figure 6-2 shows the I2C pins on the Raspberry Pi GPIO.

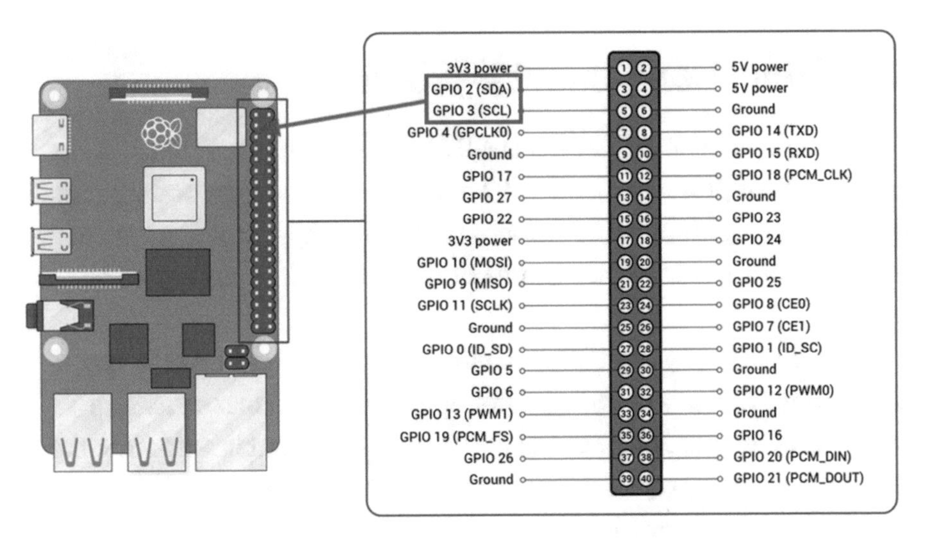

Figure 6-2. *I2C hardware interface location (Raspberry Pi)*

Note The Qwiic host adapter reroutes the I2C pins to its own circuits to support Qwiic and STEMMA QT modules.

The protocol is designed to allow the use of multiple devices (slaves) with a single master (the host board with adapter). Thus, each I2C breakout board will have its own address or identity that you will use in the driver to connect to and communicate with the device. Thus, you can connect a number of devices to the same "chain" of I2C modules. For example, the Qwiic host adapter can support over 100 modules.

Tip You can read about the details of I2C including how it works at `https://learn.sparkfun.com/tutorials/i2c/all`.

Now that we know what I2C is, let's see how easy the Qwiic Component System makes use of it.

The Qwiic Component System

Qwiic was created and released in 2017 by SparkFun (`sparkfun.com`) to speed up their prototyping efforts. They found they were spending some time doing a lot of soldering. To save them time, they created a system of connectors to modularize connections for the I2C communication protocol.

The Qwiic Component System eliminates soldering and tedious connections that most hardware examples require; it does so using a common connector. This frees you to learn more about how to write the code behind your project and thereby make more sophisticated projects with far fewer hardware issues.

The Qwiic system is a set of host adapters (also called hats, shields, carrier boards, etc.) available for a variety of platforms (Arduino, Raspberry Pi, etc.) and modules that contain small circuits that include sensors, input devices, output devices, and more. The modules come with mounting holes for permanently mounting the module in an enclosure. A nice touch.

Each host adapter contains one or more Qwiic connectors. These host adapters simply connect to your host board enabling the use of Qwiic modules without the need for additional electronics such as breadboards and discrete components or, in most cases,[2] without soldering. SparkFun has also added Qwiic connectors to many of their own Arduino and similar platforms eliminating the cost of adding a host adapter to your project.

Each module is self-contained; all of the supporting electrical components are on the module mounted on a small PCB (most come in a pretty red color in fact). All you need to do is connect the modules to your host adapter using a Qwiic wiring lead, and your hardware is done.

Capabilities

The capabilities of the Qwiic system are as elegant and simple as the products themselves and include the following:

- Modularized I2C bus.
- Easy, polarized connectors (no incorrect or reversed connections[3]).
- Modules can be daisy chained to form a linear array of modules.
- No soldering required!
- Over 100 modules can be chained together.

[2] A few host adapters may come without headers soldered.

[3] Perhaps the greatest bane of anyone working with I2C is inadvertently reversing the data and clock connections. Qwiic eliminates that guesswork entirely.

How Does It Work?

Qwiic uses a four-pin cable with JST connectors.[4] As mentioned, the cable is polarized, which means you can only connect the cable to the device one way, so you always know the connections are correct. You can even daisy chain modules together to make complex projects. Figure 6-3 shows a simple example.

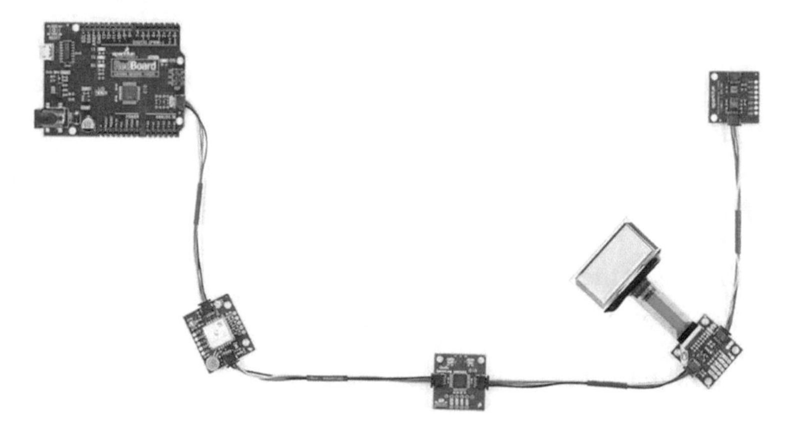

Figure 6-3. *Qwiic daisy chain (courtesy of sparkfun.com)*

The pinout for the cable and the color for each wire are as shown in Table 6-1. The connector itself is black in color (those purchased from SparkFun) or white (the STEMMA QT connector is the same pinout – more on that in the next section).

[4] `https://en.wikipedia.org/wiki/JST_connector`

Table 6-1. *Qwiic Connector Pinout*

Pin	Connection	Color
1	Ground	Black
2	3.3V	Red
3	SDA (data)	Blue
4	SCL (clock)	Yellow

Figure 6-4 shows a closeup of the female connector (the one you will plug the cable into) along with a drawing of the connector pins. Note the pins are listed from bottom to top in the drawing.

Figure 6-4. *Qwiic connector and pinout (courtesy of sparkfun.com)*

Notice the "ears" on either side of the female connector. These are offset so that you can only plug the cable in one direction. Cool! Figure 6-5 shows a typical Qwiic cable. This one is 50mm long making it one of the shorter cables. You can purchase cables in lengths 50, 100, 200, and 500mm (`www.sparkfun.com/categories/tags/qwiic-cables`).

Figure 6-5. *50mm Qwiic cable (courtesy of sparkfun.com)*

The Qwiic modules typically have two female connectors that allow you to form the daisy chain. Just plug a cable into each socket to chain modules together. The order doesn't matter because each module will "answer" at a specific address. The last module in the chain does not need to be terminated - just don't plug another cable into the last connector. Figure 6-6 shows a typical Qwiic module.

Figure 6-6. *Qwiic module (courtesy of sparkfun.com)*

Notice the two Qwiic connectors as well as additional holes at the bottom of the board. These additional holes can be used to solder a header allowing you to make use of any specialized operations. For example, notice the holes marked SCL, SDA, 3.3V, and GND. Yes, these are the I2C pins, which means you can use this module in other projects that do not have a Qwiic host adapter. Most modules have similar breakout pins.

The host adapter has one or more Qwiic connectors that you can use to form chains of modules. There are a variety of host adapters available for a growing list of host boards. This includes the Arduino (three varieties), micro:bit (three varieties), MicroMod (four varieties), Photon, Raspberry Pi (seven varieties), and Teensy (two varieties) boards. You can discover the latest offerings by visiting the following links. There's even a Qwiic hat for the Raspberry Pi 400 (`www.sparkfun.com/products/17512`):

- *Hats* (e.g., Raspberry Pi): `www.sparkfun.com/categories/tags/qwiic-hats`
- *Shields* (e.g., Arduino, Teensy): `www.sparkfun.com/categories/tags/qwiic-shields`
- *Carrier boards* (e.g., micro:bit, MicroMod): `www.sparkfun.com/categories/tags/qwiic-carrier-boards`

The hat that most will want to use is called the Qwiic Pi HAT (`www.sparkfun.com/products/14459`) that provides four Qwiic connectors as well as some most used GPIO pins broken out. It is designed to mount on the Raspberry Pi GPIO header so that the hat extends beyond the Raspberry Pi board making it easy to use the board with a case. Figure 6-7 shows a Qwiic hat for the Raspberry Pi.

Figure 6-7. *Qwiic host adapter (hat) for Raspberry Pi (courtesy of sparkfun.com)*

Notice the most used pins represented as a line of holes below the GPIO header. This is where you can solder a pin header (or wires directly) to access those GPIO pins easier. Another cool feature!

Note If your host adapter has multiple Qwiic connectors, do not form a loop by connecting the last module in the chain to the host adapter.

Now that we know what the Qwiic system is and how it works, let's examine some of the limitations.

Limitations

Like most systems, there are some limitations. Fortunately, there are few, and only the largest or most complex projects may need to heed.

Aside from the fact that the interface supports only I2C, there are few other limitations. Most notable is the system is designed for 3.3V, so platforms that use 5V will need to drop the voltage to use the modules, most likely on the host adapter.

Another limitation is the length of the daisy chain. The I2C bus was designed originally to be used on printed circuit boards rather than over longer wires. However, SparkFun reports that chains up to about 4 feet should work. If you need to use a longer cable or daisy chain, you will need to use the differential I2C breakout board (www.sparkfun.com/products/14589), which permits you to send the I2C signal over an Ethernet cable (RJ-45 connector). This means you could potentially use up to a 100-foot cable to connect a remote module. Figure 6-8 shows the differential I2C breakout board.

***Figure 6-8.** Differential I2C breakout board (courtesy of sparkfun.com)*

You may also think that if the I2C module you already own or find elsewhere doesn't have a Qwiic connector, you cannot use it with the Qwiic system. That limitation can easily be overcome by using the Qwiic Adapter (www.sparkfun.com/products/14495), which is a tiny board that you can add additional Qwiic connectors. Figure 6-9 shows the Qwiic Adapter.

Figure 6-9. *Qwiic Adapter (courtesy of sparkfun.com)*

Now that we know more about Qwiic, let's look at the STEMMA QT system.

The STEMMA QT Component System

To understand the STEMMA QT system, we must start with the original project from Adafruit from which STEMMA QT originates: STEMMA. Adafruit created the STEMMA system in 2018 and began adding connectors to some of their components.

Interestingly, the goals of this project are very similar to Qwiic: to minimize soldering during prototyping or building projects. As you can imagine, this makes the STEMMA system very similar to the Qwiic system. But there are some differences.

Rather than support a single protocol, Adafruit designed STEMMA to support I2C, pulse wave modulation (PWM), analog, and digital – just about all of the major protocols you may use for complex projects. These connectors are either three- or four-pin JST connectors. Each connector supports a different segment of these protocols as follows:

- *STEMMA four-pin JST*: 0.2mm pitch connector supporting I2C. Connectors are the same size as those used in the Grove system allowing for integration.
- *STEMMA three-pin JST*: 2.0mm pitch connector that supports PWM, analog, and digital.
- *STEMMA QT*[5] *four-pin JST*: 0.1mm pitch connector supporting I2C. Connectors are the same size as the Qwiic system making all STEMMA QT modules compatible with Qwiic host adapters.

You may be wondering why the system has three different connectors. The answer is practicality. Some boards and modules are smaller and may not have enough space for the larger connectors, making the smaller pitch connectors better suited for smaller modules and breakout boards.

While the STEMMA system supports more protocols than Qwiic, we will limit our discussion to the STEMMA QT connectors that support I2C and are pin compatible with Qwiic. Feel free to explore the other STEMMA and STEMMA QT offerings from Adafruit by visiting `www.adafruit.com/category/1005` for a complete list.

Tip If you'd like to learn more about STEMMA and the other connector types, see `https://learn.adafruit.com/introducing-adafruit-stemma-qt`.

---

[5] Comically pronounced "cutie."

Like Qwiic, the STEMMA QT system is a collection of modules that can be connected together with a simple cable to the host board. However, unlike Qwiic, there are currently only a few STEMMA QT host adapters available. Notable adapters from Adafruit include those with a color display for the Raspberry Pi (`www.adafruit.com/product/4393` and `www.adafruit.com/product/4484`), which have Qwiic connectors on the bottom. Instead, Adafruit has added STEMMA QT connectors to many of their boards including the Feather, Circuit Playground, and more.

The STEMMA QT modules are where the Adafruit products shine. You can get many different modules supporting STEMMA QT from cameras, sensors, a variety of displays, and even a speaker. Adding the fact that the STEMMA QT modules are compatible with Qwiic host adapters means choosing Qwiic and STEMMA QT enables you to choose from a long list of modules!

Capabilities

The capabilities of the STEMMA QT system are very similar to the Qwiic system and equally as elegant including the following:

- Modularized I2C bus.
- Easy, polarized connectors (no incorrect or reversed connections[6]).
- Modules can be daisy chained to form a linear array of modules.
- No soldering required!
- Over 100 modules can be chained together.

[6] Perhaps the greatest bane of anyone working with I2C is inadvertently reversing the data and clock connections. STEMMA QT eliminates that guesswork entirely.

The major difference is the STEMMA QT modules can accept 3–5V DC, whereas the Qwiic operates at 3.3V. Adafruit wanted to keep the 5V capability in order to power some of their modules that use LEDs, which operate better at the higher voltage. So one must be careful when mixing Qwiic and STEMMA QT modules. More on that later in this chapter.

How Does It Work?

Recall STEMMA QT has the same four-pin cable with JST polarized connectors as Qwiic and you can even daisy chain modules together to make complex projects. Figure 6-10 shows a simple example using a temperature and pressure sensor with Adafruit's Feather Express board (`www.adafruit.com/product/4382`).

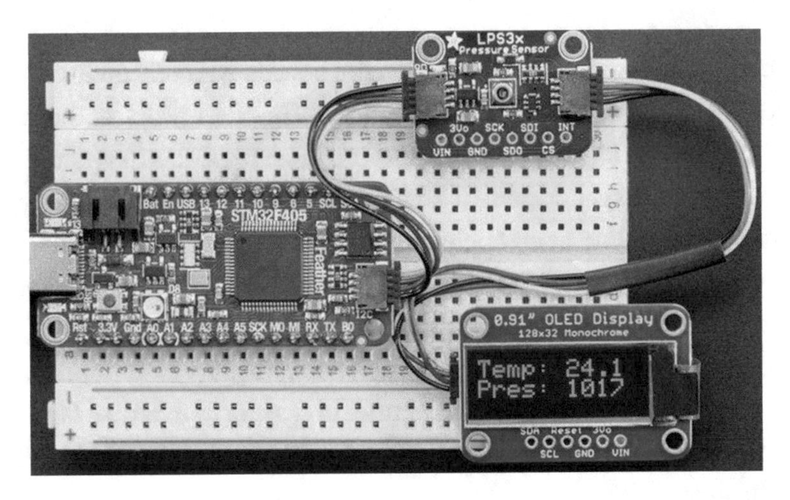

***Figure 6-10.** STEMMA QT daisy chain (courtesy of adafruit.com)*

The pinout for the cable and the color for each wire are the same as the Qwiic connector with the notable exception that STEMMA QT can support 3–5V power. Similarly, STEMMA QT cables are also the same connectors, so you can use either in your project.

Like Qwiic, STEMMA QT modules typically have two female connectors that allow you to form the daisy chain. Since these are made or sold by Adafruit, they come with a black or blue PCB. Figure 6-11 shows a typical STEMMA QT module.

Figure 6-11. *STEMMA QT module (courtesy of adafruit.com)*

Once again, like Qwiic connectors, the boards typically include a breakout section, so you can use the modules in other projects that do not have a Qwiic or STEMMA QT host adapter. Most modules have similar breakout pins. Most also come with mounting holes for permanently mounting the module in an enclosure. A nice touch.

Recall there are not that many STEMMA QT host adapters available beyond the ones for the Raspberry Pi. Fortunately, you can use the Qwiic host adapters instead. Regardless, if you want a display in your project and you're using the Raspberry Pi, you cannot go wrong with either Mini PiTFT host adapter. Figure 6-12 shows the Mini PiTFT - 135×240 Color TFT Add-on for Raspberry Pi.

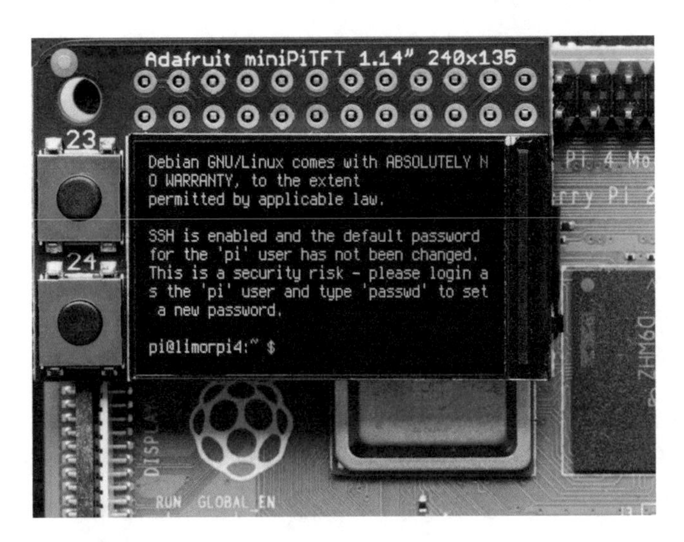

Figure 6-12. *Adafruit Mini PiTFT - 135×240 Color TFT Add-on for Raspberry Pi (courtesy of adafruit.com)*

Notice unlike the Qwiic host adapters, this one does not have any additional breakout pins because it is designed to be as compact as possible and fit within the confines of the Raspberry Pi board footprint. The programmable buttons are a nice touch. Better still, you can use this device as a console output or with a custom interface of your design.

Figure 6-13 shows the reverse side of the board with the STEMMA QT connector at the lower right.

Figure 6-13. *Mini PiTFT Color TFT Add-on reverse side (courtesy of adafruit.com)*

Notice the most used pins are represented as a line of holes below the GPIO header. This is where you can solder a pin header (or wires directly) to access those GPIO pins easier. Another cool feature!

Now that we know what the STEMMA QT system is and how it works, let's examine some of the limitations.

Limitations

The limitations of the STEMMA QT are largely the same as the Qwiic system with respect to its I2C interface. The biggest difference in the systems is the power level. Given some STEMMA QT modules may require 5V, you may not be able to use 5V STEMMA QT modules with Qwiic host adapters, but that is a small price to pay and not insurmountable as you can use a breakout board to adapt your 5V STEMMA QT module either by wiring it directly to your host board or through an adapter.

Now that we know what the component systems are and a bit about how they work, let's look at some of the components available for each system.

Components Available

There are a lot of components available for both the Qwiic and STEMMA QT systems. Since they are compatible, you can build projects that use a mix of both systems. This section highlights some of the categories of modules available for each system. We won't see everything that is available as the offerings for both systems continue to grow. Rather, we will see the more popular host adapters and modules as well as those we will use in upcoming chapters.

The Qwiic Component System

The following offers a pictorial representation of the components available from SparkFun. We omit showing the various cabling options as they aren't nearly as interesting as the host adapters and modules.

But first, let's look at some examples of development boards from SparkFun that have the Qwiic host connectors onboard.

Development Boards

There are a number of host boards (SparkFun calls them development boards) available that include the Qwiic connectors making them ideal for an IoT project where you want to minimize the number of components and perhaps cost. There are several types (product brands) of development boards available including the RedBoard products.

RedBoard products are Arduino-compatible boards that support the Arduino Uno shield header and include a number of features making them an excellent alternative to the standard Arduino boards. Plus, there are several versions that come with Qwiic connectors including the SparkFun RedBoard Qwiic (`www.sparkfun.com/products/15123`) and the RedBoard Turbo (`www.sparkfun.com/products/14812`). Figure 6-14 shows the RedBoard Qwiic. Notice the Qwiic connector located on the right side of the board. And, yes, they are red.

***Figure 6-14.** SparkFun RedBoard Qwiic (courtesy of sparkfun.com)*

There is another RedBoard variant that may interest those working with Arduino-based projects who want to mount the board in an enclosure or make user-defined operations like buttons accessible. The RedBoard Edge (`www.sparkfun.com/products/14525`) shown in Figure 6-15 provides the Qwiic connector, four LEDs, a reset button, a programmable toggle switch, and a terminal mount for power in a format that can be mounted on a panel.

***Figure 6-15.** RedBoard Edge (courtesy of sparkfun.com)*

Other development boards from SparkFun that include Qwiic connectors include the following:

- *SparkFun Thing Plus*: An Arduino-compatible board with a powerful SAMD51 microcontroller in a unique, minimal footprint that is compatible with the Adafruit Feather layout (`www.sparkfun.com/products/14713`).
- *Qwiic Pro Micro*: An Arduino-compatible board in the same format as the Arduino Micro Pro with a USB-C connector instead of the smaller, more fragile micro-USB connector as well as a host of additional features (`www.sparkfun.com/products/15795`).
- *Qwiic Micro*: A super-small board that is about the same size as a Qwiic module featuring an SAMD21 microcontroller programmable by Arduino or CircuitPython (`www.sparkfun.com/products/15423`).

If you are interested in any of these development boards for your project, be sure to visit the product website to learn more and purchase them from SparkFun.

Tip See `www.sparkfun.com/categories/399` for the complete list of Qwiic products including a number of development boards.

Host Adapters

Aside from the impressive list of modules, the variety of host adapters available from SparkFun is very impressive. Recall there are host adapters for Raspberry Pi, Arduino, micro:bit, MicroMod, Teensy, Photon, and more.

Since we are working with Arduino and Raspberry Pi in this book, let's look at versions for these platforms.

There are shields for the Arduino Uno shield header as well as one for the Arduino Nano. The SparkFun Qwiic Shield for Arduino shown in Figure 6-16 supports four Qwiic connectors as well as breakout pins for the entire Arduino as well as a prototype area that is a nice touch for those wanting to add custom circuits.

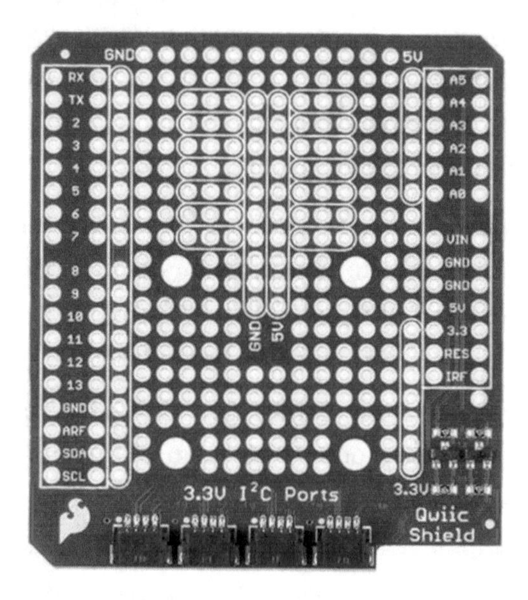

Figure 6-16. *SparkFun Qwiic Shield for Arduino (courtesy of sparkfun.com)*

There is one, small drawback for the beginner wanting to use this shield. It comes unassembled, which means you will need to solder the supplied headers onto the shield in order to use it by stacking it on top of your Arduino. That isn't too great a problem to overcome since it gives you an opportunity to learn how to solder, or you can simply find someone who does and get them to solder the headers for you.

Tip See `www.sparkfun.com/categories/tags/qwiic-shields` for the complete list of Qwiic shields available for the Arduino and other platforms.

There are a variety of hats for the Raspberry Pi that include some interesting features. The following lists the hats with other features. All of these mount to the GPIO header and have Qwiic connectors for expansion. Some also permit access to the GPIO header:

- *GPS-RTK Dead Reckoning pHAT*: GPS module (`www.sparkfun.com/products/16475`).
- *Top pHAT for Raspberry Pi*: A TFT display designed to be placed on top of other hats (`www.sparkfun.com/products/16653`).
- *Auto pHAT*: A hat designed to support robotics projects (`www.sparkfun.com/products/16328`).

Tip See `www.sparkfun.com/categories/tags/qwiic-hats` for the complete list of Qwiic hats available for the Raspberry Pi.

We've already seen the typical hat used for Qwiic – the SparkFun Qwiic Pi HAT for Raspberry Pi. That's the one you're most like going to use. However, there is another option. SparkFun also makes a tiny hat (called a shim) that supports a single Qwiic connector in a compact format. You just slip it onto the GPIO header, and off you go. It's small enough to fit inside almost any Raspberry Pi case. Figure 6-17 shows the SparkFun Qwiic SHIM for Raspberry Pi.

Figure 6-17. *SparkFun Qwiic SHIM for Raspberry Pi (courtesy of sparkfun.com)*

Modules

SparkFun offers a wide variety of modules that contain sensors, input, output, and display capabilities. So many that it is not possible to list them all here. Table 6-2 lists the categories of modules available with a link to each category for further reading. All URLs (links) begin with `www.sparkfun.com/categories/tags/`.

Table 6-2. *Categories of Qwiic Modules*

Category	Description	Category Link
Sensors	Modules that allow you to sample the world around us	qwiic-gps
Imaging	Modules that sense the light spectrum	qwiic-imaging
Distance	Modules that sense distance or proximity	qwiic-distance
Movement	Modules that sense movement of the device	qwiic-movement
Environment	Modules with sensors designed to measure the environment	qwiic-environmental
Other	Additional accessory modules for a variety of specialized operations such as RFID, capacitive touch, and more	qwiic-other

So what are the modules available in these categories? The following is a list of subcategories that help in understanding the breadth of options available:

- *Sensors*: Typically contain a single sensor that produces output (readings or values) on the I2C bus. Examples include temperature, humidity, pressure, distance, magnetometer, light, and environmental (gases) sensors.
- *Displays*: Modules that contain an output device for displaying data. Examples include OLED and LED displays.
- *Relays*: Modules that contain relays that permit you to switch higher-power devices on or off.
- *Motors*: Modules that permit you to control small electric motors.
- *Input*: Modules that contain one or more buttons, potentiometers, keypads, or switches.
- *ADC/DAC*: Modules that provide analog-to-digital conversion (ADC) or digital-to-analog conversion (DAC) that permit incorporation of other circuits into your project.
- *Accessory*: Various modules that provide handy operations such as data loggers, cryptographic operations, and even an MP3 trigger module.

Now, let's look at the Qwiic modules we will be using in the upcoming chapters as we explore how to write the code for IoT projects using the Qwiic system beginning with an output device.

The SparkFun Micro OLED Breakout (Qwiic) (`www.sparkfun.com/products/14532`) is a tiny screen we will use to present the data collected in the example projects. It is a 64×48-pixel monochrome OLED display that is only about 2cm square (about 7/8 inch). Figure 6-18 shows the module with the OLED facing side. The Qwiic connectors are located on the back of the PCB along with the I2C breakout pins. It has a nifty cutout for securing the Qwiic cables.

Figure 6-18. *SparkFun Micro OLED Breakout (Qwiic) (courtesy of sparkfun.com)*

The SparkFun Proximity Sensor Breakout - 20cm, VCNL4040 (Qwiic) module (`www.sparkfun.com/products/15177`) provides an infrared distance sensor that you can use to measure distance (technically, proximity) of objects up to 20cm away. Figure 6-19 shows the module with the sensor facing side.

Figure 6-19. *Proximity Sensor Breakout (Qwiic) (courtesy of sparkfun.com)*

The SparkFun Environmental Combo Breakout - CCS811/BME280 (Qwiic) module (`www.sparkfun.com/products/14348`) supports two sensors, the CCS811 and BME280, that measure barometric pressure, humidity, temperature, TVOCs, and equivalent CO2 (or eCO2) levels. Figure 6-20 shows the module with the sensor facing side.

Figure 6-20. *Environmental Combo Breakout (Qwiic) (courtesy of sparkfun.com)*

The SparkFun Triple Axis Magnetometer Breakout - MLX90393 (Qwiic) module (`www.sparkfun.com/products/14571`) supports a magnetometer, which is used to measure magnetic fields and can be used as a compass. Figure 6-21 shows the module with the sensor facing side.

Figure 6-21. *Triple Axis Magnetometer Breakout (Qwiic) (courtesy of sparkfun.com)*

Once again, there are many modules available. These are just a sampling of the modules available from SparkFun. A compact list of all Qwiic devices and modules is available at `www.sparkfun.com/qwiic`.

Cabling and Connectors

See `www.sparkfun.com/categories/tags/qwiic-cables` and `www.sparkfun.com/categories/tags/qwiic-connectors` for a list of cables and connectors to support the Qwiic system.

The STEMMA QT Component System

The following offers a pictorial representation of the components available from Adafruit. We omit showing the various cabling options as they aren't nearly as interesting as the host adapters and modules.

Since we are using both Qwiic and STEMMA QT modules in this book, we will use the Qwiic host adapters described previously for the projects in this book. While there aren't that many host adapters currently available other than the two for the Raspberry Pi, Adafruit provides a number of hosts (called controllers) for the STEMMA QT system.

Controllers

Adafruit offers a number of controllers that support several of their popular microcontroller boards. Most support their Feather and CircuitPython products, but there are a few that are interesting that you may want to look into once you've completed the sample projects in this book.

The first is the Adafruit QT Py – SAMD21 Dev Board with STEMMA QT (`www.adafruit.com/product/4600`), which is a very tiny board that supports the popular Arduino-compatible SAMD21 microcontroller with a USB-C connector for programming. Figure 6-22 shows the QT Py – SAMD21 Dev Board.

Figure 6-22. *QT Py – SAMD21 Dev Board (courtesy of adafruit.com)*

The board does not come with the headers soldered on, but Adafruit does supply them with the board. Better still, the board is very inexpensive. It can be used in projects where space is a premium and fewer GPIO pins are needed.

Another interesting controller is the Adafruit Feather STM32F405 Express (`www.adafruit.com/product/4382`), which runs CircuitPython natively (on-chip Python development), or you can program it with MicroPython or through the Arduino IDE (with special hardware libraries). Best of all, there are a number of accessories available for the Feather including shields that support sensors and other devices. Figure 6-23 shows the board.

***Figure 6-23.** Adafruit Feather STM32F405 Express (courtesy of adafruit.com)*

The board does not come with the headers soldered on, but Adafruit does supply them with the board.

Tip See `www.adafruit.com/category/621` for a complete list of the latest STEMMA QT controllers available from Adafruit.

While we do not use any of these controllers in the examples in this book, we will use some of the STEMMA QT modules.

Modules

Adafruit offers a wide variety of modules that contain sensors, input, output, cameras, speakers, and display capabilities. So many that it is not possible to list them all here. STEMMA QT modules cover the same range of categories as the SparkFun Qwiic modules. Please refer to the preceding Qwiic section for the categories of modules available.

Now, let's look at the STEMMA QT modules we will be using in the upcoming chapters as we explore how to write the code for IoT projects using the STEMMA QT system beginning with an output device.

The Monochrome 0.91" 128×32 I2C OLED Display - STEMMA QT / Qwiic (`www.adafruit.com/product/4440`) is a tiny OLED screen that is about 1" in size diagonal, which can be programmed to display text or graphics. It can be used with the 3.3V Qwiic host adapter. Figure 6-24 shows the module enlarged for better viewing (it is tiny - about the height of a US quarter dollar coin).

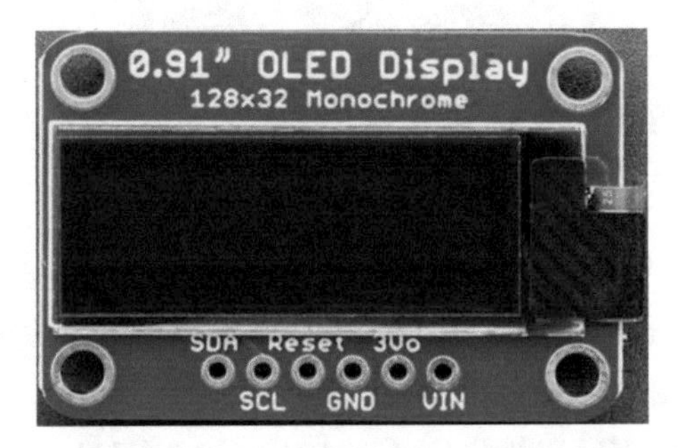

Figure 6-24. *Monochrome 0.91" 128×32 I2C OLED Display (courtesy of adafruit.com)*

One thing to consider is this is one of the few modules that have only a single STEMMA QT connector on the back, so you must place this module at the end of the daisy chain.

The Adafruit PCF8591 Quad 8-bit ADC + 8-bit DAC - STEMMA QT / Qwiic (`www.adafruit.com/product/4648`) is an interesting module that we will use to incorporate analog sensors into our projects. It provides an analog-to-digital converter that supports up to four inputs. This an excellent example of how to incorporate non-STEMMA QT (or Qwiic) components into your projects. Figure 6-25 shows the module.

Figure 6-25. *PCF8591 Quad 8-bit ADC + 8-bit DAC (courtesy of adafruit.com)*

In case you're curious and cannot wait until Chapter 10 to learn more, we will pair this with a soil moisture sensor from SparkFun (`www.sparkfun.com/products/13637`).

The Adafruit LSM6DS33 6-DoF Accel + Gyro IMU – STEMMA QT / Qwiic (`www.adafruit.com/product/4480`) allows you to add motion and orientation sense to your project. With this module, you can determine if the board is moving through six degrees of freedom (DoF). Figure 6-26 shows the board.

Figure 6-26. *LSM6DS33 6-DoF Accel + Gyro IMU (courtesy of adafruit.com)*

While not the most accurate or fastest sensor, it is a good value for the money.

Tip See `www.adafruit.com/category/620` for a complete list of the latest STEMMA QT devices and sensors available from Adafruit.

Cabling and Connectors

See `www.adafruit.com/category/619` for a list of cables and connectors to support the STEMMA QT system.

Where to Buy Qwiic and STEMMA QT Components

You can purchase Qwiic components directly from SparkFun (sparkfun.com). In addition to individual components, you can also find a variety of kits that combine Qwiic modules, cables, and a host adapter. They offer kits for developers and robotics and kits specific to the Raspberry Pi. The following are the links to the kit product pages:

- *Developer kits*: `www.sparkfun.com/categories/tags/qwiic-development-kits`
- *Robotics kits*: `www.sparkfun.com/categories/tags/qwiic-robotics-kits`
- *Raspberry Pi kits*: `www.sparkfun.com/categories/tags/qwiic-raspberry-pi`

One kit you may want to consider getting started is the SparkFun Qwiic Starter Kit for Raspberry Pi (`www.sparkfun.com/products/16841`), which includes a variety of cables, the host adapter, as well as three modules, the OLED, distance, and environmental modules, that we will use in this book.

You can purchase STEMMA QT components directly from Adafruit (adafruit.com).

Now, let's discuss how to use these systems in your projects.

Using the Components in Your Projects

Plugging your choice of Qwiic (or STEMMA QT) host adapter onto your host board and plugging the modules together with the cables is pretty easy. Recall the connectors only go one way so you can't cross-connect anything.

However, there is one important step you must do before connecting everything: power off your board. Qwiic and STEMMA QT are not hot pluggable. You cannot (and more importantly should not) connect and disconnect modules while your board is powered on. This could lead to damaging the module(s) or your host board.

Caution Do not plug or unplug Qwiic or STEMMA QT modules while your board is powered on.

Once the hardware is plugged together, the next step is to start working on the code to enable your modules and complete your project. To do so, you are likely required to load one or more software libraries.

The source for these software libraries may vary from one vendor to another and one module to another. The best place to start is to look at the documentation for the module. Both SparkFun and Adafruit are excellent at pointing you to the software and tools you need.

The following summarizes the steps necessary for the Arduino and Raspberry Pi. The following does not include all of the steps needed for all of the projects in the book; rather, the section is an overview of what you can expect to configure your PC to implement the projects. Specific details for each example are included in each chapter.

Loading Qwiic and STEMMA QT Libraries for the Arduino

Recall from Chapter 2 we can install software libraries using the Library Manager in the Arduino IDE. Simply open the Library Manager and search for Qwiic or STEMMA.

For example, in the next chapter, we will use the SparkFun Proximity Sensor Breakout - 20cm, VCNL4040 (Qwiic) module. According to the hookup guide, all we need to do is search the Library Manager for "SparkFun VCNL4040" (no quotes). Figure 6-27 shows how that would appear in the Arduino IDE.

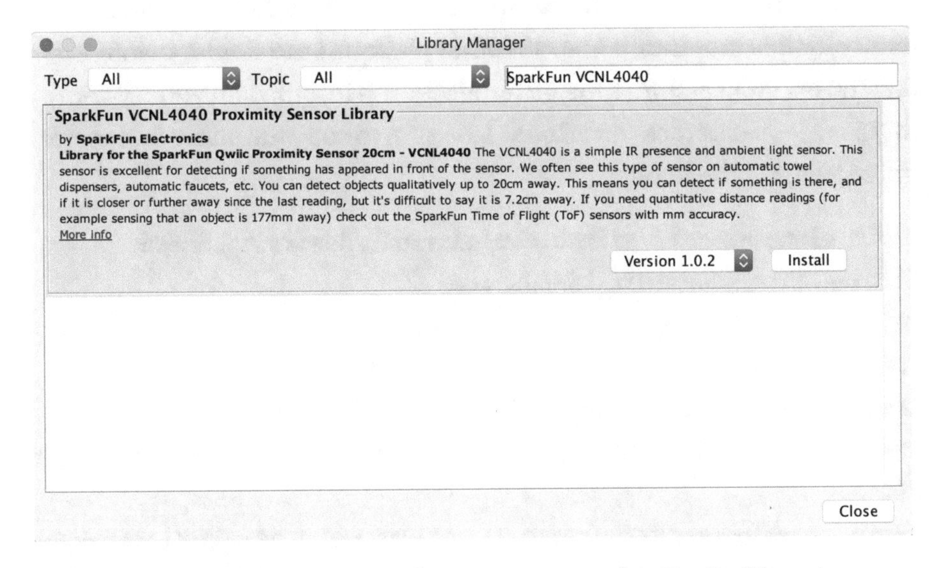

Figure 6-27. *Searching the Library Manager for Qwiic libraries*

The library is also available for downloading and manual installation via GitHub (`https://github.com/sparkfun/SparkFun_VCNL4040_Arduino_Library`).

Note Each module may require a different library, so be sure to check the product website for the latest updates.

However, for loading libraries for the STEMMA QT modules, you must use the manual method. For example, the Adafruit LSM6DS33 6-DoF Accel + Gyro IMU - STEMMA QT / Qwiic used in Chapter 11 can be downloaded from GitHub (`https://github.com/adafruit/Adafruit_LSM6DS`). Once you download the library, you will unzip the library and copy the folder to your Arduino libraries folder. In some cases, you may need to rename the folder to remove the `-master` appendix from the name.

The following shows a transcript of downloading and copying the library for this module on macOS (other platforms are similar). Note that the example assumes the git application is installed on your system (`https://git-scm.com/downloads`). Be sure to copy the entire folder, not the contents of the folder:

```
% git clone https://github.com/adafruit/Adafruit_LSM6DS
Cloning into 'Adafruit_LSM6DS'...
remote: Enumerating objects: 128, done.
remote: Counting objects: 100% (128/128), done.
remote: Compressing objects: 100% (86/86), done.
remote: Total 997 (delta 78), reused 70 (delta 41), pack-reused
869
Receiving objects: 100% (997/997), 953.47 KiB | 381.00 KiB/s, done.
Resolving deltas: 100% (649/649), done.

% cp -R Adafruit_LSM6DS ~/Documents/Arduino/libraries.
```

Tip The Arduino libraries folder location can be found on the Arduino IDE Preferences dialog.

Once you copy the library, you can open the Arduino IDE and locate the example sketches as demonstrated in Figure 6-28. If you do not see the examples, be sure to check the location of your Arduino libraries folder or restart your IDE to pick up the latest changes (newly copied library folder).

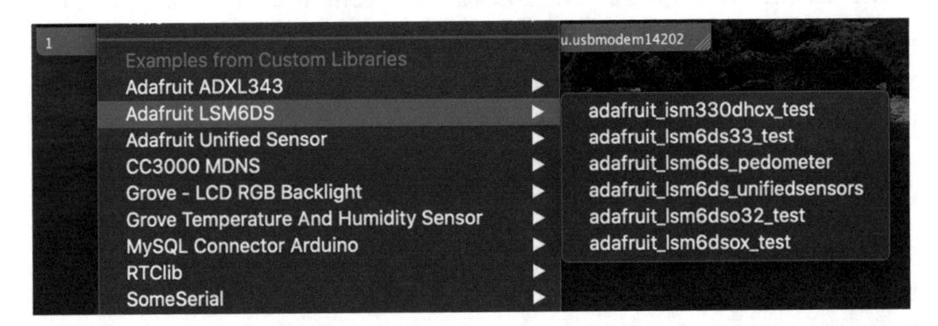

Figure 6-28. *Examples from manually installed library (Arduino IDE)*

Of course, you can do all of these steps on your PC using your desktop interface.

Loading Qwiic and STEMMA QT Libraries for the Raspberry Pi

Software libraries on the Raspberry Pi are a little different. Here, we are using Python libraries, which are installed differently. We would install them either using the `pip` command or, in some rare cases, by downloading the library and copying it to our project folder.

For example, SparkFun has created a Python library that contains all of the libraries needed for their Qwiic modules. To install it, simply enter the following command on your Raspberry Pi in a terminal. Once you do that, you're all set:

```
$ pip3 install sparkfun-qwiic
```

To learn more about the Qwiic Python library including alternative installation methods and the director layout, visit www.sparkfun.com/news/2958.

Sadly, there isn't a unified library for the Adafruit STEMMA QT modules. Few STEMMA QT modules have Python libraries available, but that doesn't mean we cannot use the modules. It means we have to find a library that will work. Fortunately, there are alternatives for some that provide a place to start.

For example, there is a Python library for the Adafruit PCF8591 Quad 8-bit ADC + 8-bit DAC – STEMMA QT / Qwiic module used in Chapter 10. To learn more about the library, you can visit https://github.com/adafruit/Adafruit_CircuitPython_PCF8591. It can be installed with the following command:

```
$ pip3 install adafruit-circuitpython-pcf8591
```

Notice the name mentions CircuitPython. Don't let that distract you. While CircuitPython is an on-chip system created by Adafruit, many of the libraries can be used on the Raspberry Pi.

Similarly, there is a Python library for the Monochrome 0.91" 128×32 I2C OLED Display – STEMMA QT / Qwiic. It can be installed with the following command:

```
$ pip3 install adafruit-circuitpython-ssd1306
```

Unfortunately, there isn't a Python library listed for the Adafruit LSM6DS33 6-DoF Accel + Gyro IMU – STEMMA QT / Qwiic module. A little creative googling results in locating a Python library for the LSM6DS33 breakout board. Thus, we can use that library as a starting point (perhaps without modification). This library can be installed with the following command:

```
$ pip3 install adafruit-circuitpython-lsm6ds
```

We will look at this library in greater detail in Chapter 10.

Integrating Additional Components

Now that we've seen what the Qwiic and STEMMA QT systems are and how they work and had a glimpse at the components available, what do you do if you want to use a specific module or circuit in your project but there isn't a module available?

Fortunately, the nice, thoughtful folks at SparkFun have got you covered here too. So long as what you want to incorporate can communicate with the I2C protocol, you can use various components designed to make almost anything usable in a Qwiic chain.

Assembling the Hardware

As an example, let's assume you have an I2C module or breakout board from another vendor or project. In this case, we have an older BMP180 module that we want to use to measure temperature and humidity. In order to use it with your Qwiic and STEMMA QT modules, you will need to add the right connectors.

Recall the Qwiic Adapter (`www.sparkfun.com/products/14495` from earlier), which is a tiny board that you can wire to your module. You can use this module to simply add the Qwiic connectors to the I2C breakout board. Figure 6-29 shows how you would wire up an I2C breakout board to the Qwiic Adapter. You simply solder wires from one to the other matching the pins as shown.

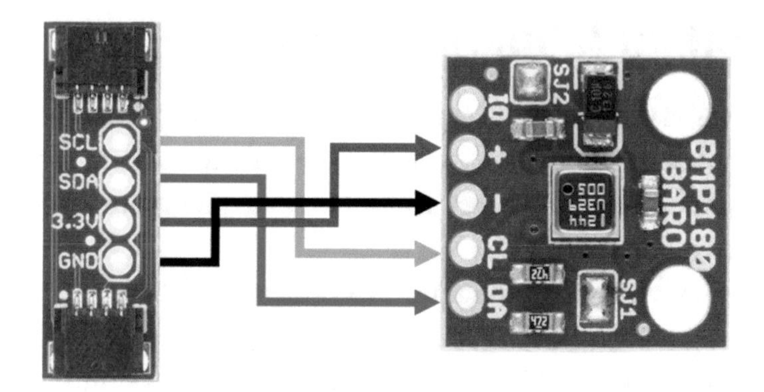

Figure 6-29. *Qwiic Adapter and BMP180 module (images courtesy of sparkfun.com)*

Another example of how to use the Qwiic Adapter is to solder right-angle headers to the adapter and the module as SparkFun suggests. Figure 6-30 shows an example. Here, we see the Qwiic Adapter used to enable Qwiic connectors on a BME280 breakout board.

Figure 6-30. *Adapting I2C modules (courtesy of sparkfun.com)*

Hardware is only part of the solution. The next step is find and use the correct software library.

Adapting Software Libraries

Note that while the wiring bits may be a challenge if you've never worked with a breadboard or don't know how to solder, another potential challenge is finding a software library that works with the module you want to use. Adding the Qwiic connectors as shown previously won't solve that for you.

In most cases, the original software library may work just fine. It's best to try it and see. If it doesn't work, the most likely issue is either the address of the module differs or there is another module in your daisy chain with the same address. If you have the source code for the library, changing the address is really easy. Plus, some libraries allow you to specify one of several alternative addresses.

Another possibility is the software library is out of date or doesn't work with your chosen host board. In those cases, it is best to check with the author of the library for assistance. But before you do so, be sure to do your homework by searching for possible solutions. It is always better to be knowledgeable of the problem instead of blaming the author when things don't work.

These challenges aside, it is not overly complicated to integrate existing I2C devices into your project using the adapter from SparkFun. There is another way to integrate analog devices into your project using the ADC described previously, but we will cover that in greater detail in Chapter 10.

Summary

The heart of building IoT projects without a lot of electronics experience and especially soldering and breadboarding is the Qwiic and STEMMA QT component systems. With a simple, no-error connector, you can wire together a very impressive set of modules to create your IoT solution. From sensors to displays, nothing is impossible to make your project successful.

Now that the hardware challenges have been nearly eliminated, we can turn our attention back to learning how to write the code for our projects. As you saw in this chapter, this may require installing software libraries to support the modules you are using or adapting existing libraries to suit your needs.

The next chapter begins a series of projects that use Qwiic and STEMMA QT components to teach you how to work with the systems for both the Arduino and Raspberry Pi.

CHAPTER 7

Keep Your Distance!

Thus far in the book, we've discovered how to program the Arduino platform using the Arduino C++-like language and how to write Python programs to run on the Raspberry Pi and had a tour of the Qwiic and STEMMA QT systems. Now it is time to start putting what we've learned to use.

In this chapter and the next four chapters, we will explore a variety of smaller projects designed to give you experience working with modules and writing the code to communicate with them. As such, we won't explore complete IoT projects because, as you will see in Chapters 17 and 18, working with the cloud is a different level of skills. It is best to become proficient in the basics before tackling the cloud element.

To balance this intentional format, each of the example project chapters includes a section entitled "Going Further," which gives you hints on how to make the project better and alter it for other uses and ideas on how to leverage cloud technologies to make it an IoT project – something you can do on your own once you finish the cloud chapters.[1]

Each chapter also includes a list of the hardware you will need to complete the project for the Arduino and Raspberry Pi. You are encouraged to use whichever platform you feel most comfortable with including implementing the code on both platforms and to read through both sections

---

[1] The books I remember and appreciate the best are those that inspired me to explore beyond the confines of the text by giving me ideas to implement with the knowledge presented rather than simply walk through a project that is limited in scope.

C. Bell, *Beginning IoT Projects*, https://doi.org/10.1007/978-1-4842-7234-3_7

for complete coverage as discussions about specifics are not repeated in both sections. After all, the hardware is the same except for the host adapter, and the code is very similar despite being written in different languages.

Let's jump into our first project that uses a distance sensor.

Project Overview

The project for this chapter is designed to demonstrate how to get started building hardware projects using sensors. The sensor chosen for this project is the proximity sensor we saw in Chapter 6. We will use the sensor to detect when something is near and display the value and a message on a small screen. Think of it as a simple object detector rather than a distance measuring device. This is because the proximity sensor is limited in range and best used to detect the presence of something close by rather than a precise distance calculation.

Interestingly, the proximity sensor can also detect ambient and white light, but we will not use those measurements in this project. Rather, we save that to the end for ideas on how to employ the sensor in other projects.

What Will We Learn?

By implementing this project, we will learn how to connect Qwiic modules to our host boards and how to form a daisy chain of Qwiic modules. We will pick up a few tips on working with the hardware along the way. Thus, the project itself is very simple and is not likely to impress, but it is well suited for learning all of the nuances of building Qwiic projects.

The programming tasks will reveal how to set up the I2C bus, read a value from the sensor, and interpret the value to make a decision on the proximity of objects. We will also learn how to write messages to the OLED module to help make the project useful on its own, that is, without having it plugged into our computer running in debug mode.

Let's see what hardware we will need.

Hardware Required

The hardware needed for this project is listed in Table 7-1. URLs for each component are included for ease of ordering including duplicate entries for alternative vendors.

Table 7-1. *Hardware Needed for the Keep Your Distance Project*

Component	URL	Qty	Cost
Proximity Sensor Breakout – 20cm, VCNL4040	www.sparkfun.com/products/15177	1	$6.95
Micro OLED Breakout	www.sparkfun.com/products/14532	1	$16.95
Qwiic cable (any length can be used)	www.sparkfun.com/products/14427	2	$1.50
SparkFun RedBoard Qwiic (Arduino Uno or compatible)	www.sparkfun.com/products/15123	1	$19.95
Raspberry Pi 3B or later	www.sparkfun.com/categories/233 www.adafruit.com/category/176	1	$35.00+
Qwiic pHAT for Raspberry Pi	www.sparkfun.com/products/15945	1	$5.95

About the Hardware

Let's discuss these components briefly. We will discover how to work with the hardware in more detail later in the chapter.

Sensor

Note that I chose the infrared sensor over the longer-range sensors due mainly to cost. The other sensors that have greater range are several times the cost of the infrared sensor.

A Note About I2C Addresses

Addresses are commonly referred to in hexadecimal values and are depicted in code with the 0x prefix, for example, 0x48, 0x60, 0x7D, etc.

One of the newest features in the latest Arduino boards is the addition of a cryptography co-processor chip, which enables a higher degree of security by providing a place to generate and store sensitive data such as keys. However, the cryptography chip is connected to the I2C bus at address 0x60. And, no, you cannot disconnect or turn it off. It's permanently part of the Arduino board.[2]

This is fine so long as no other device on your I2C bus uses address 0x60. Unfortunately, the proximity sensor chosen for this project uses address 0x60. Thus, you cannot use one of the newer Arduino boards such as the Uno WiFi Rev2, Zero, and MKR series boards.

See `www.arduino.cc/reference/en/libraries/arduinoeccx08/` for a complete list of Arduino boards that have the cryptography chip installed.

[2] For the really adventurous, you could locate the chip and unsolder it from the board. That'll show them who's really driving their I2C bus!

RESOLVING I2C ADDRESS COLLISIONS

If you encounter a situation where you have two or more modules that use the same I2C address and those modules are not part of the host board (Arduino or Raspberry Pi), you can overcome the problem by using SparkFun's Qwiic Mux Breakout – 8 Channel (TCA9548A) (`www.sparkfun.com/products/16784`) board. This board allows you to connect up to eight sensors that use the same address and communicate with them individually. For example, you may want to use several of the same module in your project. The Mux is an ingenious board that works really well. See the hookup guide for more details about how to use this board in your Arduino and Raspberry Pi (Python) projects – they've included documentation for both.

Arduino Board

The Arduino board used in this project is the SparkFun RedBoard Qwiic, which has a Qwiic connector onboard. You can use whatever Arduino board you want, but make sure it does not have a cryptography chip because it uses the same address as our proximity sensor.

If you choose to use a different Arduino without a Qwiic connector on the board, you can use the Qwiic Cable - Breadboard Jumper (`www.sparkfun.com/products/14425`) that permits you to connect the wires directly to your Arduino instead of using a shield. We will see this option in action later in the chapter.

Another option is to use the Qwiic Shield for Arduino (`www.sparkfun.com/products/14352`), but that shield only works for the older Arduino boards that use A4 and A5 for the I2C data and clock lines. The shield will not work without modification for Arduino boards that use dedicated pins for I2C.

Tip To save on shipping, see the Appendix for a complete list of all hardware used in the book.

Assemble the Qwiic Modules

Now, let's see how to connect the modules together. We will go through the connections step by step in this chapter, and later chapters present a drawing for you to follow.

You should always connect your Qwiic modules together first before connecting them to your host board and always while the board is powered off. Never attempt to connect or disconnect Qwiic modules while the board is powered on.

Caution Turn off your host board before connecting Qwiic cabling.

Since most Qwiic modules have connectors on both sides, the order in which you connect them doesn't matter or is just possibly for aesthetics or convenience, that is, placing the OLED module in the position that permits the best view. Since we're not installing the project in an enclosure, order is not important.

Begin by examining the Qwiic cable. If you look closely, you will see the black connectors have two small "ears" that are offset. These prevent the cable from being connected incorrectly. Never force a Qwiic cable to connect to a host module. The connection should require light pressure to ensure the cable is seated. Similarly, disconnecting the cables should be done with care. Using excessive force can damage the Qwiic connector.

Caution Do not force a Qwiic cable connection or disconnection. It may break the connector!

Take a moment and look at a typical Qwiic cable connector. Figure 7-1 shows the connector from two sides (enlarged for clarity). The side that shows no pins I like to call the flat or top side.

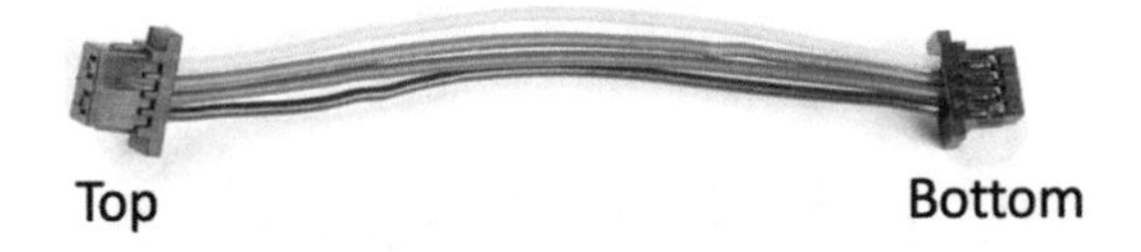

Figure 7-1. *Quick cable connector (top and bottom)*

The best way to connect the Qwiic cable to the Qwiic connector on the module is to orient the cable so that the flat side is facing up and insert it into the Qwiic connector as shown in Figure 7-2. Again, don't force the connection. If you meet resistance, try turning the cable over. It should seat easily into the connector.

Figure 7-2. *Qwiic cable connected to module*

Notice the connector looks like it isn't fully inserted, but that is not the case. If you look at the connector on the cable more closely, you will see there are several small protrusions or ridges. The connector will only seat as far as those ridges.

Now that we know how the cables are connected, let's connect the Qwiic modules for this project. Connect one Qwiic cable to the right side of the SparkFun Proximity Sensor Breakout - 20cm, VCNL4040. With the cable connected to the right side of the sensor, connect that cable to the top connector on the back side of the SparkFun Micro OLED Breakout module as shown in Figure 7-3. I recommend leaving the small plastic cover on the OLED to protect the screen. Also, there is an "Up" to the OLED, so you may want to orient it once you get the project running so you can read it better.

Figure 7-3. *Proximity and OLED module connections*

Now we have our Qwiic modules assembled and ready to connect to our host board. We'll call it our Qwiic daisy chain.

You may not be too impressed at this point since we've only connected two wires to two modules, but trust me to know the alternative of using breadboards and breakout boards to solder and wire circuits is much more difficult and time consuming.

Connecting to the Arduino

How you connect the Qwiic daisy chain to the Arduino depends on which Arduino you use. If you choose to use an Arduino board with a Qwiic connector, you can just use a Qwiic cable to connect to the daisy chain and the other end to your board, and you're done. If you are using a newer Arduino that has dedicated I2C pins, you will need to use a special cable.

Using a Shield

However, if you recall, connecting the Qwiic daisy chain to your board depends on which Arduino you are using. If you use one of the older Arduino boards including the Uno (not the WiFi version) and its variants that connect the I2C data and clock to A4 (SDA) and A5 (SCL), you can use the SparkFun Qwiic Shield (`www.sparkfun.com/products/14352`) attached to your Arduino and use another Qwiic cable to connect to the Qwiic daisy chain and connect the other end to one of the connectors on the shield. However, please be advised the shield comes unassembled, so you will have to solder or find someone to solder the headers onto the shield for you.

If you choose to use the Qwiic Shield for Arduino, your project should resemble Figure 7-4.

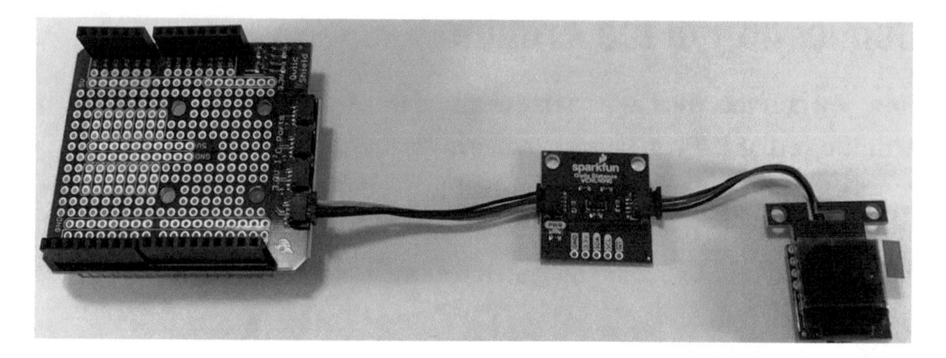

Figure 7-4. *Connecting Qwiic daisy chain to Arduino shield*

Using a Special Cable

A better solution and indeed the best solution for Arduino boards with dedicated I2C pins is to use the Qwiic Cable - Breadboard Jumper (`www.sparkfun.com/products/14425`). This cable has a Qwiic connector on one end and four male breadboard pins on the other as shown in Figure 7-5.

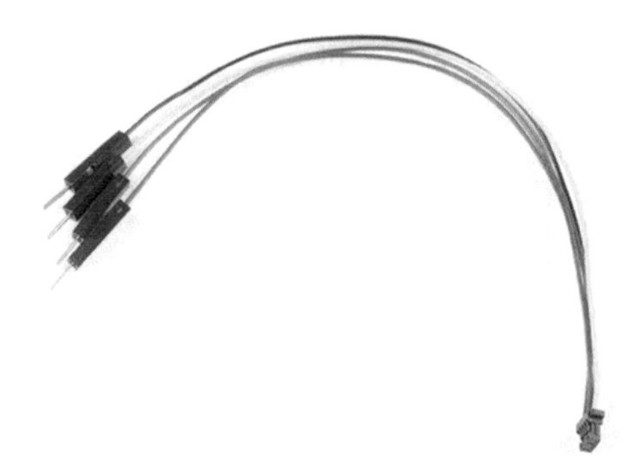

Figure 7-5. *Qwiic Cable - Breadboard Jumper (courtesy of sparkfun.com)*

This cable is a great alternative to using a shield, and it is my preferred mechanism for connecting Qwiic modules to my Arduino boards. The only downside is you must be careful connecting the pins to the Arduino to ensure you get the connections correct and the pins can pull loose if you move the project around (don't do that).

The connections are very simple. Connect the wires (male pins) to your Arduino as shown in Table 7-2.

Table 7-2. *Connecting the Qwiic Breadboard Cable to Arduino*

Wire Color	Description	Arduino Pin
Black	Ground	GND
Red	Power – 3.3V	3.3V
Blue	Data (SDA)	SDA or A4
Yellow	Clock (SCL)	SCL or A5

Take special note of the power cable. This needs to be connected to the 3.3V power pin only. If your Arduino board is not marked with such a pin, check the documentation to ensure the board operates with 3.3V power. If it does not, do not attempt to use the Qwiic modules with your board. Choose another board to use until you can find a suitable adapter/shield.

Caution Be sure to connect power to 3.3V, not VIN, 5V, or any other pin with power. Using higher voltage will damage the modules.

Notice the SDA and SCL pins are indicated as either a pin with the same name located near the USB connector or A4/A5 as part of the analog pin subheader. The older Arduino boards use A4 and A5, but newer boards have the I2C pins isolated on their own. On some specialty Arduino boards like the Leonardo, SDA is on pin 2 and SCL is on pin 3. Table 7-3 summarizes the location of the I2C pins for the more popular Arduino boards.[3]

[3] `www.arduino.cc/en/reference/wire`

Table 7-3. *Location of I2C Pins for Arduino Boards*

Arduino	I2C Pins Location
Uno, Ethernet	A4 (SDA), A5 (SCL)
Mega 2560	20 (SDA), 21 (SCL)
Leonardo	2 (SDA), 3 (SCL)
Due	20 (SDA), 21 (SCL), SDA1, SCL1

Once you've identified the correct pins on your Arduino board, you can connect the breadboard cable as shown in the preceding tables. Figure 7-6 shows a closeup of what the connections look like (and where the connections are located) on the Arduino Uno.

Figure 7-6. *Qwiic breadboard cable connections – Arduino*

Figure 7-7 shows the project once all connections have been made.

Figure 7-7. *Connecting to Arduino with breadboard breakout cable*[4]

You may be wondering what to do with the modules all flopping around.

Mounting Modules

Recall Qwiic modules have two or more mounting holes that you can use to mount the modules either to a piece of wood, an enclosure, or some other medium. However, some of the Qwiic shields and hats have matching holes, so you can mount the modules on top of the shield or hat.

For example, the Qwiic pHAT for Raspberry Pi has four holes where you can mount risers as shown in Figure 7-8.

[4] Arduino shown is for demonstration purposes. We know this project will not work on the Uno WiFi Rev2.

Figure 7-8. *Mounting Qwiic modules to a shield or hat*

You can then attach the modules to the risers as shown in Figure 7-9 (Arduino shield shown). Not only does this make a tidy package, but it also saves you some effort should you want to move the project or set it aside for another day.

Figure 7-9. *Mounting modules on Arduino shield*

Detecting I2C Devices on the Arduino

Before we move on to the Raspberry Pi, let's discuss for a moment how to detect what I2C devices are connected to your Arduino. You may think, *Yeah, I know. I can see them!* But that isn't always a good measure. Sure, you can tell the modules are connected, but what if one of them isn't working or you are trying to use a library that isn't working?

Both of these conditions can be diagnosed or partially diagnosed by running a sketch to detect what the I2C addresses are used or are visible on the bus. Listing 7-1 shows a short sketch created to scan through all I2C addresses and report which ones respond. Rather than explain the sketch in detail, I leave it to the reader as an exercise. You can find the code on the book website.

Listing 7-1. I2C Scanner Sketch (Arduino)

```
#include <Wire.h>

void setup()
{
  Serial.begin(9600);
  while (!Serial);
}

void loop()
{
  byte address = 0x00;
  byte error = 0x00;
  int row = 1;

  // Draw header
  Serial.println("\n\nI2C Address Scanner");
  Serial.println("-------------------");
```

```
  Serial.println("     0  1  2  3  4  5  6  7  8  9  a  b
  c  d  e  f");
  Serial.print("00:    ");

  // Look for addresses that respond without errors.
  for (address = 1; address < 127; address++) {
    Wire.beginTransmission(address);
    error = Wire.endTransmission();
    if ((address % 16) == 0) {
      Serial.println();
      Serial.print(row);
      Serial.print("0: ");
      row += 1;
    }
    // If no error, the address may be valid.
    if (error == 0) {
      if (address < 16) {
        Serial.print("0");
      }
      Serial.print(address, HEX);
      Serial.print(" ");
    } else {
      Serial.print("-- ");
    }
  }
  delay(3000);
}
```

When you run this sketch on an Arduino that has I2C devices connected, each address found is displayed in a grid (values are in hexadecimal) as shown in the following. The scan is repeated every three seconds:

```
I2C Address Scanner
-------------------
     0  1  2  3  4  5  6  7  8  9  a  b  c  d  e  f
00:     -- -- -- -- -- -- -- -- -- -- -- -- -- -- --
10: -- -- -- -- -- -- -- -- -- -- -- -- -- -- -- --
20: -- -- -- -- -- -- -- -- -- -- -- -- -- -- -- --
30: -- -- -- -- -- -- -- -- -- -- -- -- -- 3D -- --
40: -- -- -- -- -- -- -- -- -- -- -- -- -- -- -- --
50: -- -- -- -- -- -- -- -- -- -- -- -- -- -- -- --
60: 60 -- -- -- -- -- -- -- -- -- -- -- -- -- -- --
70: -- -- -- -- -- -- -- -- -- -- -- -- -- -- --
```

Here, we see two addresses that were identified: 0x3D (61 decimal) is the Micro OLED, and 0x60 (96 decimal) is the proximity sensor.

Whenever in doubt that your modules aren't working, use this sketch to check to see if they are wired correctly and responding to the `begin()` function call.

Connecting to the Raspberry Pi

Like the Arduino, how you connect the Qwiic daisy chain to the Raspberry Pi depends on which Qwiic hat or pHAT you use. There are several to choose from, but my favorite thus far are the Qwiic pHAT for Raspberry Pi (`www.sparkfun.com/products/15945`) for most projects and Qwiic SHIM for Raspberry Pi (`www.sparkfun.com/products/15794`) for when I want a quick, temporary connection or I want to use Qwiic with some other hat. Like the Arduino, we can also use a special cable to connect directly to the GPIO header.

And, like the Arduino, you must power down your Raspberry Pi before connecting or disconnecting any Qwiic modules.

Using a Hat

The typical Qwiic hat for Raspberry Pi mounts onto the GPIO header. Just align the pins and insert the hat onto the header and then attach your Qwiic daisy chain to one of the Qwiic connectors, and you're ready to go.

You can also use the mounting holes on the hat to mount your modules as shown in Figure 7-10.

Figure 7-10. *Connecting Qwiic daisy chain to mounts on hat on Raspberry Pi*

If you decide to use the shim, the connections will resemble Figure 7-11.

Figure 7-11. *Connecting Qwiic daisy chain to shim on Raspberry Pi*

The shim is designed to fit on top of the GPIO without soldering. Some early versions of the shim are known to be finicky and may need to be oriented at an angle to make a good connection. If you know how to solder, the shim is cheap enough that you can solder it to your GPIO and remove all such issues.

Using a Special Cable

A better solution and indeed the best solution for Arduino boards with dedicated I2C pins is to use the Qwiic Cable – Female Jumper (4-pin) (`www.sparkfun.com/products/14988`). This cable has a Qwiic connector on one end and four female breadboard pins on the other as shown in Figure 7-12.

Figure 7-12. *Qwiic Cable - Female Jumper (4-pin) (courtesy of sparkfun.com)*

This cable is a great alternative to using a hat if your GPIO header is tucked away inside a case, and it is my preferred mechanism for connecting Qwiic modules to my faster Raspberry Pi 4B boards that are mounted inside cases with fans. The only downside is you must be careful connecting the pins to the GPIO to ensure you get the connections correct and the pins can pull loose if you move the project.

The connections are very simple. Connect the wires (female pins) to your Raspberry Pi as shown in Table 7-4. You can see a complete breakdown of all the GPIO pins at `www.raspberrypi.org/documentation/usage/gpio/`.

Table 7-4. *Connecting the Qwiic Female Jumper Cable to Raspberry Pi*

Wire Color	Description	GPIO Pin
Black	Ground	6
Red	Power – 3.3V	1
Blue	Data (SDA)	3
Yellow	Clock (SCL)	5

Figure 7-13 shows the connections in detail. This example shows a header for a Raspberry Pi 4B board mounted in an Argon ONE case (`www.argon40.com/argon-one-raspberry-pi-4-case.html`).

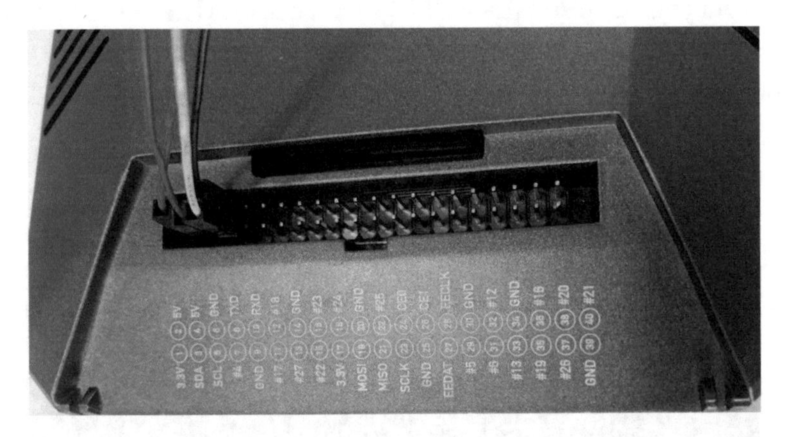

***Figure 7-13.** Qwiic female breakout cable connections – Raspberry Pi*

Notice the pins we use are the very same ones that the Qwiic SHIM used. Figure 7-14 shows the project once all connections have been made.

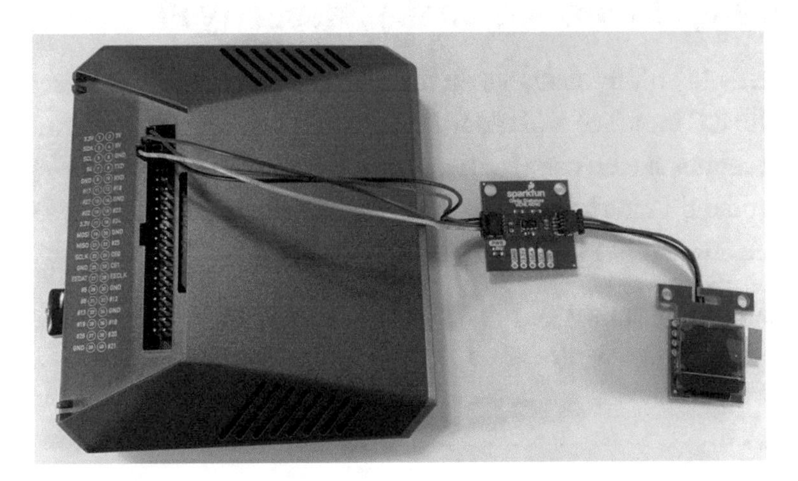

***Figure 7-14.** Connecting to Raspberry Pi with Qwiic female breakout cable*

Notice the GPIO header pins and their uses (or names) are printed on the Argon ONE case itself. This is hidden behind a removable panel and is very easy to use. Raspberry Pi does not come with the pins on the GPIO labeled, but you can purchase a GPIO Reference Card like the one from Adafruit (`www.adafruit.com/product/2263`) as shown in Figure 7-15. This card fits over the GPIO and helps you locate the pins you need quickly. It's a must-have for those working directly with the GPIO.

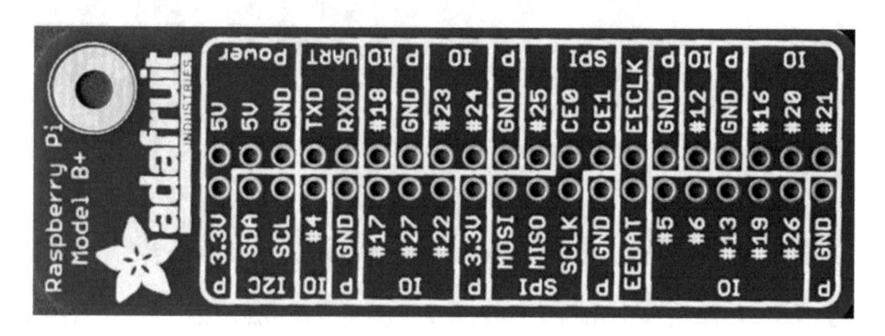

Figure 7-15. *Raspberry Pi GPIO Reference Card (courtesy of adafruit.com)*

Detecting I2C Devices on the Raspberry Pi

There is a nifty utility available on the Raspberry Pi named i2cdetect that scans the I2C interface and reports what addresses are found – similar to the i2cscanner sketch used for the Arduino. To run the utility, simply issue the command `i2cdetect -y 1` where y means do the scan and `1` is the I2C interface number as follows:

```
$ i2cdetect -y 1
     0  1  2  3  4  5  6  7  8  9  a  b  c  d  e  f
00:          -- -- -- -- -- -- -- -- -- -- -- -- --
10: 10 -- -- -- -- -- -- -- -- -- -- -- -- -- -- --
20: -- -- -- -- -- -- -- -- -- -- -- -- -- -- -- --
30: -- -- -- -- -- -- -- -- -- -- -- -- -- 3d -- --
40: -- -- -- -- -- -- -- -- -- -- -- -- -- -- -- --
```

```
50: -- -- -- -- -- -- -- -- -- -- -- -- -- -- -- --
60: 60 -- -- -- -- -- -- -- -- -- -- -- -- -- -- --
70: -- -- -- -- -- -- -- --
```

Here, we see two addresses that were identified: 0x3D (61 decimal) is the Micro OLED, and 0x60 (96 decimal) is the proximity sensor.

If you get an error when running the utility, you may need to install it with the following command:

```
$ sudo apt-get install i2ctools
```

Now that we have all of our modules connected, let's see how to write the code.

Write the Code

Writing the code for this project may seem a little strange at first, but it follows the patterns we've seen in Chapters 3 and 5 for the Arduino and Raspberry Pi (Python).

The code for this project first initializes the I2C bus, prepares the OLED for use, and then enters a loop that reads a value from the proximity sensor and then displays the value on the OLED. We will also see some logic to establish a threshold that detects when an object is too close. Recall the sensor can detect ambient light and white light levels, but we will use only the proximity reading.

The proximity sensor returns an integer value in the range of 1–65535 (in practice only about 16767), where the higher the number, the closer the object. It is important to understand that this sensor has a very low degree of accuracy and is not designed to be used to measure exact distance. Rather, it can be used to tell if something is there and if it is very close. Thus, the sensor can be used as a proximity detector such as those found in assembly lines or sorters where it is important to know when or if some object passes nearby.

Let's walk through how to prepare our computers to use the sensor and write code to read its values. We'll start with the Arduino.

Arduino

This section presents a walk-through of the sketch you will write to read values from the sensor and display them on the OLED module. But first, there are a couple of libraries we must install on our PCs.

Install Software Libraries

We will need to install the Arduino libraries for the proximity sensor and the OLED module separately. Fortunately, this is easy to do using the Library Manager. Simply open the Library Manager from the Arduino IDE menu (*Sketch* ➤ *Include Library* ➤ *Manage Libraries...*). Then search for 4040 and install the latest version of the library as shown in Figure 7-16.

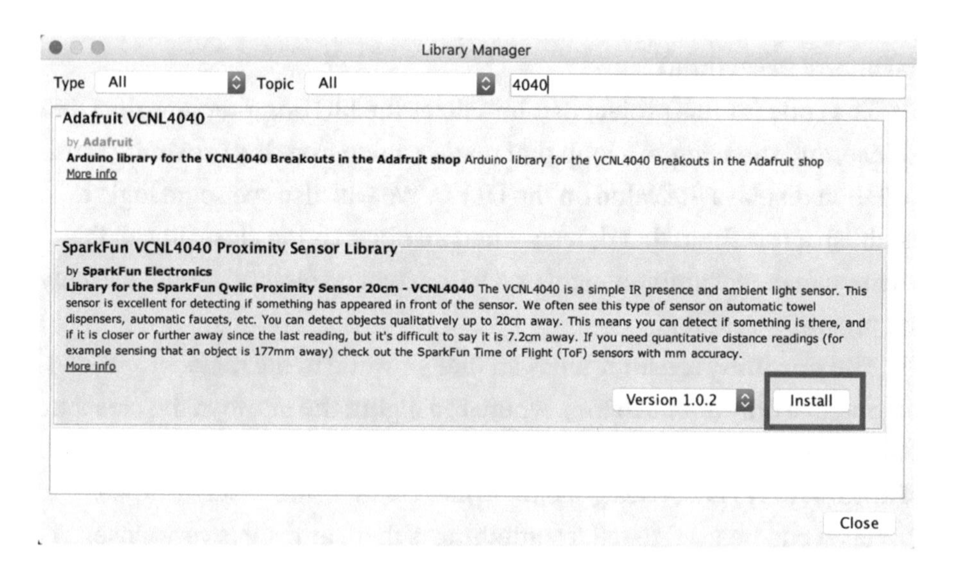

Figure 7-16. *Installing the VCNL4040 library (Arduino IDE)*

Similarly, we need to install the library for the OLED. Open the Library Manager and search for `micro OLED` and then install the latest version as shown in Figure 7-17.

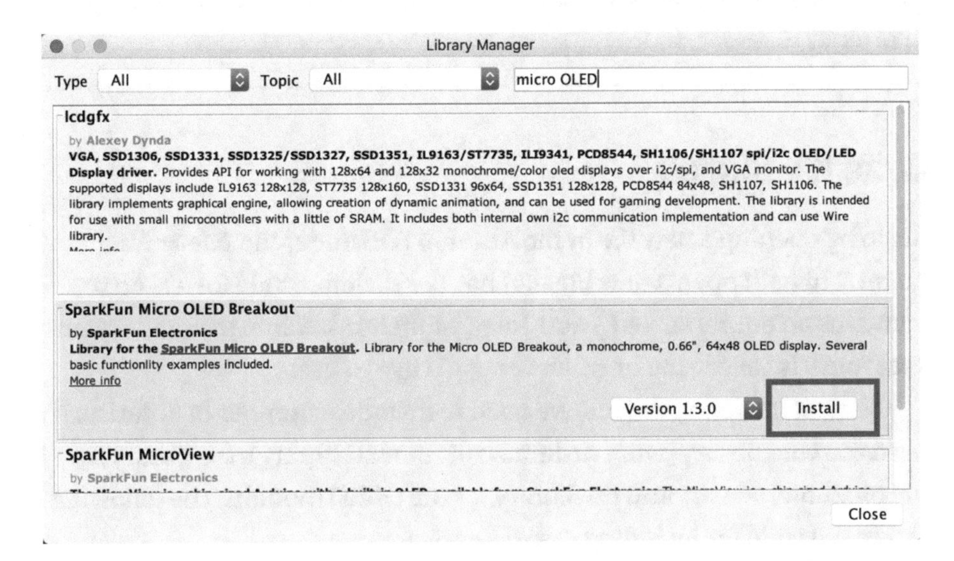

Figure 7-17. *Installing the Micro OLED library (Arduino IDE)*

Now that we have the software libraries installed, we can begin writing our sketch.

But first, take a moment to browse through the example sketches for the proximity sensor under the *File* ➤ *Examples* ➤ *SparkFun VCNL4040 Proximity Sensor Library* menu and those for the OLED under the *File* ➤ *Examples* ➤ *SparkFun Micro OLED Breakout* menu. You will find some very interesting uses of the modules in those examples all without having to change any code!

Tip Consider using the examples for any new library you install to learn more about its capabilities and to ensure you know how to write the code properly, not to mention to ensure the module is working correctly.

Write the Sketch

Begin by opening a new file in the Arduino IDE under the *File* ➤ *New* menu. This will open a new file that has the skeleton code for a sketch (contains an empty `setup()` and `loop()` function). Name the new sketch `KeepYourDistance.ino` or whatever you'd like to use.

Next, at the top of the file, we need to include a number of libraries. We need the wire library (what Arduino calls the I2C library), the library for the proximity sensor, and the library for the OLED module. The following shows the code for including these libraries:

```
#include <Wire.h>
#include <SparkFun_VCNL4040_Arduino_Library.h>
#include <SFE_MicroOLED.h>
```

You may be wondering how to know what library names to use. The answer is found in the example sketches most authors provide for you or in their documentation. SparkFun is unique by providing you a hookup guide that details all this and more for each Qwiic module. When using a module for the first time, always check these resources to know where to start.

Next, we need some constants. We need two for the OLED module to set the reset pin and DC jumper (as shown in the example sketches). The OLED is hardwired to use a reset (RST), so we only need to choose an Arduino pin to use, and any unused pin will do. In this case, we use pin 9 for `PIN_RESET`. Since we haven't modified the OLED by changing any jumpers, we can use the default of 1 for the `DC_JUMPER` (you would change to 0 if you closed the jumper on the back of the board). We also need a value to use in our check to see if something is too close. Let's start with a value of 1000 for the constant TOO_CLOSE as shown in the following:

```
#define PIN_RESET 9
#define DC_JUMPER 1
#define TOO_CLOSE 1000
```

Next, we need to initialize two variables: one for each Qwiic module as shown in the following. We will use these variables to call the functions provided by each library. Here we pass those constants defined earlier to the OLED class instance. Yes, we could have provided them directly in the constructor, but it is considered bad form to pass constants without defining them first (and it makes it easier to change them in the future):

```
MicroOLED oled(PIN_RESET, DC_JUMPER);
VCNL4040 proximitySensor;
```

OK, that's it for the initialization code. Now, we can write the code for the `setup()` function. Here, we need to start the serial monitor library, initialize/start the I2C library, and set up the OLED. Most of this code we have seen before, so we will only focus on the bits for the sensor and OLED.

We add code to check the proximity sensor to ensure it is connected and working. Note that code such as this is always a good idea to place in the `setup()` function. The following shows how to write code to check the sensor. Here, we call the `begin()` function for the sensor, and if it returns false, we print a warning message and go into an endless loop to halt the sketch:

```
  if (!proximitySensor.begin()) {
    Serial.println("ERROR: Sensor not found!");
    while (1);
  }
```

Next is the code for the OLED module, which isn't quite obvious but starts with a similar check. The following shows the code we need:

```
if (!oled.begin()) {
  Serial.println("ERROR: OLED not found!");
  while(1);
}

oled.setFontType(0);
oled.clear(PAGE);       // Clear page memory
oled.clear(ALL);        // Clear internal memory
oled.setCursor(0, 0);   // Set cursor position
oled.print("Keep Your\nDistance!");
oled.display();
```

Notice we begin with a check to ensure the OLED is present, then set the font type, clear the page memory, and then clear the internal memory. This ensures the module is ready for new output. Then, we send a test message by positioning the cursor to the top left and printing a short message. Nothing is printed (displayed) until we call the `display()` function as shown. All of these functions are called by using the `oled` object/variable we declared earlier. For example, `oled.display()` displays data sent (printed) to the OLED.

We then use a `delay()` function to pause execution for 3 seconds (3000 microseconds) to ensure all modules are initialized and ready. Listing 7-2 shows the code for the `setup()` function.

Listing 7-2. KeepYourDistance setup()

```
void setup()
{
  Serial.begin(9600);
  while (!Serial);
  Serial.println("Keep Your Distance!");
```

```
  // Initialize the i2c bus
  delay(100);
  Wire.begin();

  // Ensure the proximity sensor is connected
  // Ensure the proximity sensor is connected
  if (!proximitySensor.begin()) {
    Serial.println("ERROR: Sensor not found!");
    while(1);
  }

  // Setup the OLED and print welcome message
  if (!oled.begin()) {
    Serial.println("ERROR: OLED not found!");
    while(1);
  }

  oled.setFontType(0);
  oled.clear(PAGE);       // Clear page memory
  oled.clear(ALL);        // Clear internal memory
  oled.setCursor(0, 0);   // Set cursor position
  oled.print("Keep Your\nDistance!");
  oled.display();
  delay(3000);
}
```

Now, let's look at the code for the `loop()` function. Here, we have a bit less code. We simply get a value from the sensor using the `getProximity()` function for the proximity sensor object (variable named `proximitySensor`), display it in the serial monitor and OLED, and then determine if the object is too close by printing "TOO CLOSE!" if the sensor value is greater than our threshold or "Ok" otherwise.

Listing 7-3 shows the complete code for the loop() function. Take a moment to read through it to ensure you understand how it works. Notice the delay() at the end of the loop. This is to slow the sketch down so that there is sufficient time for the sensor to prepare the next value to be read. Some sensors will need longer delays (they have a slower refresh rate) than others.

Listing 7-3. KeepYourDistance loop()

```
void loop()
{
  // Get the proximity sensor value
  unsigned int proximity = proximitySensor.getProximity();

  // Display the value
  Serial.print("Proximity Value = ");
  Serial.println(proximity);
  oled.clear(PAGE);       // Clear page memory
  oled.setCursor(0, 0);   // Set cursor position
  oled.print("Prx = ");
  oled.print(proximity);
  oled.setCursor(0, 10);  // Set cursor position
  // Determine if the object is too close
  if (proximity > TOO_CLOSE) {
    oled.print("TOO CLOSE!");
  } else {
    oled.print("Ok");
  }
  oled.display();
  delay(1000);
}
```

OK, that's it! Listing 7-4 shows the completed sketch repeated for reference. If you've been following along writing the code as you read, you can check your code against the listing. Or, better, download the sample code from the book website if you don't want to type all of the code yourself.

Listing 7-4. KeepYourDistance Sketch (Arduino)

```
// Include the wire (i2c), OLED, and proximity sensor libraries
#include <Wire.h>
#include <SparkFun_VCNL4040_Arduino_Library.h>
#include <SFE_MicroOLED.h>

// Constants for the OLED
#define PIN_RESET 9
#define DC_JUMPER 1
#define TOO_CLOSE 1000

// Global variables (OLED and sensor)
MicroOLED oled(PIN_RESET, DC_JUMPER);
VCNL4040 proximitySensor;

void setup()
{
  Serial.begin(9600);
  while (!Serial);
  Serial.println("Keep Your Distance!");

  // Initialize the i2c bus
  delay(100);
  Wire.begin();

  // Ensure the proximity sensor is connected
```

```
  // Ensure the proximity sensor is connected
```

```
  if (!proximitySensor.begin()) {
    Serial.println("ERROR: Sensor not found!");
    while(1);
  }

  // Setup the OLED and print welcome message
  if (!oled.begin()) {
    Serial.println("ERROR: OLED not found!");
    while(1);
  }

  oled.setFontType(0);
  oled.clear(PAGE);       // Clear page memory
  oled.clear(ALL);        // Clear internal memory
  oled.setCursor(0, 0);   // Set cursor position
  oled.print("Keep Your\nDistance!");
  oled.display();
  delay(3000);
}

void loop()
{
  // Get the proximity sensor value
  unsigned int proximity = proximitySensor.getProximity();

  // Display the value
  Serial.print("Proximity Value = ");
  Serial.println(proximity);
  oled.clear(PAGE);       // Clear page memory
  oled.setCursor(0, 0);    // Set cursor position
  oled.print("Prx = ");
  oled.print(proximity);
  oled.setCursor(0, 10);   // Set cursor position
```

```
  // Determine if the object is too close
  if (proximity > TOO_CLOSE) {
    oled.print("TOO CLOSE!");
  } else {
    oled.print("Ok");
  }
  oled.display();
  delay(1000);
}
```

Compile the Sketch

The last step is to compile the sketch before uploading it to your board. It is important to do this step separately so that you can ensure you don't have any issues in the code. Not only will the compilation check the code you've written, but it will also ensure the software libraries you installed are also free of errors.

To compile the sketch, use the *Sketch* ➤ *Verify/Compile* menu or click the leftmost button in the Arduino IDE editor. You may see dozens of lines pass by in the output window, but the ones you are looking for (the last to be displayed) should resemble the following:

```
Sketch uses 14822 bytes (30%) of program storage space. Maximum
is 48640 bytes.
Global variables use 1011 bytes (16%) of dynamic memory,
leaving 5133 bytes for local variables. Maximum is 6144 bytes.
```

If you encounter any errors, be sure to fix them and recompile to ensure the sketch compiles without errors or serious warnings.

Once everything compiles, we're ready to start testing. But first, let's look at the code for the Raspberry Pi. You can skip to the "Execute the Project" section if you're curious to see how the project works (it will be the same on both platforms).

Raspberry Pi

This section presents a walk-through of the Python code you will write to read values from the sensor and display them on the OLED module. But first, there are a couple of libraries we must install on our Raspberry Pi.

Install Software Libraries

There are two actions needed to get your Raspberry Pi set up: install the Python libraries we need and enable I2C on the Raspberry Pi. Both are one-time events so you will not need to repeat them. More specifically, you do not need to install any additional software libraries for the next project to use any of the SparkFun Qwiic modules. If you use third-party Qwiic modules (such as the STEMMA QT modules), you may need to install more libraries, but we see those in later chapters if needed.

Let's begin with the SparkFun Qwiic Python software libraries. This can be done with the pip command as shown in Listing 7-5. If you haven't already, go ahead and boot your Raspberry Pi and open a terminal to enter the command shown in bold. Listing 7-5 shows an excerpt of the install for brevity.

Listing 7-5. Installing the SparkFun Python Libraries (Raspberry Pi)

```
% pip3 install sparkfun_qwiic
Collecting sparkfun_qwiic
  Downloading https://files.pythonhosted.org/packages/ef/b3/
c7f170d4e4f429f47ab0f07165a94ff7c01a9650f06129d9fc6017bfa67f/
sparkfun_qwiic-1.0.16-py2.py3-none-any.whl (219kB)
    100% |████████████████████████████████| 225kB 1.3MB/s
Collecting sparkfun-qwiic-proximity (from sparkfun_qwiic)
  Downloading https://files.pythonhosted.org/packages/ad/d4/4
01537c1113fcb278bc0b6c03221b0c28567049b7e485dd5817be9367e21/
sparkfun_qwiic_proximity-0.9.0-py2.py3-none-any.whl
```

```
...
Successfully installed pynmea2-1.15.0 smbus2-0.4.0 sparkfun-
pi-servo-hat-0.9.0 sparkfun-qwiic-1.0.16 sparkfun-qwiic-
adxl313-0.0.7 sparkfun-qwiic-bme280-0.9.0 sparkfun-qwiic-
ccs811-0.9.4 sparkfun-qwiic-dual-encoder-reader-0.0.2
sparkfun-qwiic-gpio-0.0.2 sparkfun-qwiic-i2c-0.9.11
sparkfun-qwiic-icm20948-0.0.1 sparkfun-qwiic-joystick-0.9.0
sparkfun-qwiic-keypad-0.9.0 sparkfun-qwiic-max3010x-0.0.2
sparkfun-qwiic-micro-oled-0.9.0 sparkfun-qwiic-pca9685-0.9.1
sparkfun-qwiic-proximity-0.9.0 sparkfun-qwiic-relay-0.0.2
sparkfun-qwiic-scmd-0.9.1 sparkfun-qwiic-serlcd-0.0.1 sparkfun-
qwiic-tca9548a-0.9.0 sparkfun-qwiic-titan-gps-0.1.1 sparkfun-
qwiic-twist-0.9.0 sparkfun-qwiic-vl53l1x-1.0.1 sparkfun-top-
phat-button-0.0.2 sparkfun-ublox-gps-1.1.3
```

You will see a number of libraries being installed. In fact, you will see all of the SparkFun Python libraries for all Qwiic modules and their dependencies being installed. You should see a successful message at the end as shown in the listing.

There is one more thing you need to do. You must enable the I2C interface in Raspbian. You do this by clicking the main menu and then selecting *Preferences* ➤ *Raspberry Pi Configuration*. On the *Interfaces* tab, tick the *Enable* button to the left of *Enable* for I2C as shown in Figure 7-18.

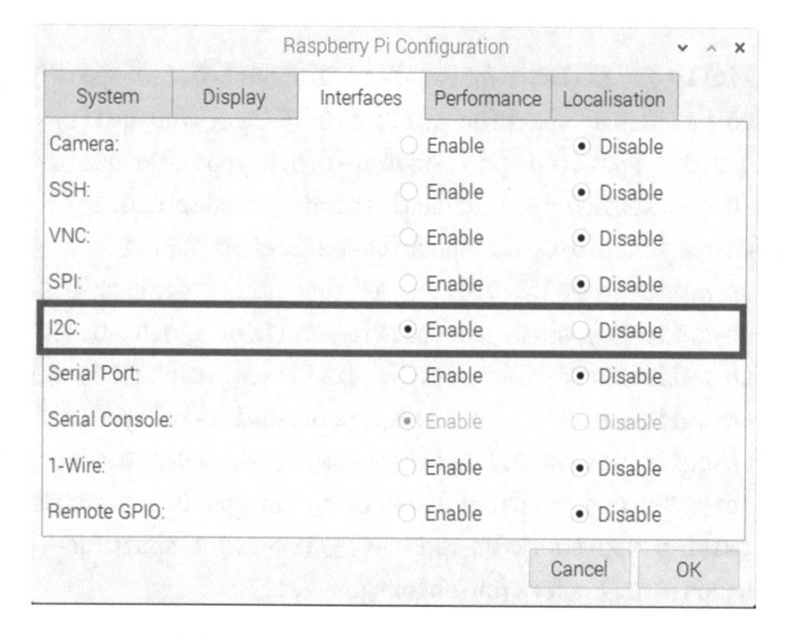

Figure 7-18. *Enabling the I2C interface (Raspberry Pi)*

Once you've set the I2C interface to Enable, click *OK* and then reboot your Raspberry Pi to ensure the changes take effect.

If you have additional devices connected to your Raspberry Pi such as those found in some enclosures and especially the pi-top products[5] (`www.pi-top.com/`), you may already have the I2C interface enabled. Or, when you do enable it, you may discover there are already I2C devices at certain addresses. If you find your code not working, try disconnecting any peripherals attached to the GPIO if possible and try again. This is another case where the `i2cdetect` utility can be helpful.

[5] I discovered my pi-top showed all I2C addresses taken until I disconnected the pi-top peripherals.

Write the Code

The code for the Python version of this project is a bit shorter than the Arduino code. We still do the same steps in the (almost) same order, but in a Python manner.

Begin by launching the Thonny Python IDE under the *Main ➤ Programming* submenu. The IDE opens with a new, blank file in the editor. Name the new file `keep_your_distance.py` or whatever you'd like to use. Or you can download the source code for the book and open the file by that name. Since the code is very similar to the Arduino sketch, we will only skim through the highlights.

Start at the top of the file. We need to import two libraries: the SparkFun Qwiic and the time library as shown in the following:

```
import time
import qwiic
```

Next, we will create a constant for our threshold to detect when something is too close:

```
TOO_CLOSE = 1000
```

Next, we create two variables that are instances of the proximity and OLED modules and then call the initialization function (`begin()`) for each:

```
prox = qwiic.QwiicProximity()
oled = qwiic.QwiicMicroOled()
prox.begin()
oled.begin()
```

Next, we clear the OLED and print a welcome message and then wait for 3 seconds using the `time.sleep()` function:

```
oled.clear(oled.PAGE)
oled.clear(oled.ALL)
oled.set_font_type(0)
```

```
oled.set_cursor(0,0)
oled.print("Keep Your Distance!")
oled.display()
time.sleep(3)
```

OK, now we are ready to get into the code proper. Since Python doesn't have a concept of the `loop()` function we saw in the Arduino sketch, we will simulate it by writing an endless `while` loop. We can use the *CTRL+C* keys on our keyboard to stop execution when we're done running the code.

Inside the loop, we read a sensor value, print it to the screen (the terminal) for debugging purposes, and then print it to the OLED module. We use the value to determine if the object sensed is too close to the sensor. Finally, we wait for one second before repeating the loop. Listing 7-6 shows the completed code for the Python program. Take some time to read through it until you are satisfied you understand it. As you will see, the Python code (to me at least) reads much easier than the Arduino code.

Listing 7-6. KeepYourDistance Code (Python)

```
# Import libraries
import time
import qwiic

# Constants for proximity sensor
TOO_CLOSE = 1000

# Create instances of the Qwiic classes (modules)
prox = qwiic.QwiicProximity()
oled = qwiic.QwiicMicroOled()

# Start I2C modules for sensor and OLED
prox.begin()
oled.begin()
```

```
# Clear the screen and print greeting
oled.clear(oled.PAGE)
oled.clear(oled.ALL)
oled.set_font_type(0)
oled.set_cursor(0,0)
oled.print("Keep Your Distance!")
oled.display()
time.sleep(3)

# Main loop
while True:
    # Read the sensor value
    proximity = prox.get_proximity()
    # Display the value to the screen and OLED
    print("Proximity = {0}".format(proximity))
    oled.clear(oled.PAGE)
    oled.set_cursor(0,0)
    oled.print("D = ")
    oled.print(proximity)
    oled.set_cursor(0,10)
    # Determine if the object is too close and print message
    if proximity > TOO_CLOSE:
        oled.print("TOO CLOSE!")
    else:
        oled.print("Ok")
    oled.display()
    # Wait for 1 second...
    time.sleep(1)
```

OK, that's it! We've written the code. Unlike the Arduino, we do not need to compile the Python code. So we're now ready to execute the project!

Execute the Project

Now that we've spent many pages exploring the Qwiic modules and writing the code to interact with them, it is time to test the project by executing (running) it.

When the project runs (executes), you will see some diagnostic message written to the serial monitor (Arduino) or the terminal (Raspberry Pi). You will also see a welcome message appear on the OLED followed by a short pause. Then sensor values will appear, and under that is the Ok or TOO CLOSE! message depending on the value read. Figure 7-19 shows an example of what you should see on the OLED.

Figure 7-19. *Executing the KeepYourDistance project*

You can affect the sensor readings by placing your hand or another solid object in the path of the IR sensor on the proximity sensor module. Be sure to face the module so the silk screen writing is facing up. If the sensor is facing another direction, it will not respond to your movements.

Executing the code depends on which platform you're using. Let's look at the Arduino first.

Sketch on the Arduino

Executing the sketch on the Arduino requires connecting our board to our PC and then uploading the sketch to the Arduino. Recall the sketch will run so long as the USB cable is connected to our PC (and the Arduino).

Execute the Sketch

To execute the sketch, be sure your Arduino is connected and you've selected the correct board under the *Tools* ➤ *Board* menu. You also need to ensure you have the correct port selected under the *Tools* ➤ *Port* menu.

Once those items are set, you can click the *Upload* button or choose *Sketch* ➤ *Upload* from the menu. The Arduino IDE will compile the sketch and then upload it to your Arduino. Once you see the `Done uploading...` message, you can open the serial monitor. You should see the output begin momentarily as shown in the following:

```
Keep Your Distance!
Proximity Value = 6
Proximity Value = 6
Proximity Value = 7
Proximity Value = 12
Proximity Value = 655
Proximity Value = 1539
Proximity Value = 1294
Proximity Value = 2035
Proximity Value = 6
Proximity Value = 6
Proximity Value = 7109
Proximity Value = 8418
Proximity Value = 16162
...
```

You should also see output on the OLED. Go ahead, and try it out!

Something Isn't Working... Now What?

If you don't see anything in the serial monitor, go back and make sure all of your Qwiic connections are tight. While it is unlikely you reversed the cables, if you're using the breadboard cable, one of the pins may have come loose. Be sure to power off (unplug the USB cable) before you start investigating your wiring.

Also, be sure that you have the correct baud rate chosen for the serial monitor. An incorrect baud rate will cause the serial monitor to display odd characters or nothing at all.

If those checks fail, you can also follow the troubleshooting tips discussed in Chapters 2 and 3.

Python Code on the Raspberry Pi

Executing the sketch on the Raspberry Pi requires running the Python code in a terminal after connecting your Qwiic daisy chain to your Raspberry Pi via a hat or the Qwiic female breakout cable. Recall the code will run until you stop it with *CTRL+C* on the keyboard.

Execute the Python Code

To run the Python code on the Raspberry Pi, you can issue the command `python3 ./keep_your_distance.py` from the same folder where the file was saved as shown in the following:

```
$ python3 ./keep_your_distance.py
Proximity = 9
Proximity = 9
Proximity = 23
Proximity = 56
Proximity = 27
Proximity = 7278
```

```
Proximity = 7304
Proximity = 7381
Proximity = 11
Proximity = 25
Proximity = 4
Proximity = 6968
Proximity = 7241
Proximity = 8689
Proximity = 10913
Proximity = 14636
Proximity = 15912
Proximity = 16176
Proximity = 16088
...
```

Something Isn't Working… Now What?

If you see a message stating there is an `OSError: [Errno 121] Remote I/O error`, this indicates the software library is not communicating or has lost communication with the I2C module. If you see this error, go back and make sure all of your Qwiic connections are tight. While it is unlikely you reversed the cables, if you're using the breadboard cable, one of the pins may have come loose. Be sure to shut down your Raspberry Pi before you start investigating your wiring.

Another possible error is `Error: Failed to connect to I2C bus 1`. This indicates the I2C bus is not available. In this case, be sure you have enabled the I2C interface in the Raspberry Pi Configuration utility.

You can also try running the `i2cdetect` utility again to ensure the module is still responding on the expected address.

If those checks fail, you can also follow the troubleshooting tips discussed in Chapters 4 and 5.

If everything worked as executed, congratulations! You've just built your first Qwiic project all without having to solder or mess with discrete components. Cool, eh?

Going Further

This example was designed to get you started with Qwiic modules. There isn't an IoT element to the project, but we have to start somewhere. The skills you learned here can help you expand the project. In this section, we will learn how we can take the basic elements of this project and expand them to other, perhaps more useful projects.

Mounting the Project in a Case

One of the things most will encounter once they get past the initial learning phase and example projects is the need to mount the project for use. More specifically, when you build a project and want to use it more than as a curiosity, you will need some way to mount it. This is where you will need to get creative.

While the Qwiic and STEMMA QT host adapters and boards have mount holes so you can mount them on a piece of wood or perhaps in a general-purpose enclosure (or on a piece of wood mounted in an enclosure), there are no (current) cases for the Arduino or Raspberry Pi that have room for mounting the modules. I am certain, with a little effort, you can come up with something. I like the wood panel idea as it allows me to mount the modules and other components easily.

Another area where you will need to use some creativity is getting power to the project. None of the projects in this book require anything more than the power for the host board, so that's not that big of an issue, but locating your project near a power source may be something else to think about.

Whatever you decide to do, the fun will be in the process of making something to mount your project. It need not be an aesthetically pleasing enclosure (but there's nothing wrong with that). There is a bit of gee whiz factor seeing your modules bolted to a plank hanging on your wall. Dare I suggest art?

Alternative Project Ideas

While it may seem a bit of a stretch, there are some ideas where you could make this project into an IoT project. Here are just a few suggestions you can try once we have learned how to take our projects to the cloud. View these as challenges or homework – things you can do to apply what you've learned rather than read another step-by-step instruction. You've got the skills now. Put them to work!

- *Greater range*: We can increase the range of the sensor by combining or replacing the infrared distance sensor with the Distance Sensor Breakout – 4 Meter, VL53L1X (Qwiic) from SparkFun (`www.sparkfun.com/products/14722`).
- *Parking aide*: Set up your device in your garage to help you get your car parked nice and tight to the wall or other obstacle.
- *Six feet, no more!*: Create a wearable version of the project and wear it on your hat or shirt to remind folks around you to keep a safe distance.
- *Cloud data:* Collect the data from your project to display how many times the sensor detected a too-close threshold or range of values read.

Summary

In this chapter, we got some hands-on experience making projects with Qwiic modules. We used a distance sensor to detect objects and displayed the calculated range values on a small OLED.

Along the way, we learned many of the fundamental aspects of working with Qwiic modules from how to plug them together, mounting the shield and/or hat, direct wiring the Qwiic daisy chain to the host board, and even programming and executing the code.

We also saw some potential to make this project better by upgrading the distance sensor as well as some ideas for how to adapt the project for practical uses.

In the next chapter, we will see another Qwiic project example that we can use to build a home-based weather station.

CHAPTER 8

How's the Weather?

Perhaps one of the ubiquitous projects that you will find on the Internet is a simple weather station. A basic weather station is a great project because it shows what is possible for sensing the environment and it's a project almost anyone can relate to. After all, the weather is often a topic in almost every polite social interaction. If you had your own weather station, particularly one you created yourself, you could contribute a bit more to the conversation!

In this chapter, we will create a basic weather station. It isn't a full weather station because it doesn't include some of the more frequent sensors such as wind speed and rainfall totals (but you could add them). Rather, we're going to start with the basic observations for temperature, humidity, and barometric pressure.

Note The details having to do with connecting the Qwiic modules and alternatives to using shields or hats on your host board have been omitted in this and the next three project chapters for brevity. Please refer to Chapter 7 to review any of these details.

Like the last chapter, we will see how to implement this project on the Arduino and Raspberry Pi. Let's get started.

C. Bell, *Beginning IoT Projects*, https://doi.org/10.1007/978-1-4842-7234-3_8

Project Overview

The project for this chapter is designed to demonstrate how to get started building a weather station with a single Qwiic environmental sensor. We will use the sensor to read the current temperature, relative humidity, and barometric pressure displaying the data on a small screen. Once again, this is a very basic weather station and is small enough to be both interesting and useful.

Interestingly, the environmental sensor can also calculate altitude, but we will not use such measurement in this project. The primary reason is it may not be very interesting to present because the values change from one reading to the next and, more importantly, the measurement is not very accurate (but can be more so with some adjustments).

What Will We Learn?

While we won't see anything new with the hardware other than a different sensor, by implementing this project, we will reinforce what we learned from Chapter 7, specifically how to connect Qwiic modules to our host boards and how to form a daisy chain of Qwiic modules.

The challenges for this project are in the programming tasks, which are similar to the last project except we will see how to use functions to make the code easier to maintain. We will also encounter an interesting problem with the software libraries and see how to solve the problem. The goal is to see how we can adapt and overcome problems not of our own making – perhaps more importantly, how to know when it is a problem with a software library and not our own code.

Let's see what hardware we will need.

Hardware Required

The hardware needed for this project is listed in Table 8-1. URLs for each component are included for ease of ordering including duplicate entries for alternative vendors.

Table 8-1. *Hardware Needed for the Weather Station Project*

Component	URL	Qty	Cost
Environmental Combo Breakout - BME280	www.sparkfun.com/products/15440	1	$14.95
Micro OLED Breakout	www.sparkfun.com/products/14532	1	$16.95
Qwiic cable (any length can be used)	www.sparkfun.com/products/14427	2	$1.50
SparkFun RedBoard Qwiic (Arduino Uno or compatible)	www.sparkfun.com/products/15123	1	$19.95
Raspberry Pi 3B or later	www.sparkfun.com/categories/233 www.adafruit.com/category/176	1	$35.00+
Qwiic pHAT for Raspberry Pi	www.sparkfun.com/products/15945	1	$5.95

Tip There are a number of Qwiic environmental sensors available. See `www.sparkfun.com/categories/tags/qwiic-environmental` for a complete list.

About the Hardware

Let's discuss these components briefly. We will discover how to work with the hardware in more detail later in the chapter.

Sensor

The SparkFun BME280 Atmospheric Sensor Breakout is very easy to use with a minimal amount of setup and code. The sensor can measure temperature, relative humidity, barometric pressure, and altitude.

The sensor, like most Qwiic modules, is small and can be mounted in small enclosures. In fact, the sensor can be used either indoors or outdoors. So you can measure environmental conditions inside your home, office, dog house, or barn or even on your terrace or patio.

As mentioned, the sensor can also be used as an altimeter. The sensor can use the current sensor readings including the barometric pressure to calculate the current altitude. But this requires a bit more information to get any reasonable accuracy.

The piece of information you need to provide is the current barometric pressure at sea level for your current position. You can get this by going to a weather website and searching for your town or city. Look specifically for the value at sea level. For example, on the east coast, the barometric pressure at sea level is approximately 1013.25 hectopascal (hPa) units.[1]

---

[1] `https://en.wikipedia.org/wiki/Pascal_(unit)`

Most libraries for the sensor will provide a means for you to provide this data.

Unfortunately, you will need to use data that is current and measured to the nearest hour of when you take samples with the sensor because stale data can reduce the accuracy considerably. If you want to try it out yourself, we will see how to make the adjustment (but it is not needed to read temperature, humidity, and barometric pressure).

OLED

The OLED module we will use is the same from Chapter 7. If you'd like to experiment with other output devices, you can, but it is recommended to use the Micro OLED so that the example code works without modification. We will learn more about how to adapt our code to changes in modules in later chapters.

Tip To save on shipping, see the Appendix for a complete list of all hardware used in the book.

Assemble the Qwiic Modules

Recall from Chapter 7 we can use a single Qwiic cable to connect our BME280 sensor to the OLED module and then another to attach to the host adapter on our host board. Figure 8-1 shows an example of how you should connect your modules to form a Qwiic daisy chain.

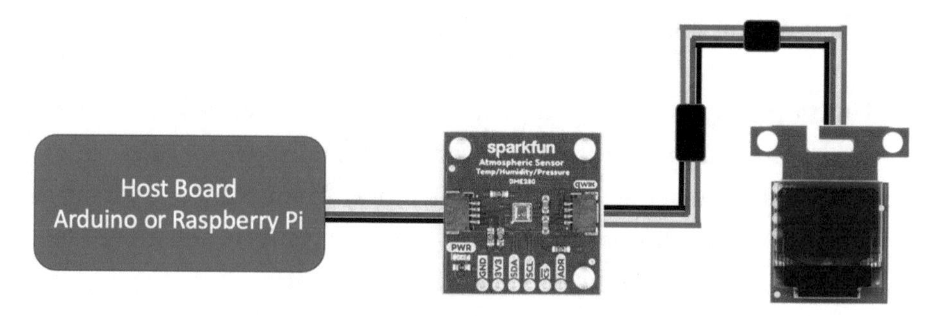

Figure 8-1. *Weather project Qwiic connections*

Tip Refer to Chapter 7 for specifics of how to attach the Qwiic daisy chain to your host board.

Now that we have all of our modules connected, we can now learn how to write the code.

However, before we jump into how to write the code for the Arduino and Raspberry Pi, let's take a short detour and discuss how you would go about researching sensors and modules to learn how to use them.

Researching the Hardware

You may be wondering how you will be able to use any Qwiic (or STEMMA QT) modules on your own after reading this book. After all, we will cover only a very few of the possible modules available. So how do you get started using a new module?

There are several steps I go through whenever I consider or encounter a new module, sensor, or device I want to use. The following outlines the steps I typically take to get an idea of how to use the module and start forming my own code. As you will see, there is no special knowledge needed other than a well-developed curiosity and a good dose of perseverance.

Read the Documentation!

Yes, this is quite the cliché. However, you would be surprised how many people skip this step. The "some assembly required" warning is sadly interpreted as "when all else fails, read the instructions." Don't do that!

Vendors almost always have a documentation section, page, blog, or article for every module they sell. For example, SparkFun includes a description of each module along with a list of links to documentation. Figure 8-2 shows an example of the list of documentation pages for the BME280 sensor we are using in this chapter. I simply navigated to the page and then clicked the *Documents* tab.

Figure 8-2. *BME280 module Documents page (courtesy of sparkfun.com)*

Notice in the list of documents there are links for the schematics, files for electronic computer-aided design (CAD) (eagle files), a hookup guide, a manufacturer datasheet, the software libraries for Arduino and Python, and the hardware repository on GitHub (`github.com`).

The one item you should zero in on and read thoroughly is the hookup guide.[2] Start with the hookup guide and read it from start to end at least once before you attempt to use the samples in the software libraries. Not only will you see examples of how to use the module, but you will also learn key information about the hardware itself.

For example, the hookup guide for the BME280 module contains information about every aspect of the board (`https://learn.sparkfun.com/tutorials/qwiic-atmospheric-sensor-bme280-hookup-guide`). This includes information about how to connect the module using the breakout pins and how to set the jumpers on the back and the technical details about the sensor such as accuracy, I2C address, and more!

Some modules allow you to change the I2C address with a jumper, and the BME280 module is one of them. You would never know about this by looking at the board (unless you know what to look for on the tiny screen print). Specifically, this module can be addressed at 0x77 (the default) or 0x76 by setting the ADR jumpers on the back. Figure 8-3 shows a detailed view of the jumpers.

[2] While a bit too old-school for some, I like to download the .pdf version and either store it on my iPad for later reading or print it out (if I'm traveling). OK, so I'm weird like that.

Figure 8-3. *Setting address jumpers on BME280 (courtesy of sparkfun.com)*

If you look closely, you will see a small joint between the leftmost pins (bars) on the jumper. To change the address, you would break the connection and make a new connection from the center pin to the right pin. Yes, this requires a small bit of soldering to make it work.

Changing the address is a big deal in complex projects because the more modules you have, the more likely you are to encounter a case where two modules use the same address. You may also encounter a situation where you want to use a different software library but that library is written to use a different address. At least this way, you have a possibility of resolving these conflicts.

This an excellent example of why you should take the time to read these documents. You'd never know about this feature without doing so, unless, of course, you've had lots of experience and know how to spot it. But why take the chance that it works the way you think it may?

Other vendors, like Adafruit, often include similar well-written articles to the hookup guide from SparkFun. For example, the article for their BME280 module can be found at `https://learn.adafruit.com/adafruit-bme280-humidity-barometric-pressure-temperature-sensor-breakout`.

Tip If you encounter a module that doesn't have documentation, do not despair. Most vendors sold versions of the sensor prior to packaging it for Qwiic, so you can look at the non-Qwiic product and use that documentation. In most cases, the software libraries are the same.

For those who want to see how the sausage is made,[3] you can read the manufacturer's datasheet and look at the schematics. For the novice, this may not add much knowledge, but if you want to learn how the sensor works and thus how the software library is written to read the data, the datasheet will provide those answers (but they may not be easy to read).

Install the Software Libraries

Once you have read all that you can about the module in the documentation, you should then install the software libraries. Most will be available for the Arduino through the Library Manager or installed on Raspberry Pi with the `pip3` command.

However, it is also a good idea to click the links for the software libraries to visit the repository (typically on `github.com`). You will encounter additional documentation there along with a place to go if you get stuck. We'll see an example of this in a later section.

---

[3] Gah! Another cliché.

The repository will contain the latest version of the code, so if you need to use the latest features, you can use the repository to download and install the library. While this chapter and the others like it show you how to install the libraries, the documentation should be used as a guide as well. For example, it is possible a new release of the library is introduced that isn't part of the Arduino Library Manager or available from Pip. In those cases, you should follow the documentation for manually installing the latest library code.

Explore the Sample Code

We've already learned this valuable lesson. Recall we discussed using the sample sketches and example code vendors include as a way to learn how to use the module. Not only does this help you understand how the code you want may be written, but it also helps you test the module in a controlled, known good mechanism that will remove doubt as to whether your hardware is working correctly.

By loading the examples, you will be able to run them as tests to ensure your module and the wiring are correct. Plus, some of the examples can be fun.

Write Your Own Code

Once you've followed all of these steps, you now know enough to start writing your own code. In fact, you should have been able to answer most of what you want to do and what you can do with the module by this point.

You may even find most of what you need in the examples provided by the vendor. This isn't always the case as we will see in the next chapter, but these exploratory skills will be an advantage in those cases as well.

Since this is a beginner's book, we will learn how to form our own code by following the example projects. It may take you some practice, but you should be able to start writing code for you own projects after you finish this book. Of course, the examples from the software libraries are a great place to start!

Now, let's see how to write the code for this project.

Write the Code

The code for this project reveals the start of a pattern of how the code to work with IoT projects is typically structured. There are other ways to construct the code (mainly to do with order of operations and modularization), but the basics are still the same.

We begin by initializing the I2C bus and preparing the OLED for use and then executing a loop that reads values from the sensor and then displays the values on the OLED. More specifically, we will read the current temperature, relative humidity, and barometric pressure. We will present the temperature in degrees Celsius, the relative humidity as a percentage, and the barometric pressure in hectopascal (hPa) units.

Let's walk through how to prepare our computers to use the sensor and write code to read its values. We'll start with the Arduino.

Arduino

This section presents a walk-through of the sketch you will write to read values from the sensor and display them on the OLED module. But first, there are a couple of libraries we must install on our PCs.

Install Software Libraries

We will need to install the Arduino libraries for the BME280 sensor and the OLED module separately. Fortunately, this is easy to do using the Library

Manager. Simply open the Library Manager from the Arduino IDE menu (*Sketch* ➤ *Include Library* ➤ *Library Manager...*). Then search for BME280 and install the latest version of the SparkFun Qwiic BME280 library as shown in Figure 8-4.

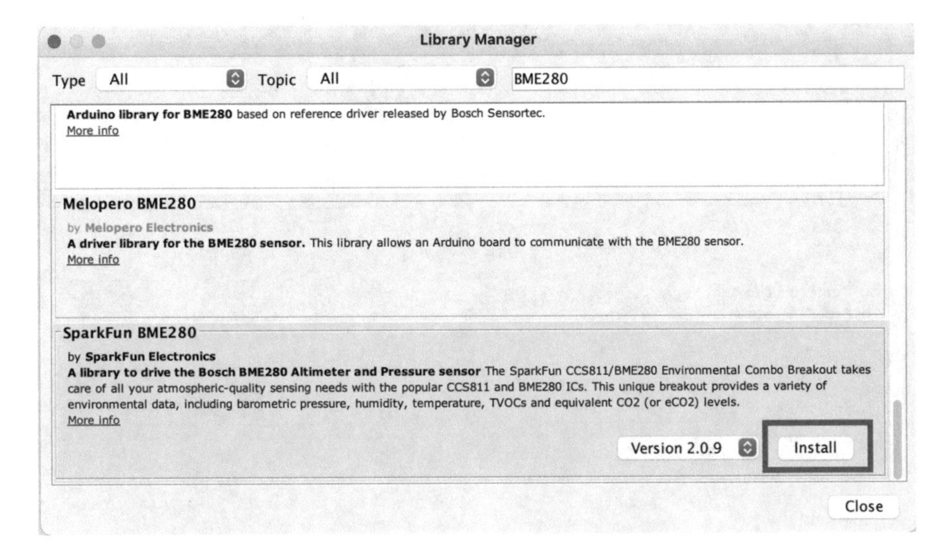

Figure 8-4. *Installing the SparkFun BME280 library (Arduino IDE)*

When you complete the search, you may be surprised to see a long list of libraries for the BME280 sensor. This is a good indication the sensor is popular and likely a good choice for your project. It also means there is more than one choice for libraries you can use. While it is always best to use the library provided by the vendor of the module, sometimes it may be possible or even necessary to use a different library.

For example, if you discover anomalous readings and have verified your wiring is correct, using a different library to test your module will help you determine if you have defective hardware or there is a defect in the vendor's supplied library or more likely you're not using the library correctly. Viewing examples from other libraries will help you determine which is the likely cause of the anomalous readings.

Similarly, we need to install the library for the OLED. If you haven't already loaded the library when you completed the project in Chapter 7, open the Library Manager and search for `micro OLED` and then install the latest version as shown in Figure 8-5.

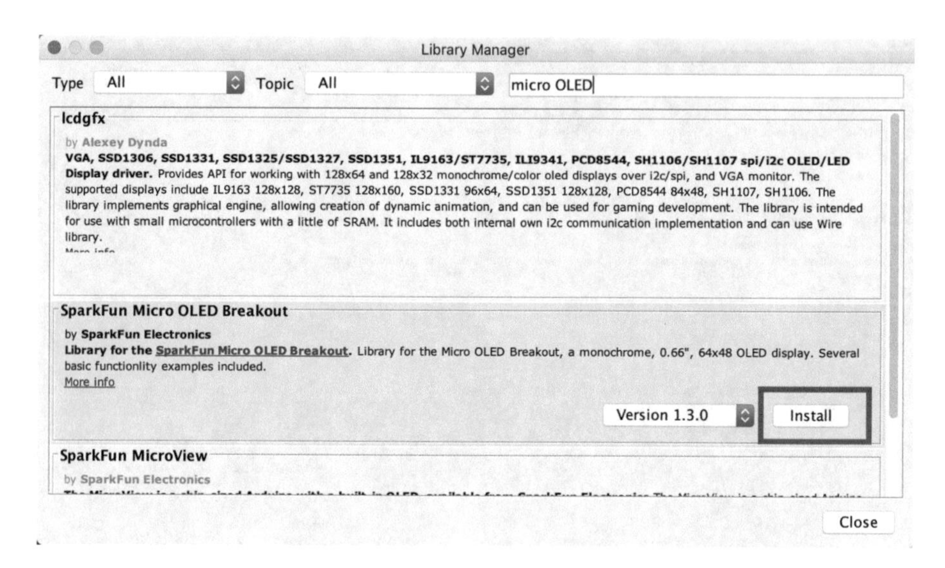

Figure 8-5. *Installing the Micro OLED library (Arduino IDE)*

Now that we have the software libraries installed, we can begin writing our sketch.

Write the Sketch

Begin by opening a new file in the Arduino IDE under the *File ➤ New* menu. This will open a new file that has the skeleton code for a sketch (contains an empty `setup()` and `loop()` method). Name the new sketch `Weather.ino` or whatever you'd like to use.

Next, at the top of the file, we need to include a number of libraries. We need the wire library (what Arduino calls the I2C library), the library for the BME280 sensor, and the library for the OLED module. The following shows the code for including these libraries:

```
#include <Wire.h>
#include <SFE_MicroOLED.h>
#include "SparkFunBME280.h"
```

Next, we need some constants. We need two for the OLED module to set the reset pin and DC jumper (as shown in the example sketches). The OLED is hardwired to use a reset (RST), so we only need to choose an Arduino pin to use, and any unused pin will do. In this case, we use pin 9 for `PIN_RESET`. Since we haven't modified the OLED by changing any jumpers, we can use the default of 1 for the `DC_JUMPER` (you would change to 0 if you closed the jumper on the back of the board). We also include a constant for the barometric pressure at sea level for our location as shown in the following:

```
#define PIN_RESET 9
#define DC_JUMPER 1
#define SEALEVEL_REFERENCE (101325.00)
```

The value for the barometric pressure at sea level is not strictly needed for this project. It is only needed if you want to read the altitude value from the sensor because the sensor needs the value for your location to calculate the altitude. It is included in the sketch to show you how to do it and as an example of where to place similar operations for future projects.

Next, we need to initialize two variables: one for each Qwiic module as shown in the following. We will use these variables to call the methods provided by each library. Here we pass those constants defined earlier to the OLED class instance:

```
MicroOLED oled(PIN_RESET, DC_JUMPER);
BME280 bme280;
```

The next thing we're going to do is a bit different than the last project. We will use functions (or methods if you prefer) to set up the BME280 and OLED modules and display information to the serial monitor and

OLED. This is an exercise to demonstrate how you can move parts of your code to functions to make your code easier to read.

First is a function to set up the BME280 as shown in Listing 8-1. We will return a Boolean, which we can use to determine if the sensor is set up properly. We name the function `setupBME280()`.

Listing 8-1. setupBME280 (Arduino)

```
bool setupBME280()
{
  bme280.settings.commInterface = I2C_MODE;
  bme280.settings.I2CAddress = 0x77;
  return bme280.begin();
}
```

Here, we set up the settings for the software library including the communication interface. This is needed because this module can communicate with I2C as well as SPI. In this case, we want the I2C mode. We also assign the address. Recall the default is 0x77, but we could change it to 0x76 by altering the jumpers on the back of the board. Finally, we call the `begin()` function and return the result. This way, we can check the result of our function, which is the same result as the library function. Nice.

Next, we create a function to set up the OLED module moving the initialization code we saw in Chapter 7 to the function as shown in Listing 8-2. We name the function `setupOLED()`.

Listing 8-2. setupOLED (Arduino)

```
bool setupOLED()
{
  if (!oled.begin()) {
    return false;
  }
```

```
  oled.setFontType(0);
  oled.clear(PAGE);      // Clear page memory
  oled.clear(ALL);       // Clear internal memory
  oled.setCursor(0, 0);  // Set cursor position
  oled.print("What's theweather?");
  oled.display();
  return true;
}
```

Notice this code is written a bit differently where we call the `begin()` function at the top of the function returning `false` if the initialization fails or true after we've cleared the screen and displayed the welcome message.

Next, we create a function to display the values read to the serial monitor. We name the function `printDiagnostics()` because this function is useful only while the project is running and the serial monitor is open and connected. Listing 8-3 shows the new function to print the values to the serial monitor.

Listing 8-3. Print Diagnostics (Arduino)

```
void printDiagnostics(float temp, float humid, float pres)
{
  Serial.print("BME280 values: ");
  Serial.print(temp, 2);
  Serial.print("C, ");
  Serial.print(humid, 2);
  Serial.print("%rh, ");
  Serial.print(pres / 100.00, 2);
  Serial.println(" hPa");
}
```

Notice we do not return anything (return type is void), but we have a comma-separated list of parameters for the temperature, humidity, and pressure of type float. This means to use this function, we will need to pass (include) these values in a parameterized list.

Next is a similar function to print the values read to the OLED module. We name this function showDataOLED(), and it also takes the same parameters as shown in Listing 8-4. Thus, we will need to also include the values as parameters to call the function.

Listing 8-4. Show Data on the OLED (Arduino)

```
void showDataOLED(float temp, float humid, float pres)
{
  oled.clear(PAGE);          // Clear page memory
  oled.setCursor(0, 0);      // Set cursor position
  oled.print(temp);
  oled.print(" C");
  oled.setCursor(0, 15);     // Set cursor position
  oled.print(humid);
  oled.print(" %rh");
  oled.setCursor(0, 30);     // Set cursor position
  oled.print(long(pres / 100.00));
  oled.print(" hPa");
  oled.display();
}
```

OK, all of the initialization, setup, and helper functions are defined. Now we can see how this affects our setup() function. Listing 8-5 shows the completed setup() function.

Listing 8-5. Setup Function (Arduino)

```
void setup()
{
  Serial.begin(115200);
  Serial.println("How's the weather?");
  Serial.println("------------------");
  Wire.begin();
  if (!setupBME280()) {
    Serial.println("The sensor did not respond. Please check
    wiring.");
    while(1); //Freeze
  }
  if (!setupOLED()) {
    Serial.println("ERROR: OLED not found!");
    while(1);
  }
  bme280.setReferencePressure(SEALEVEL_REFERENCE);
  delay(3000);
}
```

Notice we simply include calls to the functions we wrote to break out the setup routines for the sensor and OLED. We also see the call to set the reference pressure as noted previously.

Finally, we can write the code for the `loop()` function. In this case, we need to simply read the values, then call the `printDiagnostics()` and `showDataOLED()` functions, and then wait a couple of seconds as shown in Listing 8-6.

Listing 8-6. Loop Function (Arduino)

```
void loop()
{
  // Read the sensor
  float temperature = bme280.readTempC();
  float humidity = bme280.readFloatHumidity();
  float pressure = bme280.readFloatPressure();

  // Display data to serial monitor
  printDiagnostics(temperature, humidity, pressure);

  // Display data to OLED
  showDataOLED(temperature, humidity, pressure);
  delay(2000);
}
```

Now, let's see all of this code in context. Listing 8-7 shows the completed sketch with all functions included.

Listing 8-7. Weather Sketch (Arduino)

```
// Include the wire (i2c), OLED, and environment sensor
libraries
#include <Wire.h>
#include <SFE_MicroOLED.h>
#include "SparkFunBME280.h"
// Constants for the OLED
#define PIN_RESET 9
#define DC_JUMPER 1
#define SEALEVEL_REFERENCE (101325.00)

// Global variables (OLED and sensor)
MicroOLED oled(PIN_RESET, DC_JUMPER);
BME280 bme280;
```

```
// Initialize the BME280 sensor
bool setupBME280()
{
  bme280.settings.commInterface = I2C_MODE;
  bme280.settings.I2CAddress = 0x77;
  return bme280.begin();
}

// Setup the OLED and print welcome message
bool setupOLED()
{
  if (!oled.begin()) {
    return false;
  }
  oled.setFontType(0);
  oled.clear(PAGE);       // Clear page memory
  oled.clear(ALL);        // Clear internal memory
  oled.setCursor(0, 0);   // Set cursor position
  oled.print("What's theweather?");
  oled.display();
  return true;
}

// Print values to the serial monitor
void printDiagnostics(float temp, float humid, float pres)
{
  Serial.print("BME280 values: ");
  Serial.print(temp, 2);
  Serial.print("C, ");
  Serial.print(humid, 2);
  Serial.print("%rh, ");
```

```
  Serial.print(pres / 100.00, 2);
  Serial.println(" hPa");
}
// Display values on OLED screen
void showDataOLED(float temp, float humid, float pres)
{
  oled.clear(PAGE);        // Clear page memory
  oled.setCursor(0, 0);    // Set cursor position
  oled.print(temp);
  oled.print(" C");
  oled.setCursor(0, 15);   // Set cursor position
  oled.print(humid);
  oled.print(" %rh");
  oled.setCursor(0, 30);   // Set cursor position
  oled.print(long(pres / 100.00));
  oled.print(" hPa");
  oled.display();
}

void setup()
{
  Serial.begin(115200);
  Serial.println("How's the weather?");
  Serial.println("------------------");
  Wire.begin();
  if (!setupBME280()) {
    Serial.println("The sensor did not respond. Please check
    wiring.");
    while(1); //Freeze
  }
```

```
  if (!setupOLED()) {
    Serial.println("ERROR: OLED not found!");
    while(1);
  }
  bme280.setReferencePressure(SEALEVEL_REFERENCE);
  delay(3000);
}

void loop()
{
  // Read the sensor
  float temperature = bme280.readTempC();
  float humidity = bme280.readFloatHumidity();
  float pressure = bme280.readFloatPressure();

  // Display data to serial monitor
  printDiagnostics(temperature, humidity, pressure);

  // Display data to OLED
  showDataOLED(temperature, humidity, pressure);
  delay(2000);
}
```

As you can see, the code is a bit easier to read, but it is a bit longer. That is often the trade-off for making your code easier to read and maintain.

Compile the Sketch

The last step is to compile the sketch before uploading it to your board. It is important to do this step separately so that you can ensure you don't have any issues in the code. Not only will the compilation check the code you've written, but it will also ensure the software libraries you installed are also free of errors.

To compile the sketch, use the *Sketch* ➤ *Verify/Compile* menu or click the leftmost button in the Arduino IDE editor. You may see dozens of lines pass by in the output window, but the ones you are looking for (the last to be displayed) should resemble the following:

```
Sketch uses 17784 bytes (55%) of program storage space. Maximum
is 32256 bytes.
Global variables use 1128 bytes (55%) of dynamic memory,
leaving 920 bytes for local variables. Maximum is 2048 bytes.
```

If you encounter any errors, be sure to fix them and recompile to ensure the sketch compiles without errors or serious warnings.

Once everything compiles, we're ready to start testing. But first, let's look at the code for the Raspberry Pi. You can skip to the "Execute the Project" section if you're curious to see how the project works (it will be the same on both platforms).

Raspberry Pi

This section presents a walk-through of the Python code you will write to read values from the sensor and display them on the OLED module. But first, there are a couple of libraries we must install on our Raspberry Pi.

Locating Alternative Software Libraries

This is where things get interesting. In testing the code for this book, I made a startling discovery. The values returned by the SparkFun Qwiic BME280 library for barometric pressure were incorrect. While it is true the sensor can produce values that vary a bit, the values returned by the library were too far off to be minor variations. You may wonder how I knew the values were off. Simply put, I knew what values to expect, and I had the Arduino sketch results to compare. So what do you do in this case?

There are several possibilities. Everyone approaches a problem differently, but I like to first attempt to validate the issue. In this case, the output of the Arduino sketch helps, but it wasn't enough because it was executing on a different platform in a different programming language.

Fortunately, as I mentioned, the BME280 is a popular sensor, and there are other libraries available. My backup plan includes using libraries from vendors I trust such as SparkFun and Adafruit. If one is giving me problems, I look to the other to validate my observations. If neither has a compatible library, I then look for others, but in this case Adafruit had exactly what I needed.

Adafruit's BME280 library is written for their version of Python called CircuitPython[4] that competes with MicroPython,[5] but in this case, the code for the library works perfectly with Python 3 on the Raspberry Pi. You can find the Adafruit BME280 library at `https://github.com/adafruit/Adafruit_CircuitPython_BME280`.

Once I installed the Adafruit library and altered the code to work with the different library, I was able to produce values in the expected ranges. This led me to conclude there is a defect in the SparkFun BME280 library.

Whenever you discover something like this and, most importantly, you can reproduce the problem, you should contact the vendor to let them know there may be a problem. In fact, you can visit the GitHub site for the library and open an issue like I did. Be sure to communicate the problem as succinctly and as accurately as you can and, if possible, present example output and the reason(s) why you think it is a defect. You may not get an immediate response, but at least you will be on the record as having reported the issue.

[4] `https://learn.adafruit.com/welcome-to-circuitpython`
[5] `https://micropython.org/`

WHAT IF YOU KNOW WHAT'S WRONG AND HOW TO FIX IT?

There may be a case where you find a defect and you know how to fix it. In this case, you can send the potential fix to the vendor by describing it in the issue. Be sure to show your work and example test runs. No one is going to accept a change if it isn't justified and proven to fix the problem. This is the best way to provide the fix to the vendor if you are not confident in using GitHub and coding. However, if you are an experienced developer, creating a branch of the repository and a pull request is the fastest way to get your fix into the production code.

So the lesson to learn here is if you encounter problems with a software library, be sure to check other vendors' libraries for a compatible substitute. What you should look for is if the vendor has a similar module with the same interface (Qwiic, STEMMA QT) and the same sensor. Be sure to try out their examples to verify the library will work with your module. In this case, the Adafruit library solved the problem. Let's see how to install it.

Install Software Libraries

Since we are using a different library than what is included in the SparkFun Qwiic Python pip package, we must install additional libraries, in this case the Adafruit BME280 library and its supporting libraries.

Fortunately, we can do all of this with a single command as follows. Note that it could install a number of libraries that are needed to support the library:

```
$ pip3 install adafruit_circuitpython_bme280
Looking in indexes: https://pypi.org/simple, https://www.
piwheels.org/simple
Collecting adafruit_circuitpython_bme280
```

```
  Using cached https://www.piwheels.org/simple/adafruit-
circuitpython-bme280/adafruit_circuitpython_bme280-2.5.1-py3-
none-any.whl
Collecting Adafruit-Blinka (from adafruit_circuitpython_bme280)
  Downloading https://www.piwheels.org/simple/adafruit-blinka/
Adafruit_Blinka-5.10.0-py3-none-any.whl (140kB)
    100% |████████████████████████████████| 143kB 511kB/s
Collecting adafruit-circuitpython-busdevice (from adafruit_
circuitpython_bme280)
  Downloading https://www.piwheels.org/simple/adafruit-
circuitpython-busdevice/adafruit_circuitpython_busdevice-5.0.3-
py3-none-any.whl
Collecting Adafruit-PlatformDetect>=2.18.1 (from Adafruit-
Blinka->adafruit_circuitpython_bme280)
...
Installing collected packages: Adafruit-PlatformDetect,
Adafruit-PureIO, Adafruit-Blinka, adafruit-circuitpython-
busdevice, adafruit-circuitpython-bme280
Successfully installed Adafruit-Blinka-5.10.0 Adafruit-
PlatformDetect-2.25.0 Adafruit-PureIO-1.1.8 adafruit-
circuitpython-bme280-2.5.1 adafruit-circuitpython-
busdevice-5.0.3
```

Tip If you haven't installed the SparkFun Qwiic Python libraries, you must install them to run this project (`pip3 install sparkfun_qwiic`).

Now we're ready to write the code.

Note If you did not implement the project in Chapter 7, please refer to the "Raspberry Pi" section in Chapter 7 to enable the I2C interface on your Raspberry Pi.

Write the Code

The code for the Python version of this project is a bit shorter than the Arduino code. We still do the same steps in the (almost) same order, but in Python. We will also implement the code as functions like we did with the Arduino sketch, but we won't need as many functions. Also, we will see how to write the code in a Python-friendly manner using a "main" function that we will call when execution begins.

Begin by launching the Thonny Python IDE under the *Main* ➤ *Programming* submenu. The IDE opens with a new, blank file in the editor. Name the new file `weather.py` or whatever you'd like to use. Or you can download the source code for the book and open the file by that name. Since the code is very similar to the Arduino sketch, we will only skim through the highlights.

Start at the top of the file. We need to import several libraries. We need the supporting libraries for the Adafruit BME280 library (as taken from the example code) and the system, time, and SparkFun Qwiic libraries as shown in the following:

```
import sys
import time
import board
import busio
import adafruit_bme280
import qwiic
```

We will also create two constants to store a formatting string we will use repeatedly in the code to save us some typing and remove risk of typing errors. The first ensures Python will print a floating-point number with only two decimals, and the second will print a floating-point number with no decimals. These are used by the Python `format()` function as you will see:

```
TWO_DECIMALS = "{0:.2f}"
NO_DECIMALS = "{0:.0f}"
```

Next, we will create some global variables: one for the BME280 sensor and another for the OLED module as shown in the following. We also need to initialize the BME280 I2C interface when we create an instance to the object. This is different than how the SparkFun library works, so there won't be a setup function for the BME280:

```
i2c = busio.I2C(board.SCL, board.SDA)
bme280 = adafruit_bme280.Adafruit_BME280_I2C(i2c)
oled = qwiic.QwiicMicroOled()
```

Like we did with the Arduino sketch, we will use functions to set up the OLED module and display information to the terminal and OLED. This is an exercise to demonstrate how you can move parts of your code to functions to make your code easier to read.

First, we create a function to set up the OLED module moving the initialization code we saw in Chapter 7 to the function as shown in Listing 8-8. We name the function `setup_oled()`.

Listing 8-8. setup_oled (Python)

```
def setup_oled():
    if oled.begin() is False:
        return False
```

```
    # Clear the screen and print greeting
    oled.clear(oled.PAGE)
    oled.clear(oled.ALL)
    oled.set_font_type(0)
    oled.set_cursor(0, 0)
    oled.print("How's theWeather?")
    oled.display()
    time.sleep(3)
    return True
```

Notice this code is written a bit differently where we call the begin() function at the top of the function returning false if the initialization fails or true after we've cleared the screen and displayed the welcome message.

Next, we create a function to display the values read to the terminal. We name the function print_diagnostics() because this function is useful only while the project is running and the terminal is open and connected. The following shows the new function to print the values to the terminal (command window):

```
def print_diagnostics(temp, humid, press):
    str_format = "BME280 values: {0:.2f} C, {1:.2f} %rh,
    {2:.2f} hPa"
    print(str_format.format(temp, humid, press))
```

Notice we do not return anything (return type is void), but we have a comma-separated list of parameters for the temperature, humidity, and pressure of type float. This means to use this function, we will need to pass (include) these values in a parameterized list.

Next is a similar function to print the values read to the OLED module. We name this function show_data_oled(), and it also takes the same parameters as shown in Listing 8-9. Thus, we will need to also include the values as parameters to call the function.

Listing 8-9. Show Data on the OLED (Python)

```
def show_data_oled(temp, humid, press):
    oled.clear(oled.PAGE)   # Clear page memory
    oled.set_cursor(0, 0)   # Set cursor position
    oled.print(TWO_DECIMALS.format(temp))
    oled.print(" C")
    oled.set_cursor(0, 15)  # Set cursor position
    oled.print(TWO_DECIMALS.format(humid))
    oled.print(" %rh")
    oled.set_cursor(0, 30)  # Set cursor position
    oled.print(NO_DECIMALS.format(press))
    oled.print(" hPa")
    oled.display()
```

Finally, we create a `main()` function that contains the code we want to execute to start (run) the project. We will see why we do this in a moment. The code is as you'd expect; we read the sensor at the top, print the diagnostics, show the data on the OLED, and repeat. The only new thing is once again a placeholder for setting the barometric pressure at sea level for reading the altitude. Listing 8-10 shows the `main()` function.

Listing 8-10. Main Function (Python)

```
def main():
    print("\nHow's the Weather?")
    print("------------------")
    if not setup_oled():
        print("ERROR: The OLED module is not found. Please
        check your connections!")
        sys.exit(1)

    # Compensate for reference pressure
    bme280.sea_level_pressure = 1013.25
```

```
    while True:
        # Read the sensor data
        temperature = bme280.temperature
        humidity = bme280.humidity
        pressure = bme280.pressure
        # Display diagnostics
        print_diagnostics(temperature, humidity, pressure)
        # Show data on OLED
        show_data_oled(temperature, humidity, pressure)
        time.sleep(3)
```

OK, so how do we use this `main()` function? When a Python script is read, the code is executed as it is read from top to bottom. Functions are declared, but they are not called until they are explicitly called. When you use a `main()` function in this manner, you can place special code at the bottom of the script (typically) that, if the script is executed, calls the `main()` function as shown in the following:

```
if __name__ == '__main__':
    try:
        main()
    except (KeyboardInterrupt, SystemExit) as err:
        print("\nbye!\n")
```

Here, we see the special condition (if statement) that checks to see if the reserved variable `__name__` is equal to `__main__`, which indicates the script is being executed. If true, we use a `try` block to call the main function. The `try` block is set up to capture a keyboard interrupt signal from the operating system. This means we can stop the program gracefully by pressing *CTRL+C*, and the code will exit. Cool, eh? This is a very common construct in Python scripts.

Now, let's see the code all together. Listing 8-11 shows the completed script for this project.

Listing 8-11. Weather Script (Python)

```
# Import libraries
import sys
import time
import board
import busio
import adafruit_bme280
import qwiic

# Constants
TWO_DECIMALS = "{0:.2f}"
NO_DECIMALS = "{0:.0f}"

# Create instances of the Qwiic classes (modules)
i2c = busio.I2C(board.SCL, board.SDA)
bme280 = adafruit_bme280.Adafruit_BME280_I2C(i2c)
oled = qwiic.QwiicMicroOled()

def setup_oled():
    """setup_oled"""
    if oled.begin() is False:
        return False

    # Clear the screen and print greeting
    oled.clear(oled.PAGE)
    oled.clear(oled.ALL)
    oled.set_font_type(0)
    oled.set_cursor(0, 0)
    oled.print("How's theWeather?")
    oled.display()
    time.sleep(3)
    return True
```

```
# Print values to the terminal
def print_diagnostics(temp, humid, press):
    """print_diagnostics"""
    str_format = "BME280 values: {0:.2f} C, {1:.2f} %rh,
{2:.2f} hPa"
    print(str_format.format(temp, humid, press))

# Display values on OLED screen
def show_data_oled(temp, humid, press):
    """show_data_oled"""
    oled.clear(oled.PAGE)   # Clear page memory
    oled.set_cursor(0, 0)   # Set cursor position
    oled.print(TWO_DECIMALS.format(temp))
    oled.print(" C")
    oled.set_cursor(0, 15)  # Set cursor position
    oled.print(TWO_DECIMALS.format(humid))
    oled.print(" %rh")
    oled.set_cursor(0, 30)  # Set cursor position
    oled.print(NO_DECIMALS.format(press))
    oled.print(" hPa")
    oled.display()

def main():
    """main"""
    print("\nHow's the Weather?")
    print("------------------")
    if not setup_oled():
        print("ERROR: The OLED module is not found. Please
check your connections!")
        sys.exit(1)

    # Compensate for reference pressure
    bme280.sea_level_pressure = 1013.25
```

```
    while True:
        # Read the sensor data
        temperature = bme280.temperature
        humidity = bme280.humidity
        pressure = bme280.pressure
        # Display diagnostics
        print_diagnostics(temperature, humidity, pressure)
        # Show data on OLED
        show_data_oled(temperature, humidity, pressure)
        time.sleep(3)

if __name__ == '__main__':
    try:
        main()
    except (KeyboardInterrupt, SystemExit) as err:
        print("\nbye!\n")
sys.exit(0)
```

OK, that's it! We've written the code. Unlike the Arduino, we do not need to compile the Python code. So we're now ready to execute the project!

Execute the Project

Now that we've spent many pages exploring the Qwiic modules and writing the code to interact with them, it is time to test the project by executing (running) it.

When the project runs (executes), you will see some diagnostic message written to the serial monitor (Arduino) or the terminal (Raspberry Pi). You will also see a welcome message appear on the OLED followed by a short pause. Then sensor values will appear. Figure 8-6 shows an example of what you should see on the OLED.

Figure 8-6. *Executing the weather project*

You can affect the sensor readings by placing your hand or another solid object in the path of the IR sensor on the proximity sensor module. Be sure to face the module so the silk screen writing is facing up. If the sensor is facing another direction, it will not respond to your movements.

Executing the code depends on which platform you're using. Let's look at the Arduino first.

Sketch on the Arduino

Executing the sketch on the Arduino requires connecting our board to our PC and then uploading the sketch to the Arduino. Recall the sketch will run so long as the USB cable is connected to our PC (and the Arduino).

Execute the Sketch

To execute the sketch, be sure your Arduino is connected and you've selected the correct board under the *Tools* ➤ *Board* menu. You also need to ensure you have the correct port selected under the *Tools* ➤ *Port* menu.

Once those items are set, you can click the *Upload* button or choose *Sketch* ➤ *Upload* from the menu. The Arduino IDE will compile the sketch and then upload it to your Arduino. Once you see the `Done uploading...` message, you can open the serial monitor. You should see the output begin momentarily as shown in the following:

```
How's the weather?
------------------
BME280 values: 20.13C, 35.25%rh, 1020.18 hPa
BME280 values: 20.10C, 35.22%rh, 1020.19 hPa
BME280 values: 20.08C, 35.21%rh, 1020.11 hPa
BME280 values: 20.08C, 35.23%rh, 1020.15 hPa
BME280 values: 20.04C, 35.19%rh, 1020.14 hPa
BME280 values: 20.00C, 35.33%rh, 1020.14 hPa
BME280 values: 19.99C, 35.27%rh, 1020.17 hPa
BME280 values: 19.60C, 35.75%rh, 1020.08 hPa
BME280 values: 19.58C, 35.85%rh, 1020.08 hPa
BME280 values: 19.56C, 35.96%rh, 1020.06 hPa
...
```

You should also see output on the OLED. Go ahead, and try it out!

If something isn't working, check your connections or refer to Chapter 7 for troubleshooting tips.

Python Code on the Raspberry Pi

Executing the sketch on the Raspberry Pi requires running the Python code in a terminal after connecting your Qwiic daisy chain to your Raspberry Pi via a hat or the Qwiic female breakout cable. Recall the code will run until you stop it with *CTRL+C* on the keyboard.

Execute the Python Code

To run the Python code on the Raspberry Pi, you can issue the command `python3 ./weather.py` from the same folder where the file was saved as shown in the following:

```
$ python3 ./weather.py
How's the Weather?
------------------
BME280 values: 17.99 C, 39.84 %rh, 1020.21 hPa
BME280 values: 18.00 C, 39.81 %rh, 1020.15 hPa
BME280 values: 18.01 C, 39.91 %rh, 1020.15 hPa
BME280 values: 18.01 C, 39.66 %rh, 1020.16 hPa
BME280 values: 18.03 C, 39.83 %rh, 1020.12 hPa
BME280 values: 18.04 C, 39.91 %rh, 1020.13 hPa
BME280 values: 18.04 C, 39.76 %rh, 1020.15 hPa
bye!
```

If everything worked as executed, congratulations! You've just built your second Qwiic project.

If something isn't working, check your connections or refer to Chapter 7 for troubleshooting tips.

Going Further

While we didn't discuss them in this chapter, there are some ideas where you could make this project into an IoT project. Here are just a few suggestions you can try once we have learned how to take our projects to the cloud. Put your skills to work!

- *MyWeather portal*: Build an enclosure and place the project outside and record your measurements in the cloud.
- *Test your HVAC*: Build several of these projects and place them throughout your house, workshop, etc. to see how well your HVAC system keeps you comfortable. If you see large variances in temperature, for example, it could mean the HVAC isn't configured (called balanced) correctly.
- *More weather*: Add rainfall and wind speed sensors to the project.
- *How high?*: Add the altitude reading by using the correct barometric pressure at sea level for your location.

Summary

In this chapter, we got more hands-on experience making projects with Qwiic modules. We used an environmental sensor to read temperature, relative humidity, and barometric pressure and displayed the values on a small OLED.

Along the way, we learned more about how to work with Qwiic modules including how to research the module to discover its capabilities through the documentation and example sketches and code. We also learned how to mitigate problems with software libraries by researching and using alternative software libraries.

We also saw some potential to make this project better as well as some ideas for how to adapt the project for practical uses.

In the next chapter, we will see another Qwiic project that demonstrates how to use multiple sensors that have the same address. We will use the Qwiic Mux Breakout – 8 Channel (TCA9548A) to read a series of Qwiic soil moisture sensors and an ADC module (ADS1015) to read a series of analog soil moisture sensors.

CHAPTER 9

Digital Gardener

One of the most common projects for the home is a simple plant monitoring station. Such projects can have all manner of components, but the most basic is determining if your plants need to be watered. Yes, that means all of those folks who have terrible luck with plants can add a tool to help them earn their green thumbs!

In this chapter, we will create a basic plant monitoring station. It isn't a full plant monitoring station because it doesn't include mechanisms to automatically water plants and record temperature, humidity, etc. (but you could add them). Rather, we're going to read the moisture level of the soil to determine if our plants need water. That alone is a great start to a professional-grade monitoring system.

Like the last chapter, we will see how to implement this project on the Arduino and Raspberry Pi but with an interesting twist. Let's get started.

Project Overview

The project for this chapter is designed to demonstrate how to get started building a plant monitoring station with a set of soil moisture sensors. We will use the sensors to read the moisture as an analog value and use that to determine if the plant (soil) needs more water. In fact, we can use the values to determine a status of dry, OK, or wet to categorize the condition of the soil. As you will see, the soil moisture sensor is one of many that require a bit of calibration to work correctly. And, naturally, we will need plants to monitor. Small potted plants make an excellent medium to start.

C. Bell, *Beginning IoT Projects*, https://doi.org/10.1007/978-1-4842-7234-3_9

What Will We Learn?

By implementing this project, we will explore more Qwiic modules including how to use multiples of the same sensor in a single project. Recall sensors have predefined addresses, and while some modules allow you to change the address, most will not. Thus, we will need to read from several sensors using the same address.

The challenges for this project are in the programming tasks, which are similar to the last project except we will increase the level of sophistication by using classes to make the code easier to maintain.

We will also encounter an interesting problem with the software libraries and see how to solve the problem. More specifically, the sensor we will use doesn't have a Python-equivalent software library, so we will have to use different hardware for our Python version of the project.

Let's see what hardware we will need.

Hardware Required

This chapter is unique in that the hardware needed for the Arduino version differs from the Python version. As mentioned, this is due to the lack of a Python library for the Qwiic Soil Moisture Sensor. Thus, there are two shopping lists. The only difference is the soil moisture sensor and an I2C Mux for the Arduino and an analog-to-digital (ADC) module for the Python version.

The hardware needed for the Arduino version of this project is listed in Table 9-1. URLs for each component are included for ease of ordering including duplicate entries for alternative vendors.

Table 9-1. *Hardware Needed for the Digital Gardener Project (Arduino)*

Component	URL	Qty	Cost
Qwiic Soil Moisture Sensor	www.sparkfun.com/products/17731	1*	$8.50
20×4 SerLCD - RGB Backlight (Qwiic)	www.sparkfun.com/products/16398	1	$24.95
Qwiic Mux Breakout - 8 Channel	www.sparkfun.com/products/16784	1	$11.95
Qwiic cable (any length can be used)	www.sparkfun.com/products/17259	3**	$1.50
Qwiic cable kit (optional)	www.sparkfun.com/products/15081	1**	$7.95
SparkFun RedBoard Qwiic (Arduino Uno or compatible)	www.sparkfun.com/products/15123	1	$19.95
	www.sparkfun.com/categories/233	1	$35.00+
	www.adafruit.com/category/176		
	www.sparkfun.com/products/15945	1	$5.95

Tip You should consider using the longer Qwiic cables for the soil moisture sensors so that you can place them in your potted plants and have length to place the host board nearby.

The hardware needed for the Python version of this project is listed in Table 9-2. URLs for each component are included for ease of ordering including duplicate entries for alternative vendors.

Table 9-2. *Hardware Needed for the Digital Gardener Project (Python)*

Component	URL	Qty	Cost
Soil Moisture Sensor (with Screw Terminals)	www.sparkfun.com/products/13637	1*	$6.95
20×4 SerLCD - RGB Backlight (Qwiic)	www.sparkfun.com/products/16398	1	$24.95
Qwiic 12 Bit ADC - 4 Channel	www.sparkfun.com/products/15334	1	$10.50
Qwiic cable (any length can be used)	www.sparkfun.com/products/17259	3**	$1.50
Jumper Wires Premium 12" M/M	www.sparkfun.com/products/9387	1	$4.50
(Arduino Uno or compatible)	www.sparkfun.com/products/15123	1	$19.95
Raspberry Pi 3B or later	www.sparkfun.com/categories/233 www.adafruit.com/category/176	1	$35.00+
Qwiic pHAT for Raspberry Pi	www.sparkfun.com/products/15945	1	$5.95

**Use as many sensors as you want, but you will need at least one.*

***You will need a minimum of two (2) cables plus one for each soil moisture sensor.*

About the Hardware

Let's discuss these components briefly. We will discover how to work with the hardware in more detail later in the chapter. Once again, some of the hardware is used in only one version. If you implement the Python version, be sure to get a set of male-to-male jumper wires so you connect the analog soil moisture sensors to the ADC module.

Qwiic Soil Moisture Sensor

The Qwiic Soil Moisture Sensor is an excellent example of how an analog sensor (a sensor that generates an analog value measured typically in volts) can be packaged into a Qwiic module whose value can be read as a digital value over I2C. In that respect, it makes working with analog signals very easy.

Unfortunately, there isn't a Python version of the library to read this sensor. It is oddly missing from the SparkFun pantheon of Qwiic Python libraries. While we could write one ourselves, we are going to solve this problem by using alternative hardware for the Python version. Figure 9-1 shows the Qwiic Soil Moisture Sensor.

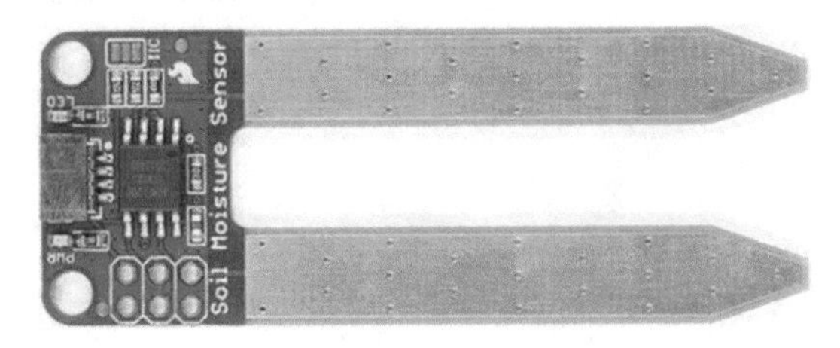

Figure 9-1. *Qwiic Soil Moisture Sensor (courtesy of sparkfun.com)*

One interesting aspect of soil moisture sensors is the coating on the arms. Most soil moisture sensors have a coating that can deteriorate over time. In fact, it is highly recommended that these sensors be powered on only when you need to read a value. Fortunately, the SparkFun soil moisture sensors (Qwiic and non-Qwiic) have a gold coating that permits the sensor to have a much longer life. That said, prolonged use of these sensors may require a maintenance cycle where you replace the sensors periodically.

Tip See `https://learn.sparkfun.com/tutorials/soil-moisture-sensor-hookup-guide` for more information about how to manage soil moisture sensors.

Qwiic I2C Mux

The SparkFun Qwiic Mux Breakout – 8 Channel (TCA9548A) is a very versatile module that everyone should have in their toolbox. It solves the address collision problem when using multiples of the same sensor. In fact, you can use up to eight of the same sensor and reference each at the same address. Better still, you can adjust the address on the I2C Mux module to use up to eight I2C Mux boards on the same project. With eight sensors per board, that means you can host up to 64 of the same sensor![1] Figure 9-2 shows the I2C Mux module.

[1] Theoretically, of course. I've not tried this myself, and I suspect there may be a soft limit to the number of sensors you can host in regard to the memory available on the host board or perhaps a limit in the cumulative length of Qwiic cables.

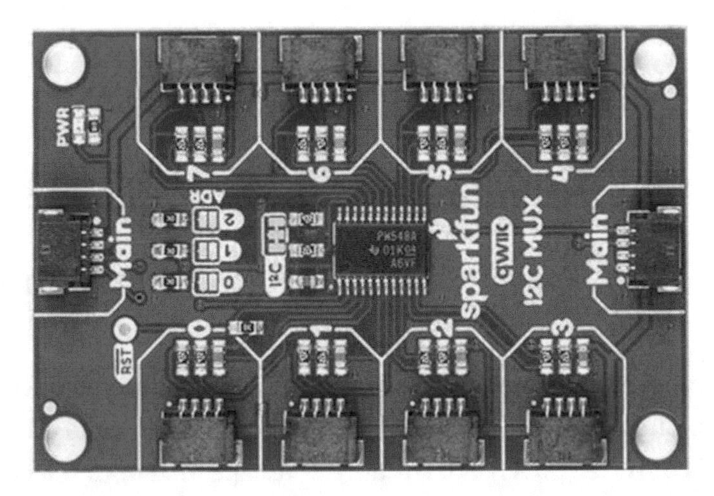

Figure 9-2. *Qwiic Mux Breakout (courtesy of sparkfun.com)*

The code for this module may seem a bit strange. It involves making an array of instances of the library class for your sensor and using the Qwiic I2C Mux library to pass the data through the Mux back to your own code. That may seem a little confusing, but once you see it in action, it will become clear.

Like most modules from SparkFun, you can change several features of the board including the I2C address. See the hookup guide for more details about the module including how to change the I2C address (`https://learn.sparkfun.com/tutorials/qwiic-mux-hookup-guide`).

Soil Moisture Sensor (with Screw Terminals)

The Soil Moisture Sensor (with Screw Terminals) is one of the normal forms of this sensor that you will encounter. Several vendors make these sensors in a variety of forms. Some have through-hole headers (soldered or not), others have a four- or three-pin connector, but the easiest to use is the screw terminal version from SparkFun. Figure 9-3 shows the Soil Moisture Sensor (with Screw Terminals).

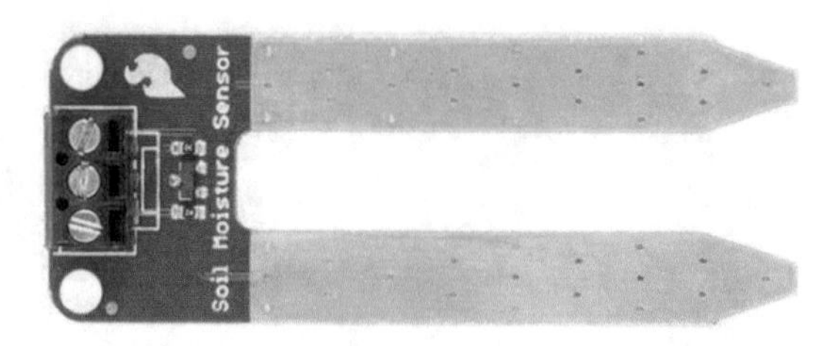

Figure 9-3. *Soil Moisture Sensor (with Screw Terminals) (courtesy of sparkfun.com)*

It is important to note that this is an analog sensor, so you cannot connect it directly to your Qwiic cabling. In fact, you will need an analog-to-digital converter (ADC) to translate the voltage read from the sensor to a number (integer) that you can use in your code.

Fortunately, we can use the ADC from SparkFun to connect this sensor.

Qwiic 12 Bit ADC

The Qwiic 12 Bit ADC from SparkFun allows you to connect up to four analog sensors (analog components) via a screw terminal header. Typically, analog sensors use three wires: ground, power, and signal. In fact, the preceding soil moisture sensor has three screw terminals marked as such.

The ADC module has a screw terminal header with six screw terminals. From the left, we see ground, power (VCC), and A0–A3, which represent the four analog signals. To hook the sensors to the board, you can connect all four of the ground wires to the GND terminal, all power wires to the VCC terminal, then the signal wires to A0–A4. We will see an example of how to do this in a later section. Figure 9-4 shows the ADC module.

Figure 9-4. *Qwiic 12 Bit ADC (courtesy of sparkfun.com)*

Note While the 12 Bit ADC module can host only four sensors, you can use the preceding Mux to host up to eight of the ADC modules giving you 32 sensors on a single project. That's a lot of plants to water!

Like most modules from SparkFun, you can change several features of the board including the I2C address. See the hookup guide for more details about the module including how to change the I2C address (`https://learn.sparkfun.com/tutorials/qwiic-12-bit-adc-hookup-guide`). For this chapter, you can use the module without modification.

LCD

The LCD module is our output module of choice. It is a character-driven LCD with four lines of 20 characters each. Its physical size is several times that of the Micro OLED, but it is easy to read and easier to mount on a panel.

It is also has a backlight that you can adjust the contrast and color (hence the RGB moniker). The text is shown in black over whatever backlight color you choose. Figure 9-5 shows the LCD panel.

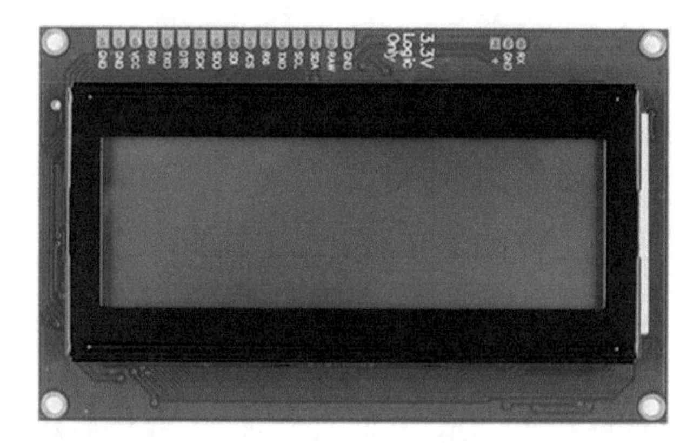

Figure 9-5. *20×4 SerLCD - RGB Backlight (courtesy of sparkfun.com)*

Note that the LCD has only a single Qwiic connector, so you will need to connect this module at the end of a daisy chain.

But this module is more than a Qwiic plug-and-play device. It has a header that you can use to connect this module to non-Qwiic host boards. For more information about the module including how to use the header pins, see `https://learn.sparkfun.com/tutorials/avr-based-serial-enabled-lcds-hookup-guide`.

Assemble the Qwiic Modules

Since we have two versions of the hardware, we will look at each hardware configuration. Note that how you wire the sensors may need to be adjusted depending on where your plants are located. More specifically, you need to insert the soil moisture sensors in soil for them to work correctly! So make sure you have enough cable to set up your project.

That said, you can assemble the project hardware without inserting the sensors in soil, but you can expect to get odd values that won't permit your code to determine the correct moisture boundaries for dry or wet soil. That's OK so long as you're developing the code, but you will need soil to calibrate the sensors. More on that in a moment.

Figure 9-6 shows an example of how you should connect your modules for the Arduino version using a Mux and the Qwiic soil moisture sensors. In this example, we use only two Qwiic soil moisture modules connected to the 0 and 1 ports on the Qwiic I2C Mux. You should use the longer Qwiic cables to connect the sensors to the Mux.

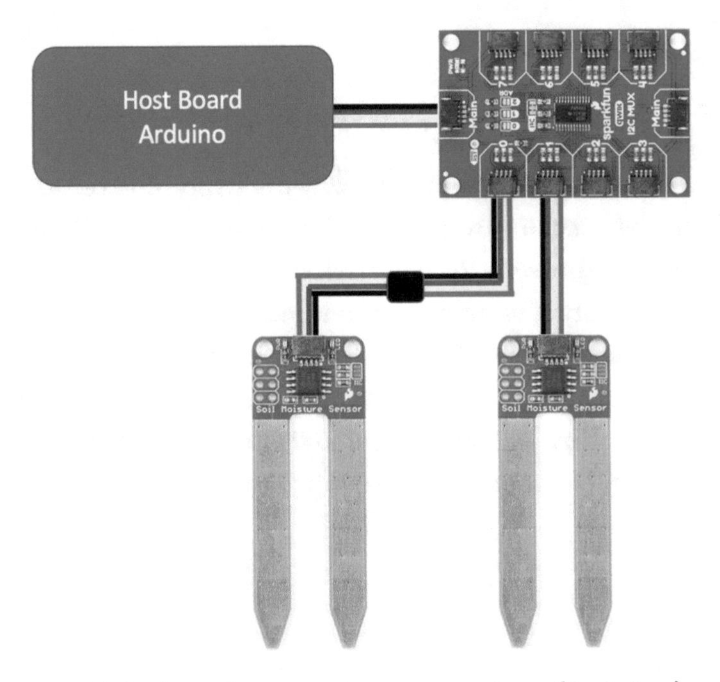

Figure 9-6. *Digital gardener project connections (Arduino)*

Figure 9-7 shows an example of how you should connect your modules for the Python version using a single ADC and the soil moisture sensors with screw terminals.

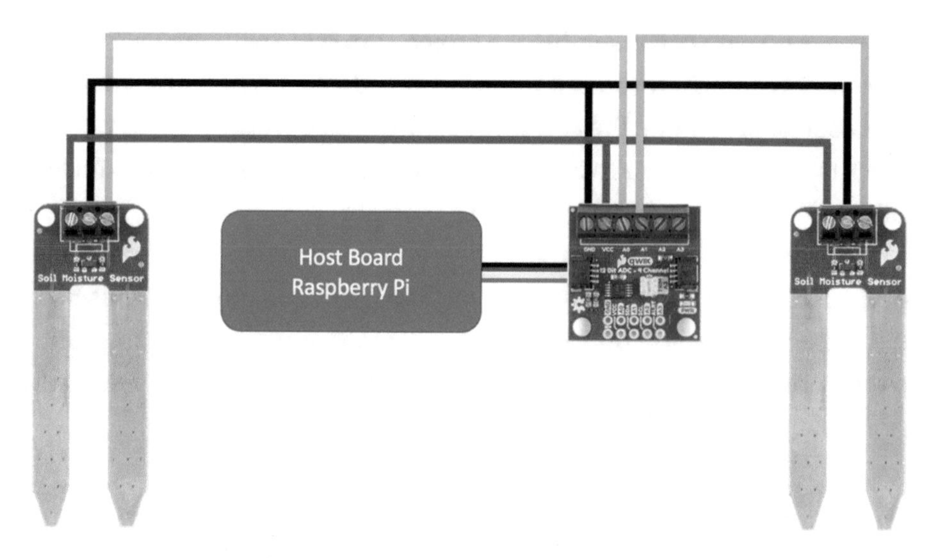

Figure 9-7. *Digital gardener project connections (Python)*

Here we see that we have connected two soil moisture sensors to the ADC module by attaching the ground (GND) and power (VCC) wires from the soil moisture sensors to the GND and VCC terminals on the ADC. There is plenty of room in the screw terminal header for solid wires like the recommended male jumper wires. The signal wires from each of the soil moisture sensors are attached to A0 and A1 as shown. Be sure to use a small flat (bladed) screwdriver to tighten the terminals. Do not overtighten them![2] Just tighten them enough so that they won't come loose.

[2] Overtightening screw terminals can break the plastic terminal housing and cause damage to the PCB.

POSITIONING THE PROJECT

This project requires the use of potted plants to insert the soil moisture sensors. As mentioned, you can write and test the code without placing the sensors in the soil, but you will need to do so for calibration and execution of the project.

Now is a good time to consider where you will set up your project to run. Choose a place where you can position your test subjects (plants) that can be reached by the wiring you've chosen and so you can place your host board nearby. You can move the project away from your computer once you have the code running, but you will still need a power source and all necessary cables to power your device (Arduino or Raspberry Pi). Careful consideration ahead of time can help avoid damaging your project wiring and make for a better experience.

Before we jump into the code, let's discuss calibration. While we have not discussed the code, it is important to know how to calibrate the sensors.

Calibrating the Sensors

The soil moisture sensor is the first of several sensors used in this book that require calibration. In this case, it is to find the thresholds for determining if the soil is dry, OK, or wet. Soil moisture sensors can vary in readings from one vendor to another, and to some degree the same model can vary from one instance to another (but with less variance).

The soil moisture sensor works by measuring the resistance between the probes returning a voltage value. In the case of the Qwiic version, the onboard ADC returns a value in the range 0–1024 where the higher value represents more moisture. Conversely, the screw terminal version connects to our ADC module and returns a value from 0 to 32572. That's quite a difference! Thus, you must tailor your code to match the characteristics of the sensors used. Furthermore, you must test the sensors either in the same environment or as close as you can get to determine their effective values - more specifically, what values represent soil that is dry, OK, or wet. To do that, we must perform a short calibration test.

To calibrate a soil moisture sensor, you will need to prepare two soil samples: one that represents soil you consider too dry for the plants you plan to monitor and another that is too wet. Dry soil may require some preparation either by removing the moisture manually by heating it (carefully) in an oven or by searching your planting area for a particularly dry area. If you cannot find any dry soil, try mixing some sand with the driest soil you have. For best results, you should choose samples of the same soil you use for your plants. A small 4"-tall container such as a small planter should be sufficient. Once you have the samples prepared, you can place the soil moisture sensor in one of the samples and run the completed project and note the values.

For example, when preparing this chapter, I used two soil moisture sensors: one in a container with dry soil and another with overly wet soil. I ran the project and noted the values for each sensor. I then used an average from the dry soil to determine the low value and an average from the wet soil to determine the high value. The Qwiic soil moisture sensors produced values of 400 for dry conditions and 800 for wet. Conversely, the screw terminal soil moisture sensors attached to the ADC produced values of 4000 for dry and 12000 for wet. Be sure to execute this test to determine the values for your soil.

Caution You must calibrate the sensors as described to determine the proper values to use in the code. Failing to do so may result in erroneous determination of the soil condition.

Now that we know more about the hardware for this chapter, let's write the code!

Write the Code

The code for this project follows what should be a familiar pattern. We start by declaring some variables, initializing the I2C bus, and preparing our modules for use. Then we execute a loop that reads values from the sensor and then displays the values on the LCD. More specifically, we will read the current soil moisture and determine if the soil is dry, OK, or wet based on the value returned from the sensor.

However, this time we're going to use a class to contain the code for working with the sensor. As you will see, the class for the Arduino version is a bit lower level than you're used to seeing, which will reveal some of the secrets about how a library can be written (the ones we've been downloading and installing). The Python version also has a class, but in this case the code is more simplified because we are using the library for the ADC module to read values from one of the analog pins.

Let's walk through how to prepare our computers to use the sensor and write code to read its values. We'll start with the Arduino.

Arduino

This section presents a walk-through of the sketch you will write to read values from the sensor and display them on the LCD module. But first, there are a couple of libraries we must install on our PCs.

Install Software Libraries

We will need to install the Arduino libraries for the I2C Mux and the LCD module separately. Fortunately, this is easy to do using the Library Manager. Simply open the Library Manager from the Arduino IDE menu (*Sketch* ➤ *Include Library* ➤ *Library Manager...*). Then search for `mux` and install the latest version of the SparkFun Qwiic I2C Mux library as shown in Figure 9-8.

Library Manager

Type All Topic All mux

Collection of Multiplexer and Hardware Controllers. Allows you to read one or more 74HC165, 74HC595, 74HC4067, 74HC4051, MCP23017 & MCP23018 also includes utilities to read buttons, encoders, pots & more!.
More info

SparkFun I2C Mux Arduino Library
by **SparkFun Electronics**
Library to control I2C multiplexers including the TCA9548/PCA9548. I2C multiplexers are useful for connecting multiple I2C devices that have only one address. This library makes it easy to work with the 8-channel TCA9548/PCA9548 I2C multiplexer but also works with smaller 4 and 2 bit multiplexers. This library support daisychaining multiple muxes so that you can get up to 64 devices on one I2C bus! It also supports generic Wire ports (Wire1, myWire, etc). Checkout the Qwiic Mux for more information.
More info
Version 1.0.1 Install

uMuxOutputLib
by **Naguissa**
Arduino, ESP8266 and STM32 7-segment multiplexed outputs library Supports Arduino AVR, STM32 and ESP8266 microcontrollers
More info

Close

Figure 9-8. *Installing the SparkFun Qwiic I2C Mux library (Arduino IDE)*

Similarly, we need to install the library for the LCD. Open the Library Manager and search for `serlcd` and then install the latest version as shown in Figure 9-9.

Library Manager

Type All Topic All serlcd

SparkFun SerLCD Arduino Library
by **Gaston R. Williams and Nathan Seidle**
Library for I2C, SPI, and Serial Communication with SparkFun SerLCD Displays An Arduino Library to allow simple control of 16x2 and 20x4 character SerLCDs from SparkFun. Includes RGB backlight control, display scrolling, cursor movement, and custom characters all over I2C, SPI, or Serial.
More info
Version 1.0.9 Install

Close

Figure 9-9. *Installing the SerLCD library (Arduino IDE)*

You may be wondering which library we need to install to use the Qwiic Soil Moisture Sensor. Unfortunately, there isn't a library for this sensor. This is likely because the board was a collaboration between SparkFun and Zio Smart-Prototyping. However, there are examples on the Zio Smart-Prototyping page that we can use as a template for writing our own class to read data from the sensor. See `https://github.com/sparkfun/Zio-Qwiic-Soil-Moisture-Sensor/tree/master/Firmware/Qwiic%20Soil%20Moisture%20Sensor%20Examples` for more details.

Now that we have the software libraries installed, we can begin writing our sketch. Since this is not our first Arduino sketch, we will discuss the code at a high level and skip the line-by-line details focusing on the mechanics of how the code works. You can study the code at your leisure to ensure you understand the sketch in more detail.

Write the Sketch

Recall we are going to use a class to read the sensor, which will be in the form of a separate file. Rather than write the main sketch file first and then add the class, we will write the class header first and then the main sketch and finally complete the class code. This is typically how programmers develop code with class modules (but not always). By creating the header for the class first, we can understand how to use the class making writing the main sketch easier.

The class will be named `QwiicSoilMoisture` and stored in two files: a header file named `QwiicSoilMoisture.h` and a source file named `QwiicSoilMoisture.cpp`. Effectively, we are moving the functions we would normally include in the main sketch to a class to make it easier to write, maintain, and understand.

However, since the Arduino IDE manages sketches, we will need to create the bare sketch file and folder first and then manually add the files to the project. There is no way (currently) to create and add new files to a sketch (but you can add existing files by clicking *Sketch* ➤ *Add File...*).

Open a new sketch and name it `gardener.ino` or whatever you'd like to use. Save the file and then close the project in the Arduino IDE.

To create the class files, navigate with your File Explorer (Finder) to the folder where you stored your main sketch (`gardener.ino`). Then, use your File Explorer or a text file editor to create two new files named `QwiicSoilMoisture.h` and `QwiicSoilMoisture.cpp`. Or you can use a terminal to navigate to the folder and issue these commands to create the empty files:

```
gardener % touch QwiicSoilMoisture.h
gardener % touch QwiicSoilMoisture.cpp
```

Now you can open the project in the Arduino IDE and see all three files in the project as shown in Figure 9-10.

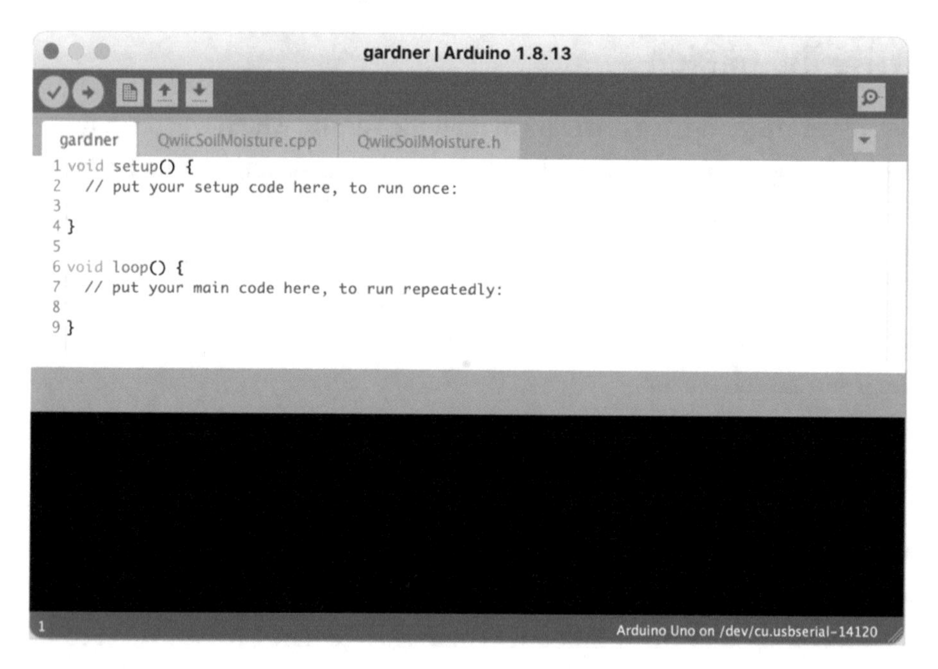

***Figure 9-10.** Starting a bare sketch with class header and code files (Arduino IDE)*

Now, let's see the code for each file starting with the header file.

Class Header File

Click the tab named `QwiicSoilMoisture.h` to open the blank file. Here, we will add the header or blueprint for the class. Recall the header file simply defines the class. We will use the `QwiicSoilMoisture.cpp` file to add the code for the functions in the class. Let's discuss those first.

The module provides a user-triggered LED we can use to indicate we are reading from the sensor. We will need a function to turn the LED on and off. We will name the function simply `led()` and pass a Boolean to indicate whether the LED should be turned on or off.

We also need a function to retrieve a value from the sensor. We will name this function `getValue()` and have it return the value read from the sensor.

Aside from those functions, we will also create a constructor so we can pass in an instance of the TwoWire class from the wire library. This is typical of how some Arduino class libraries written for I2C (Qwiic) modules are written (but there are others). Finally, we will need a function named `begin()` to initialize the sensor.

At this point, you may be wondering how such functions can be written and how we can control the module. After reading the example from the Zio Smart-Prototyping GitHub page, we learned there are three basic commands we can issue. These are represented as values that we write to the module to control it and are as follows:

```
#define COMMAND_LED_OFF         0x00
#define COMMAND_LED_ON          0x01
#define COMMAND_GET_VALUE       0x05
```

The first two simply tell the module to turn the LED on or off. The third tells the module to read a value from the sensor. Thus, reading the value requires first a command to tell the module to read and then another command to read the value. This will become clearer when we see the code, but for now we will include these definitions in the header file.

Let's look at the completed code for the header file. Listing 9-1 shows the file.

Listing 9-1. Qwiic Soil Moisture Sensor Header File

```
#import <Wire.h>

#define COMMAND_LED_OFF        0x00
#define COMMAND_LED_ON         0x01
#define COMMAND_GET_VALUE      0x05

class QwiicSoilMoisture {
public:
  QwiicSoilMoisture(TwoWire &i2c, bool diag=false) {
    wire = &i2c;
    diagnostics = diag;
  }
  bool begin();
  unsigned int getValue();

private:
  TwoWire *wire;
  int qwiicAddress = 0x28;      // Default Address
  bool diagnostics;

  void led(bool on=true);
};
```

The code should be easy to read, but there are a couple of things you can discover. First, notice how we store a variable for the I2C class that we get from the constructor as well as a Boolean we can use to turn on diagnostics, which are nothing more than `Serial.print()` statements in the class code. This is another common technique you can employ to control verbosity in your code. It also helps if there is a problem to turn on the diagnostics so you can see what is going on inside.

Finally, there is one very special and vitally important variable: the address of the module. We need this to pass to functions we call from the wire library so that we can communicate with the module over the I2C bus.

OK, let's return to the main sketch to see how we can use this class.

Main Sketch

Now click the `gardener.ino` tab to return to the main sketch. Let's begin with the preamble or top of the file. Recall here is where we include libraries we need, declare variables and constants, etc. Listing 9-2 shows the code for the main sketch preamble. As you can see, we include the wire, SerLCD, Mux, and our new class header.

Listing 9-2. Main Sketch Preamble

```
#include <Wire.h>
#include <SerLCD.h>
#include <SparkFun_I2C_Mux_Arduino_Library.h>
#include "QwiicSoilMoisture.h"
// Constants
#define NUMBER_OF_SENSORS 2  // Set number of sensors here
#define DRY_THRESHOLD 250    // Low threshold for dry soil
#define WET_THRESHOLD 400    // High threshold for wet soil

// Global Variables
QWIICMUX myMux;    // Mux
SerLCD lcd;        // Serial LCD

// Create pointer to an array of pointers to the sensor class
QwiicSoilMoisture **soilMoistureSensors;
```

Notice we also define the lower and upper thresholds for the dry and wet conditions. Anything between those values is considered OK. Refer to the "Calibrating the Sensors" section to see how we obtain these values through a calibration test.

Finally, notice we create an array of the new class. The declaration may seem a little strange, so let's examine it in more detail as repeated in the following. Notice we have declared a variable named `soilMoistureSensors` of type `QwiicSoilMoisture`. The double * means we are creating a pointer to a pointer or simply an array of `QwiicSoilMoisture` class instances. Why? Because the library for the I2C Mux requires an instance of the class for each module attached to the I2C Mux. Since we are using two, we will create an array of two instances of our new class:

```
QwiicSoilMoisture **soilMoistureSensors;
```

Next, we will code the `setup()` function, which includes the sort of initialization code we've seen in other projects. Specifically, we will set up the LCD, create the array of soil moisture sensors, initialize the I2C Mux, and then call the `begin()` function for each of the soil moisture instances. Note that we can control how many sensors are created by changing the value of the `NUMBER_OF_SENSORS` constant. Listing 9-3 shows the complete code for the `setup()` function. Read through it to ensure you understand all of the code included.

Listing 9-3. Main Sketch setup()

```
void setup()
{
  Serial.begin(9600);
  Serial.println("Digital Gardener!");

  Wire.begin();

  // Now, setup the LCD
  lcd.begin(Wire);
  lcd.clear();                      // Clear the display
  lcd.setBacklight(255, 255, 255); // Set backlight to bright white
```

```
lcd.setContrast(5);                    // Set contrast. 0<- for
                                          higher contrast
//          01234567890123456789   - Max characters we can
                                          display
lcd.print("Digital Gardener!");

// Create set of pointers for instantiated soil moisture
   classes
soilMoistureSensors = new QwiicSoilMoisture *[NUMBER_OF_
SENSORS];

// Instantiate the instances of the class
for (int x = 0; x < NUMBER_OF_SENSORS; x++)
  soilMoistureSensors[x] = new QwiicSoilMoisture(Wire);

if (myMux.begin() == false) {
  Serial.println("ERROR: Mux not detected. Freezing...");
  while (1);
}

// Initialize all the sensors
bool initSuccess = true;

for (byte x = 0; x < NUMBER_OF_SENSORS; x++) {
  myMux.setPort(x);
  if (!soilMoistureSensors[x]->begin()) {
    Serial.print("Sensor ");
    Serial.print(x);
    Serial.println(" did not initialize! Check wiring?");
    initSuccess = false;
  }
}
```

```
  if (initSuccess == false) {
    Serial.print("Freezing...");
    while (1);
  }
  Serial.println("Mux ready...");

  delay(3000);
}
```

Finally, we have the loop() function. At the highest level, we want to loop through all of the sensors attached to the I2C Mux, read their values, and display them on the LCD (as well as in the serial monitor). We will also use the value read to determine if the soil is dry, OK, or wet.

To read values through the I2C Mux, we simply use the setPort() function to "turn on" whichever port we want and then use the array of sensor class instances to read the values. In this case, we use the port number to correspond with the index of the array. In other words, the sensor on port 0 on the Mux is index [0] in the array. Nice! The I2C Mux clearly makes using multiple sensors at the same address really easy.

Listing 9-4 shows the completed code for the loop() function. Go ahead and read through the code to ensure you understand how it works. There should not be any surprises or new techniques.

Listing 9-4. Main Sketch loop()

```
void loop()
{
  unsigned int moisture[NUMBER_OF_SENSORS];

  lcd.clear();
  for (byte x = 0; x < NUMBER_OF_SENSORS; x++) {
    // Connect to the specific port on the mux
```

```
if (x > 4) {
  myMux.setPort(x-4);
} else {
  myMux.setPort(x);
}

// Get the sensor value from the port specified.
moisture[x] = soilMoistureSensors[x]->getValue();
Serial.print("Sensor #");
Serial.print(x);
Serial.print(" value = ");
Serial.print(moisture[x]);
Serial.print(" ");

// Display on the LCD, page if necessary
lcd.setCursor(0, x);
lcd.print("#");
lcd.print(x+1);
lcd.print(": ");
lcd.print(moisture[x]);
lcd.print(" = ");
// Determine if soil is dry, wet, or Ok
if (moisture[x] <= DRY_THRESHOLD) {
  lcd.print("TOO DRY");
  Serial.println("TOO DRY");
} else if (moisture[x] >= WET_THRESHOLD) {
  lcd.print("TOO WET!");
  Serial.println("TOO WET!");
} else {
  lcd.print("Ok");
  Serial.println("Ok");
}
```

```
    // Scroll if there are more than (4) sensors connected.
    if (NUMBER_OF_SENSORS > 4) {
      delay(3000);
      lcd.clear();
    }
  }
  delay(5000);
}
```

Did you notice the code at the end of the function? This solves the problem of how to display the status of more than four sensors on an LCD that only has four lines. This is a primitive form of paging. Cool, eh?

Now we can write the final portion of our project – the code for the class.

Class Code File

Click the tab named `QwiicSoilMoisture.cpp` to open the blank file. Here, we will add the code for the class. There are only three functions to write.

The `begin()` function does a quick check of the module by calling the `beginTransmission()` function passing in the address of the module. We then call the `endTransmission()` function, and if it returns anything other than 0, we return false to indicate the initialization failed. We also turn the LED on briefly to show a successful initialization.

The `getValue()` function writes the read command to the module using a series of function calls including the `beginTransmission()` function to start the protocol, then the `write()` function to write the get value command, and the `endTransmission()` function to complete the protocol. We then use the `requestFrom()` function to request 2 bytes from the module and read them inside a loop that loops while the `available()` function returns true. This way, we are sure to read all of the bytes requested. Finally, we read the status value from the module using the `read()` function.

The led() function uses a similar mechanism of function calls including the beginTransmission() function to start the protocol, then the write() function to write the led on or off command, and the endTransmission() function to complete the protocol.

Listing 9-5 shows the completed code for the class (documentation omitted for brevity). Since this is a much lower level than the code we're used to seeing, be sure to study it to understand how it works. While the protocol (how we talk to the module) is very simple, the functions we use are uncommon since we normally have libraries written for us.

Listing 9-5. Qwiic Soil Moisture Sensor Code File

```
#include <Arduino.h>
#include "QwiicSoilMoisture.h"

bool QwiicSoilMoisture::begin() {
  wire->beginTransmission(qwiicAddress);
  // Check here for an ACK from the slave.
  // If no ACK, return false.
  if (wire->endTransmission() != 0) {
    return false;
  }
  led(true);
  delay(1000);
  led(false);
  return true;
}

unsigned int QwiicSoilMoisture::getValue() {
  unsigned int status {0};
  unsigned int value {0};
```

```
  led(true);                              // turn LED on
  wire->beginTransmission(qwiicAddress); // start transmission
  wire->write(COMMAND_GET_VALUE);         // status command
  wire->endTransmission();                // stop transmitting

  wire->requestFrom(qwiicAddress, 2);     // request 2 bytes
                                             from device
  // Loop until the byte is available and tell ADC to prepare
     to read
  while (wire->available()) {
    uint8_t value_low = wire->read();  // lower values
    uint8_t value_high = wire->read();  // upper values
    // Reassemble the bytes
    value=value_high;
    value<<=8;
    value|=value_low;
    if (diagnostics) {
      Serial.print("ADC = ");
      Serial.print(value);
      Serial.print(" ");
    }
  }
  status = wire->read();             // read the status value
  if (diagnostics) {
    Serial.print("Status = ");
    Serial.print(status);
    Serial.print(" ");
  }
  delay(500);
  led(false);                             // turn off LED
  return value;
}
```

```
void QwiicSoilMoisture::led(bool on) {
  wire->beginTransmission(qwiicAddress);
  if (on) {
    wire->write(COMMAND_LED_ON);
  } else {
    wire->write(COMMAND_LED_OFF);
  }
  wire->endTransmission();
}
```

As you can see, the code is a bit more complicated, but by moving it to a class we can move the complexity out of the main sketch making that much easier to read. While this class is a very primitive example of what a class would look like to interact with an I2C module, it is a good glimpse into what those libraries contain as well as their complexity. That should make you aware of the necessity of such libraries so that we can write more powerful code without having to learn a great deal more than we would normally want to know.

Compile the Sketch

The last step is to compile the sketch before uploading it to your board. If you encounter any errors, be sure to fix them and recompile to ensure the sketch compiles without errors or serious warnings.

Once everything compiles, we're ready to start testing. But first, let's look at the code for the Raspberry Pi. You can skip to the "Sketch on the Arduino" section if you're curious to see how the project works. While the code will execute the same on both platforms, the values differ due to the differences in how the sensors are read (the range of values differs).

Raspberry Pi

This section presents a walk-through of the Python code you will write to read values from the sensor and display them on the LCD module. But first, there are a couple of libraries we must install on our Raspberry Pi.

Install Software Libraries

We need only two more software libraries to install to use the Python code in this chapter. While we are using an ADC from SparkFun, we will use the ADC code from Adafruit instead. We need the Adafruit CircuitPython ads1x15 and the SparkFun SerLCD libraries as shown in the following. Once you've run those commands, you can work on the code:

```
# pip3 install adafruit-circuitpython-ads1x15
# pip3 install sparkfun-qwiic-serlcd
```

Tip If you haven't installed the SparkFun Qwiic Python libraries, you must install them to run this project (`pip3 install sparkfun_qwiic`).

Now we're ready to write the code.

Write the Code

The code for the Python version of this project is a bit shorter than the Arduino code. We will create a class for the soil moisture sensor, but it won't be as complicated since we will be using the Adafruit ads1x15 library instead of communicating directly with the sensor (recall the soil sensors in this example are analog, now Qwiic enabled). Thus, we will encapsulate all of the code that interacts with the ADC (and the sensor) in this class making it hide more of the complexity. As you will see, it is shorter and less complex than the Arduino example. Let's get started!

Once again, we will not dive into every line of code. We will explore the code at a higher level and discuss the more complex or important parts in detail. You can read through the code and learn more about how it works at your leisure.

Like the Arduino example, we will use a class to contain the code to read the sensor. However, unlike the Arduino IDE, you can use any editor to create the class and main script. We will name the main script `gardener.py` and the class file `soil_moisture.py`. Let's start with the class.

Soil Moisture Class

Open the Thonny Python IDE under the *Main ➤ Programming* submenu. Create a new file `soil_moisture.py`. We will name the class `SoilMoisture,` and we will need a few functions. In fact, we need only the constructor and a function named `read_sensor()` to read the value.

The constructor accepts a single value that specifies the number of sensors to read. Since the ADC contains only four analog pins, we will ignore values greater than four.

The `read_sensor()` function is more complicated because we have to determine which analog pin to read. We will use a utility class from the Adafruit pantheon called `AnalogIn` that will have us read analog values from the ADC in the form of a tuple. More specifically, it creates a tuple that contains the value from the analog pin as well as the voltage read from the ADC for that pin, which could be helpful for diagnosing problems with the sensor.

However, to read the value on that pin, we must pass a specific constant. So we must use a longer `if elsif` sequence to initialize the `AnalogIn` class correctly. Once we have the class initialized correctly, we can return the values.

Listing 9-6 shows the complete code for the class with documentation removed for brevity. Take a few moments to read through the code so that you understand all of the parts of the code. As you will see, it is not nearly as complicated as the Arduino class, thanks to the helpful class library and utility class from Adafruit.

Listing 9-6. Soil Moisture Class (Python)

```
# Import libraries
import board
import busio
import adafruit_ads1x15.ads1015 as ADS
from adafruit_ads1x15.analog_in import AnalogIn

class SoilMoisture:
    """Soil Moisture Class"""
    number_of_sensors = 0
    i2c = busio.I2C(board.SCL, board.SDA)
    soil_moisture_sensor = ADS.ADS1015(i2c)

    def __init__(self, num_sensors):
        self.number_of_sensors = num_sensors
        if num_sensors > 4:
            self.number_of_sensors = 4

    def read_sensor(self, sensor_number):
        """Read Sensor"""
        channel = None
        if sensor_number == 0:
            channel = AnalogIn(self.soil_moisture_sensor, ADS.P0)
        elif sensor_number == 1:
            channel = AnalogIn(self.soil_moisture_sensor, ADS.P1)
        elif sensor_number == 2:
            channel = AnalogIn(self.soil_moisture_sensor, ADS.P2)
        elif sensor_number == 3:
            channel = AnalogIn(self.soil_moisture_sensor, ADS.P3)
```

```
        if not channel:
            return (None, None)
        return (channel.value, channel.voltage)
```

Now we can write our main script.

Main Script (Python)

Open the Thonny Python IDE under the *Main* ➤ *Programming* submenu. Create a new file `gardener.py`. There is nothing new in this code as it follows the same flow as the Arduino example but simplified. Specifically, we need only one instance of our new class since that class can read up to four soil moisture sensors on the ADC.

Once the initialization is complete, we will read through the number of sensors as specified in the constant `NUMBER_OF_SENSORS`. We also employ the main function concept we saw in the last chapter. Listing 9-7 shows the complete code for the main script for this project. You can read through it to see how all of the code works.

Listing 9-7. Main Script (Python)

```
# Import libraries
import time
import sys
import qwiic_serlcd

from soil_moisture import SoilMoisture

# Constants
NUMBER_OF_SENSORS = 4    # Set number of sensors here
DRY_THRESHOLD = 4000     # Low threshold for dry soil
WET_THRESHOLD = 12000    # High threshold for wet soil
```

```
def main():
    """Main function to run the digital gardener example."""
    lcd = qwiic_serlcd.QwiicSerlcd()
    soil_moisture_sensor = SoilMoisture(NUMBER_OF_SENSORS)

    # Use the serial LCD
    print("\nDigital Gardener!")
    if not lcd.connected:
        print("The Qwiic SerLCD device isn't connected to "
              "the system. Please check your connection",
              file=sys.stderr)
        sys.exit(1)

    lcd.setBacklight(255, 255, 255) # Set backlight to bright
                                      white
    lcd.setContrast(5)              # set contrast
    lcd.clearScreen()               # clear the screen
    lcd.print("Digital Gardener!")
    lcd.setCursor(0, 1)
    lcd.print("Getting ready")
    for i in range(0, 5):
        lcd.print(".")
        time.sleep(2) # wait sec for system messages to complete

    while True:
        lcd.clearScreen()
        for i in range(0, NUMBER_OF_SENSORS):
            value, voltage = soil_moisture_sensor.read_sensor(i)
            if value > WET_THRESHOLD:
                condition = "Too WET!"
            elif value < DRY_THRESHOLD:
                condition = "Too DRY!"
```

```
            else:
                condition = "Ok"
            msg = "#{0}: {1:5} {2}".format(i, value, condition)
            print(msg)
            lcd.setCursor(0, i)
            lcd.print(msg)
            time.sleep(0.5)
        time.sleep(5)

if __name__ == '__main__':
    try:
        main()
    except (KeyboardInterrupt, SystemExit) as err:
        print("\nbye!")
        sys.exit(0)
```

Notice the values for the thresholds for dry and wet soil differ from the Arduino version. Recall this is because we're using an ADC to read a different form of the soil moisture sensor.

OK, that's it! We've written the code. Unlike the Arduino, we do not need to compile the Python code. So we're now ready to execute the project!

Execute the Project

Now that we've spent many pages exploring the Qwiic modules and writing the code to interact with them, it is time to test the project by executing (running) it.

When the project runs (executes), you will see some diagnostic message written to the serial monitor (Arduino) or the terminal (Raspberry Pi). You will also see a welcome message appear on the LCD followed by a short pause. Then sensor values will appear one on each line of the screen. Figure 9-11 shows an example of what you should see on the LCD. In this case, we are seeing the screen running with four soil moisture sensors on the Raspberry Pi. Notice two of them are OK, but two have been overwatered (too wet).

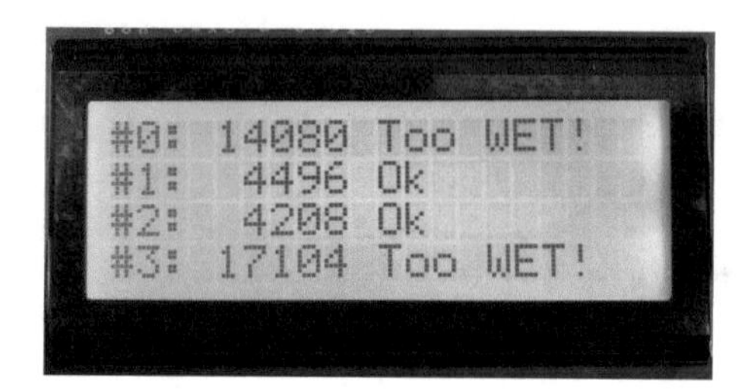

Figure 9-11. *Executing the digital gardener project*

Executing the code depends on which platform you're using. Let's look at the Arduino first.

Sketch on the Arduino

Executing the sketch on the Arduino requires connecting our board to our PC and then uploading the sketch to the Arduino. Recall the sketch will run so long as the USB cable is connected to our PC (and the Arduino).

Execute the Sketch

To execute the sketch, be sure your Arduino is connected and you've selected the correct board under the *Tools* ➤ *Board* menu. You also need to ensure you have the correct port selected under the *Tools* ➤ *Port* menu.

Once those items are set, you can click the *Upload* button or choose *Sketch* ➤ *Upload* from the menu. The Arduino IDE will compile the sketch and then upload it to your Arduino. Once you see the `Done uploading...` message, you can open the serial monitor. You should see the output begin momentarily that is the same as that on the LCD. Go ahead, and try it out! You should see values similar to the following:

```
Digital Gardener!
#0:   957 Too WET!
#1:   522 Ok
...
```

If something isn't working, check your connections or refer to Chapter 7 for troubleshooting tips.

Python Code on the Raspberry Pi

Executing the sketch on the Raspberry Pi requires running the Python code in a terminal after connecting your Qwiic daisy chain to your Raspberry Pi via a hat or the Qwiic female breakout cable. Recall the code will run until you stop it with *CTRL+C* on the keyboard.

Execute the Python Code

To run the Python code on the Raspberry Pi, you can issue the command `python3 ./gardener.py` from the same folder where the file was saved as shown in the following. You should get results similar to the following depending on the number of sensors you're using:

```
$ python3 ./gardener.py
Digital Gardener!
#0:  3010 Too DRY!
#1:  4224 Ok
#2:  4336 Ok
#3: 17104 Too WET!
...
```

If everything worked as executed, congratulations! You've just built your third Qwiic project.

If something isn't working, check your connections or refer to Chapter 7 for troubleshooting tips.

Going Further

While we didn't discuss them in this chapter, there are some ideas where you could make this project into an IoT project. Here are just a few suggestions you can try once we have learned how to take our projects to the cloud. Put your skills to work!

- *MyGarden portal*: You can display the values of the sensor readings on a web page to allow you to check on your plants from anywhere in the world.
- *Go outdoors*: Build several of these projects installing them in weather-proof enclosures and place them throughout your garden to monitor your plants. Or, if you're a produce gardener, you can build a system to help you grow better vegetables than your rival gardener down the block.
- *Alternative hardware*: Implement the hardware used in the Python version in Arduino.
- *Sensor power management*: Add a relay to turn the power on and off to the soil moisture sensors to extend their life.

Summary

In this chapter, we got more hands-on experience making projects with Qwiic modules and overcoming obstacles, which plague most when learning to build IoT projects. We used a set of soil moisture sensors to check the moisture levels and displayed the values on a small LCD.

Along the way, we learned more about how to work with Qwiic modules including how to write our own classes for managing the sensors. We also learned how to use alternative software libraries in our project. Finally, we saw some potential to make this project better as well as some ideas for how to adapt the project for practical uses.

In the next chapter, we will see another project that demonstrates how to use a STEMMA QT accelerometer to build a digital spirit level.

CHAPTER 10

Balancing Act

Now that we've learned how to use Qwiic modules, we can go a bit further by learning how to write our code using more common constructs in the language. For example, we may need to make a class to keep certain functions or pieces of data, pass variables among functions, and so on. We will see more of these in this chapter as we continue to learn to make more sophisticated projects.

In this chapter, we will learn how to use a multiple degrees of freedom (DoF[1]) module to build a spirit level, sometimes called a bubble level or digital level. We will also see how we can seamlessly combine Qwiic and STEMMA QT modules.

Of course, we will see how to implement this project on the Arduino and Raspberry Pi. Let's get started.

[1] https://en.wikipedia.org/wiki/Degrees_of_freedom

C. Bell, *Beginning IoT Projects*, https://doi.org/10.1007/978-1-4842-7234-3_10

WHAT IS A SPIRIT OR BUBBLE LEVEL?

A bubble level, sometimes called a spirit level, uses a tube or cylinder with colored liquid partially filled so that an air bubble is in the container. For tube shapes, rings are painted on the outside to allow the use of the bubble to move between the lines to indicate a level or plumb position. Tubes are commonly used in construction levels for a single axis in the vertical or horizontal position.

The version we will mimic in this chapter uses a concave glass cylinder with rings painted on the top so that the bubble can be centered to indicate level. The advantage of this form is that you can level a surface in both X and Y directions.

So, if you want to make a post level, a spirit level with tubes (carpenter level) can help you by checking for level in a vertical position on two adjoining sides (in this case, the Z axis). However, if you want to make your desk level, a level with a concave cylinder can help you level the desk in both X and Y horizontal axes simultaneously.

Project Overview

The project for this chapter is designed to demonstrate how to build a spirit level with an accelerometer sensor. We will use the sensor to read the values of the X and Y coordinates to determine if the sensor is level. We will use the data to display a set of circles on our trusty Micro OLED to help you determine if the sensor is level. And, like the last chapter, there is a bit of calibration needed to ensure the sensor is ready for use.

For those with access to a 3D printer, we will also see a simple enclosure you can print to install the modules to protect the modules and make them easier to handle. In fact, any enclosure will make the calibration sequence easier.

Interestingly, the module we will use also has a gyroscope sensor, which would allow you to build all manner of motion sensing projects, but we will stick with a project most can relate easy. After all, almost everyone has wanted to make something more level - a table, picture frame, your 3D printer, etc.

What Will We Learn?

While we won't see anything new with the hardware other than a different sensor, by implementing this project, we will reinforce what we learned from the previous chapters, specifically how to connect Qwiic and STEMMA QT modules to our host boards.

The challenges for this project are in the programming tasks, which are a step-up in complexity from the last project. Not only will we use a class to make the code easier to maintain, but we will also learn how to work with structures and mapping two coordinate systems (and the math that entails[2]).

Let's see what hardware we will need.

Hardware Required

The hardware needed for this project is listed in Table 10-1. URLs for each component are included for ease of ordering including duplicate entries for alternative vendors.

[2] Don't worry. It's basic arithmetic here.

Table 10-1. Hardware Needed for the Balancing Act Project

Component	URL	Qty	Cost
Adafruit LSM6DS33 6-DoF Accel + Gyro IMU - STEMMA QT / Qwiic	www.adafruit.com/product/4480	1	$5.95
Micro OLED Breakout	www.sparkfun.com/products/14532	1	$16.95
Qwiic cable (any length can be used)	www.sparkfun.com/products/17259	2	$1.50
Qwiic cable kit (optional)	www.sparkfun.com/products/15081	1**	$7.95
SparkFun RedBoard Qwiic (Arduino Uno or compatible)	www.sparkfun.com/products/15123	1	$19.95
Raspberry Pi 3B or later	www.sparkfun.com/categories/233 www.adafruit.com/category/176	1	$35.00+
Qwiic pHAT for Raspberry Pi	www.sparkfun.com/products/15945	1	$5.95

Tip There are quite a few Qwiic accelerometers from SparkFun and STEMMA QT accelerometers from Adafruit. Most can easily be adapted to use in this project, so feel free to substitute a different module should you need to substitute (or just want to experiment).

About the Hardware

Let's discuss these components briefly. We will discover how to work with the hardware in more detail later in the chapter.

Sensor

The Adafruit LSM6DS33 6-DoF Accel + Gyro IMU – STEMMA QT / Qwiic is a low-cost module that permits you to detect motion and orientation. It permits six degrees of freedom (notated as 6-DoF) or six different parameters that can be monitored. As mentioned, it has both an accelerometer and a gyroscope chip. The accelerometer measures direction or orientation in three axes with respect to the earth (through gravity). The gyroscope can tell us how fast the change is happening and can detect if the module is rotated.

Perhaps the best part of this module is its low cost, so it makes purchasing the module for a permanent project a bit more economical. That is, if you decide to build the project in a permanent enclosure, you won't be spending a lot of money to build it.

OLED

The OLED module we will use is the same from the previous chapters. If you'd like to experiment with other output devices, you can, but it is recommended to use the Micro OLED so that the example code works without modification. That said, the code for the Arduino example is written so that it can be used with different OLED sizes. We'll see those details when we examine the code.

Assemble the Qwiic Modules

Recall from Chapter 7 we can use a single Qwiic cable to connect our LSM6DS33 sensor to the OLED module and then another to attach to the host adapter on our host board. Figure 10-1 shows an example of how you should connect your modules to form a Qwiic daisy chain.

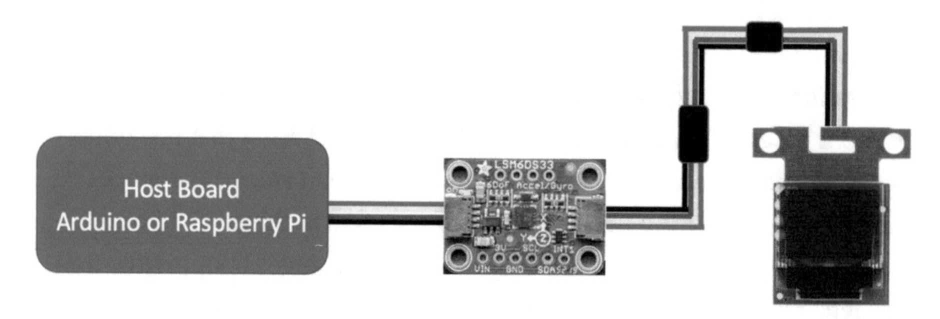

Figure 10-1. *Balancing act project Qwiic/STEMMA connections*

Using an Enclosure

Projects like the one we're about to create can be very cumbersome because we will need to move the module in a number of orientations during the calibration routine. You can achieve this more easily by using a long Qwiic cable (or join two Qwiic cables using the Qwiic breakout module), but even then it leaves the OLED flapping around.

So what can we do? If our host adapter provides mounts for mounting Qwiic modules, we could use those, but since we have a STEMMA QT module in the project, that won't work because they have a different layout (the holes are smaller and closer together). Of course, we could use a piece of wood to mount the modules, but there is a more elegant solution.

If you have your own or access to a 3D printer, you can print an enclosure. The source code for this chapter includes the 3D printing files you need to create a simple enclosure to house the accelerometer and OLED screen. Figure 10-2 shows the two parts of the enclosure with the bottom on the left.

Figure 10-2. *3D enclosure design for the spirit level project*

Notice we have a cutout in the bottom for the Qwiic cable as well as two sets of holes: one for mounting the STEMMA QT module and the other to attach the cover. The top also has holes for the OLED module as well as a cutout for the display.

You will need a few pieces of hardware to assemble the enclosure. Fortunately, SparkFun has most of what we need. The following lists the hardware needed. You may be able to find these at electronics or hobby stores:

- (2) ¾" standoffs or equivalent (`www.sparkfun.com/products/11796`)
- (4) #40 screws or equivalent (`www.sparkfun.com/products/10453`)

- (2) 2.5×10mm bolts or equivalent
- (2) 2.5mm nuts or equivalent

To print the files, you should use supports so that the recessed areas do not collapse. These are used to hide the heads of the bolts and screws.

To assemble the enclosure, begin by mounting the STEMMA QT module to the bottom as shown in Figure 10-3. Use two 2.5×10mm bolts and two 2.5mm nuts to secure the module to the base.

Figure 10-3. *Mounting the STEMMA QT board to the enclosure base*

Next, attach a short Qwiic cable to the OLED and mount it on the top using a ¾" (19mm) spacer from SparkFun as shown in Figure 10-4. Use two #40 screws to attach the OLED to the spacer as shown.

Figure 10-4. *Mounting the Micro OLED board to the enclosure top cover*

Finally, attach the Qwiic cable from the Micro OLED to the STEMMA QT on one side and another longer cable on the side with the cutout. Then, carefully press the bottom onto the top (or vice versa) and attach with two #40 screws. Figure 10-5 shows a completed example of the enclosure.

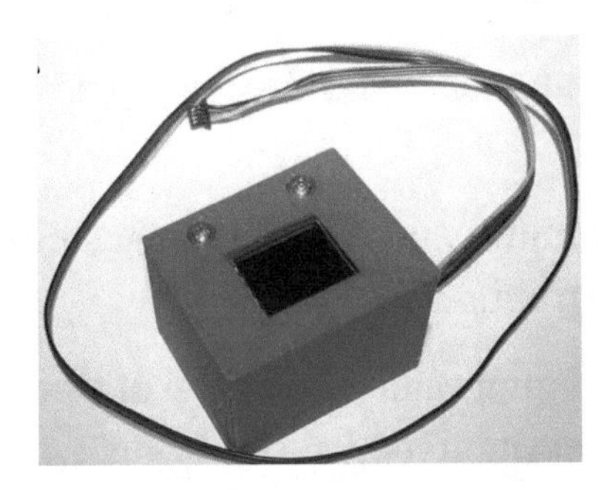

Figure 10-5. *3D printed enclosure for the spirit level project*

If you have experience creating 3D models for printing, feel free to experiment with creating your own enclosure - perhaps one that also includes a battery and a small form factor host board.

Tip Refer to sections in Chapter 8 that discuss researching modules if you'd like to use a different accelerometer for the project. You will need to learn what library to use and how it differs from how the LSM6DS33 reads data (the range of values, sensitivity, etc.).

Calibrating the Sensor

The accelerometer must be calibrated to ensure we achieve three goals: 1) the values of X and Y at rest in a known level position will be recorded and used as a bias to ensure we detect proper orientation and to compensate for enclosure mounting (board not level in the enclosure), 2) we need to know the maximum values of the X and Y axes, and 3) we need to know the minimum values of the X and Y axes. The minimum and maximum values will ensure we calculate the location of the pointer in the OLED. That is, we draw the pointer so that it shows level when positioned in the center.

To achieve this calibration, we will perform five steps as follows.

1. Obtain the values of X and Y when placed in a level position.

2. Get the maximum value of X by placing the sensor in a vertical position.

3. Get the minimum value of X by placing the server in the vertical position rotated 180 degrees.

4. Get the maximum value of Y by placing the sensor in a vertical position.

5. Get the minimum value of Y by placing the server in the vertical position rotated 180 degrees.

Caution You must calibrate the sensor as described to determine the correct values for X and Y at rest and the min/max values for each. Failing to do so may result in the spirit level not centering when the sensor is level.

Figure 10-6 shows a series of positions of a cube (to represent our preceding 3D printed enclosure) for each step.

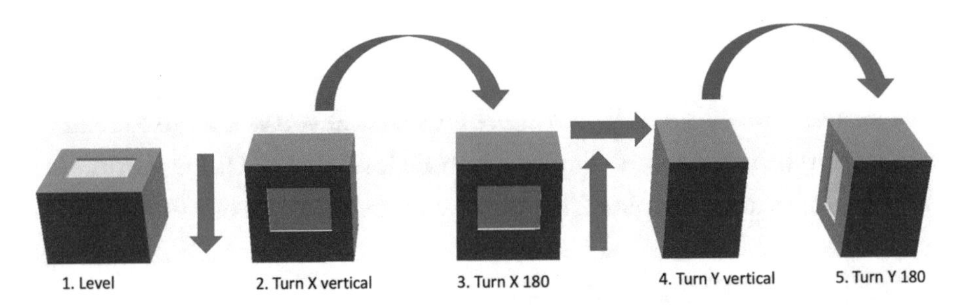

Figure 10-6. *Calibration positions*

Now that we know more about the hardware for this chapter, let's write the code!

Write the Code

The code for this project follows the pattern of code layout we've learned. There are other ways to construct the code (mainly to do with order of operations and modularization), but the basics are still the same.

We begin by initializing the I2C bus and preparing the sensor and OLED for use and then executing a loop that reads values from the sensor and then displays the values on the OLED. More specifically, we will read the X and Y axis values from the accelerometer.

However, the scale or range of values for the accelerometer may vary, so we will need to determine those ranges. We also need to calculate the at-rest values in order to factor out any orientation or environmental effects. We perform both of these steps in a calibration routine that we will execute at the start of execution.

Let's walk through how to prepare our computers to use the sensor and write code to read its values. We'll start with the Arduino.

Arduino

This section presents a walk-through of the sketch you will write to read values from the sensor and display a bubble level on the OLED module. But first, there are a couple of libraries we must install on our PCs.

Install Software Libraries

We will need to install the Arduino libraries for the Adafruit accelerometer and Micro OLED module separately. Fortunately, this is easy to do using the Library Manager. Simply open the Library Manager from the Arduino IDE menu (*Sketch* ➤ *Include Library* ➤ *Library Manager...*). Then search for `LSM6DS` and install the latest version of the Adafruit LSM6DS library as shown in Figure 10-7.

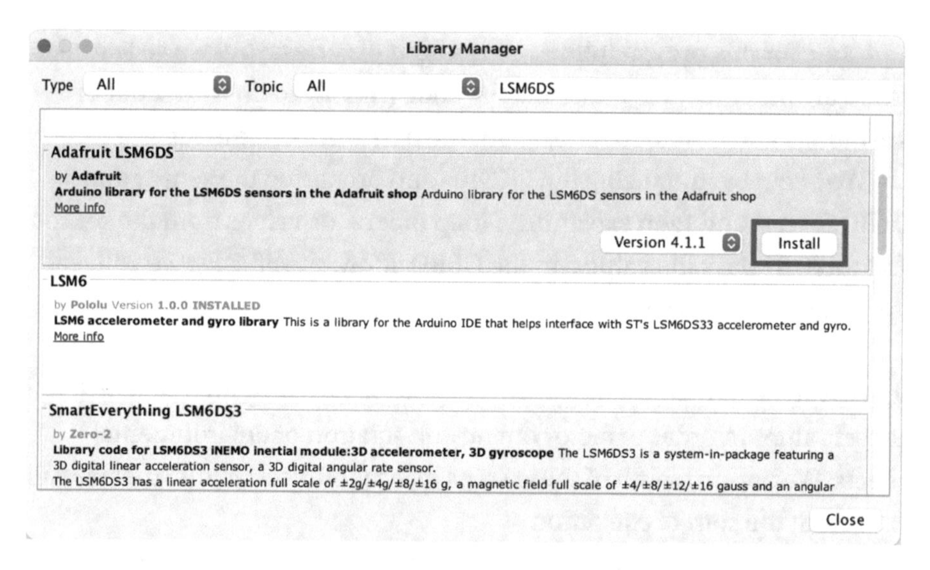

Figure 10-7. *Installing the Adafruit LSM6DS library (Arduino IDE)*

Similarly, we need to install the library for the OLED. Open the Library Manager and search for `micro OLED` and then install the latest version as shown in Figure 10-8.

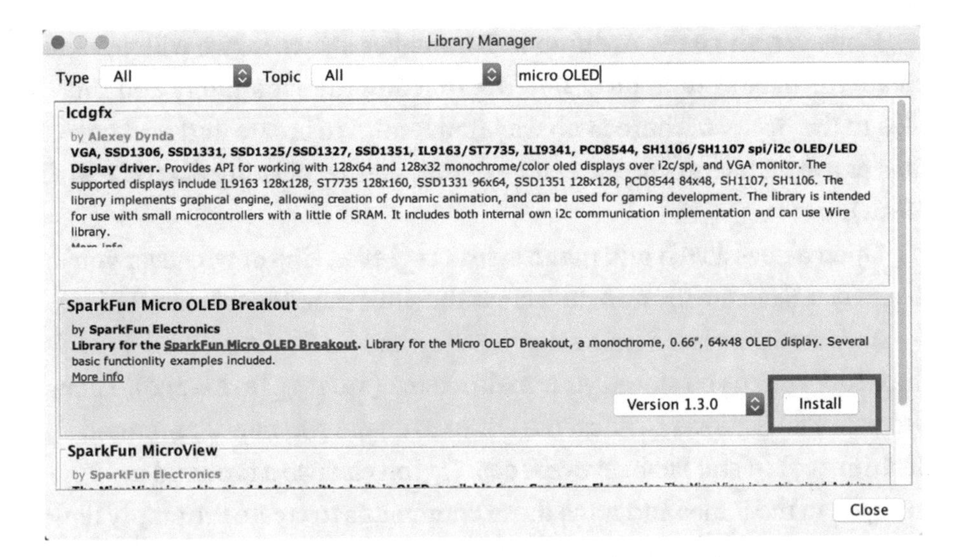

Figure 10-8. *Installing the Micro OLED library (Arduino IDE)*

Now that we have the software libraries installed, we can begin writing our sketch. Since this is not our first Arduino sketch, we will discuss the code at a high level and skip the line-by-line details focusing on the mechanics of how the code works. You can study the code at your leisure to ensure you understand the sketch in more detail.

Write the Sketch

Recall we are going to use a class to emulate the bubble level, which will be in the form of a separate file. Rather than write the main sketch file first and then add the class, we will write the class header first and then the main sketch and finally complete the class code like we did in Chapter 9.

The class will be named `BubbleLevel` and stored in two files: a header file named `BubbleLevel.h` and a source file named `BubbleLevel.cpp`. Effectively, we are moving the functions we would normally include in the main sketch to a class to make it easier to write, maintain, and understand.

However, since the Arduino IDE manages sketches, we will need to create the bare sketch file and folder first and then manually add the files to the project. There is no way (currently) to create and add new files to a sketch (but you can add existing files by clicking *Sketch* ➤ *Add File...*).

Open a new sketch and name it `spirit_level.ino` or whatever you'd like to use. Save the file and then close the project in the Arduino IDE.

To create the class files, navigate with your File Explorer (Finder) to the folder where you stored your main sketch (`spirit_level.ino`). Then, use your File Explorer or a text file editor to create two new files named `BubbleLevel.h` and `BubbleLevel.cpp`. Or you can use a terminal to navigate to the folder and issue these commands to create the empty files:

```
spirit_level % touch BubbleLevel.h
spirit_level % touch BubbleLevel.cpp
```

You can then open the project in the Arduino IDE and see all three files in the project. Let's see the code for each file starting with the header file.

Class Header File

Click the tab named `BubbleLevel.h` to open the blank file. Here, we will add the header or blueprint for the class. Recall the header file simply defines the class. We will use the `BubbleLevel.cpp` file to add the code for the functions in the class. Let's discuss those first.

This class is a bit different than the class from the last chapter. Instead of just managing the sensor, we will be using a class to manage the bubble level. That is, we'll be putting all of the code that interacts with the Micro OLED to create the bubble level in this class. Reading the sensor will be done from the main sketch instead.

Thus, we will need a number of data that is used inside the class and a few functions for use outside the class. Recall we can have a private and public section in the class. We place the data and functions used inside the class in the private section and those used outside the class in the public section. Let's start with the public section.

If you think about how we used the Micro OLED in the past, we need a setup or initialize function as well as a function for drawing the bubble level and positioning the pointer (the bubble) on the screen.

For drawing the bubble level and the pointer, we will need to know the minimum and maximum values for both the X and Y axes so that we can calculate a scale to use when positioning the pointer. For example, if the Y values can range from -8 to 8 and our screen is 38 pixels wide (with a center position of 17), we calculate a ratio of the pixels in the horizontal direction of the screen to the range of Y values or 38/16 = 2.376. So, if the Y value is -3.45, the horizontal position of the center of the pointer is (2.376 * -3.45) + 17 = 8.08 or simply pixel 8. We'll see how this works later when we write the code for the class.

Additionally, we want to be able to change the sensitivity of the bubble level making it more or less sensitive to movement. You would want it more sensitive if you are trying to level something that can be positioned with small increments or less sensitive if you're interested in getting something "close enough." We will collect this information in the constructor.

We will also create a function to allow the caller to print a message on the screen. This will be used during the calibration sequence to give the user instructions for positioning the sensor.

The functions we will create include the following:

- *Constructor*: We need a constructor for the class to accept the sensitivity and size of the center ring.
- *begin()*: Allow the caller to initialize the class and the Micro OLED.

- *setScale()*: Allow the caller to set the scale for the X and Y axes providing the minimum and maximum values for each.
- *printMessage()*: Allow the caller to write a message (text) onto the screen. This is used in the calibration sequence.
- *drawBubble()*: Draw the bubble level and position the pointer at the indicated X and Y coordinates.

One of the reasons for creating a class to mimic the bubble level is so that we can use any screen we want. More specifically, the preceding functions can still be used without changes for a different size screen. Thus, we would not have to change our main sketch to change the screen. Cool, eh?

To make that possible, we will need to store a number of data items in the private section of our class including a variable for the instance of the Micro OLED class, center position for X and Y, the sensitivity value, and the scale.

Finally, we will need some constants including the same ones we used for the Micro OLED, but we will also create a constant for X and Y of 0 and 1, respectively, to make indexing a bit easier. This is another common technique. Placing the constants in the header file means we can use them in the main sketch as well as the code file for the class.

Let's look at the completed code for the header file. Listing 10-1 shows the file.

Listing 10-1. Bubble Level Header File

```
#include <SFE_MicroOLED.h>

// Defines for OLED
#define PIN_RESET 9
#define DC_JUMPER 1
```

```
#define X 0
#define Y 1

class BubbleLevel {
public:
  BubbleLevel(float sens_ratio = 1.0, int center_ring = 6);
  bool begin();
  void setScale(float min_x, float min_y, float max_x,
  float max_y);
  void printMessage(const char *message);
  void drawBubble(int raw_x, int raw_y);

private:
  MicroOLED *oled; // OLED
  int center_x = 0;
  int center_y = 0;
  int inner_ring_size = 0;
  int oled_scale[2] = {0.0, 0.0};
  float sensitivity = 1.0;
};
```

The code should be easy to read, and there are no surprises other than the details described previously. Let's return to the main sketch to see how we can use this class.

Main Sketch

Now click the `spirit_level.ino` tab to return to the main sketch. Let's begin with the preamble or top of the file. Recall here is where we include libraries we need, declare variables and constants, etc. We will also create functions to execute the calibration sequence, which is a bit more complex than what we are used to seeing, so we will walk through those functions separately.

We will look at the preamble first and then the setup() and loop() functions. Listing 10-2 shows the preamble code for the main sketch. As you can see, we include the wire, Serial LCD, Mux, and our new class headers.

Listing 10-2. Main Sketch Preamble

```
#include <Adafruit_LSM6DS33.h>
#include "BubbleLevel.h"

// Structure to save calibration data
typedef struct {
  float bias[2] = {999.0, 999.0};
  float min_values[2] = {0.0, 0.0};
  float max_values[2] = {0.0, 0.0};
  bool calibrated = false;
} calData_ptr;

// Calibration operations enumeration
enum CAL_TESTS {
  AVERAGES,     // Get stable values at rest
  MIN_X,        // Get minimum values for X axis
  MAX_X,        // Get maximum values for X axis
  MIN_Y,        // Get minimum values for Y axis
  MAX_Y         // Get maximum values for Y axis
};

// Global variables
Adafruit_LSM6DS33 lsm6ds33;             // accelerometer
BubbleLevel bubbleLevel(1.25);
calData_ptr calData;                    // calibration data
```

Notice we define a structure to contain the calibration data and an enumeration set to establish a series of calibration tests. We use the structure to capture the calibration data including the bias from the first step in the calibration sequence and the minimum and maximum values of the X and Y axes from the remaining four steps in the sequence.

Finally, we create variables for the accelerometer, our bubble level class, and the structure. We create a variable for the structure so we can pass it to our functions for updating the data in the structure.

Next, we will code the `setup()` function, which includes the sort of initialization code we've seen in other projects. Specifically, we will set up the accelerometer and the bubble level. Listing 10-3 shows the complete code for the `setup()` function. Read through it to ensure you understand all of the code included. Notice we also have some of the optional tuning methods for the accelerometer (from the Adafruit examples) in case you want to tune the sensitivity of the sensor.

Listing 10-3. Main Sketch setup()

```
void setup() {
  Serial.begin(115200);
  while (!Serial);
  Serial.println("\nSpirit level");

  if (!lsm6ds33.begin_I2C()) {
    Serial.println("ERROR: LSM6DS33 module not found");
    while (1);
  }

  Serial.println("LSM6DS33 ready!");

  // Optionally change sensitivity settings of the module
  // lsm6ds33.setAccelRange(LSM6DS_ACCEL_RANGE_2_G);
  // lsm6ds33.setGyroRange(LSM6DS_GYRO_RANGE_250_DPS);
  // lsm6ds33.setAccelDataRate(LSM6DS_RATE_12_5_HZ);
  // lsm6ds33.setGyroDataRate(LSM6DS_RATE_12_5_HZ);
```

```
  if (!bubbleLevel.begin()) {
    Serial.println("ERROR: OLED not found!");
    while(1);
  }
  Serial.println("Readings:");
}
```

Next, we have the `loop()` function. We simply read the axis values from the sensor and then call the `drawBubble()` function for our new class. We will also check to see if the sensor is calibrated. While we could have put this in the `setup()` function, placing it in the `loop()` function permits us to modify the code to run the calibration whenever needed. See the "Going Further" section for hints on how to do this.

Once the calibration is done, we call the `setScale()` function of our new class to set the scale for the pointer. So where are these calibration functions? We haven't written them yet! Read on.

Listing 10-4 shows the completed code for the `loop()` function. Go ahead and read through the code to ensure you understand how it works. There should not be any surprises or new techniques.

Listing 10-4. Main Sketch loop()

```
void loop() {
  sensors_event_t accel;
  // We need variables for the gyroscope and temperature
  // readings, but we do not use them.
  sensors_event_t gyro;
  sensors_event_t temp;
  float x = 0.0;
  float y = 0.0;

  // Check to ensure sensor is calibrated.
  if (!calData.calibrated) {
```

```
    calibrate(&calData);
    bubbleLevel.setScale(calData.min_values[X],
    calData.max_values[X],
                         calData.min_values[X], calData.max_
                         values[Y]);
  }
  // Read the accelerometer data
  lsm6ds33.getEvent(&accel, &gyro, &temp);
  // Adjust with bias from calibration
  x = accel.acceleration.x - calData.bias[X];
  y = accel.acceleration.y - calData.bias[Y];
  Serial.print("X=");
  Serial.print(x);
  Serial.print("\tY=");
  Serial.println(y);

  bubbleLevel.drawBubble(x, y);
  delay(100);
}
```

Now, let's look at the calibration functions. We will use two functions: one to manage the calibration process named `calibrate()` and another to execute the calibration tests named `runCalibrationTests()`. Since much of the code is similar, writing a function to run all of the tests is a better option. While it does increase the complexity of the code, it avoids a lot of repetitive code.[3]

[3] Eliminating repetitive code is always a good idea as it removes risk of missing changes should you have to modify the code. That is, fixing five of the six occurrences will lead to some strange behavior and some hairpulling to get it sorted and fixed.

Let's look at the calibration tests function first (runCalibrationTests()). Recall there are five calibration steps we need to execute. This function is written to execute the specific code needed for each of the five calibration steps.

The calibration data is passed by reference as a parameter named calData. Any data we collect from the test is saved in the appropriate field of that structure. We also use a parameter named operation to determine which test to run and another named iterations for the number of times to run the test. Recall we want to run the same test many times to ensure we have a good average to use.

As you can imagine, the function that calls this function will pass the calibration data and specify a test to run as well as the number of times to run the test. Aside from that, there isn't anything complicated in the code. Listing 10-5 shows the completed function. Take some time and read through the code until you understand how it works. Notice there are two parts: one to execute the test and another to collect the data. Each is written using a switch statement that routes execution based on the operation specified.

Listing 10-5. Main Sketch Calibration Tests Function

```
void runCalibrationTests(calData_ptr* calData, CAL_TESTS
operation, int iterations) {
  sensors_event_t accel;
  sensors_event_t gyro;
  sensors_event_t temp;
  float min_value = 32767.00;
  float max_value = -32767.00;
  float sum_values[2] = {0.0, 0.0};
  float x = 0.0;
  float y = 0.0;
```

```
// Loop reading sensor values recording data for specific test
for (int i = 0; i < iterations; i++) {
  lsm6ds33.getEvent(&accel, &gyro, &temp);
  x = accel.acceleration.x;
  y = accel.acceleration.y;
  switch (operation) {
    case AVERAGES:
      sum_values[0] += x;
      sum_values[1] += y;
      break;
    case MIN_X:
      min_value = min(x, min_value);
      break;
    case MAX_X:
      max_value = max(x, max_value);
      break;
    case MIN_Y:
      min_value = min(y, min_value);
      break;
    case MAX_Y:
      max_value = max(y, max_value);
      break;
  }
  Serial.print(".");
  if ((i % 100) == 99) {
    Serial.println();
  }
  delay(10);
}
// Complete calculations
switch (operation) {
  case AVERAGES:
```

```
      calData->bias[X] = sum_values[X] / (float)iterations;
      calData->bias[Y] = sum_values[Y] / (float)iterations;
      break;
    case MIN_X:
      calData->min_values[X] = min_value;
      break;
    case MAX_X:
      calData->max_values[X] = max_value;
      break;
    case MIN_Y:
      calData->min_values[Y] = min_value;
      break;
    case MAX_Y:
      calData->max_values[Y] = max_value;
      break;
  }
  Serial.println("done.");
}
```

Finally, we look at the last function to execute the calibration tests (`calibrate()`). Like the last function, we pass the calibration data by reference so that we can store the changes in the structure for use in the main portion of the code.

There is also nothing else complicated about this function. As you will see, it serves as a driver for the preceding calibration tests function. In fact, the bulk of the code is about printing diagnostic messages to the serial monitor and writing short instructions to the Micro OLED.

Listing 10-6 shows the completed code for the calibration function. Take some time and read through it to ensure you understand how it works.

Listing 10-6. Main Sketch Calibration Function

```
void calibrate(calData_ptr* calData) {
  bubbleLevel.printMessage("Begin\nCalibrate\nsequence");
  Serial.println("\nStarting Calibration");
  delay(2000);
  // Calibrate at rest
  Serial.println("\nStep 1: Place on level surface. Do not tilt.");
  bubbleLevel.printMessage("Step 1:\nPlace on\nlevel\nsurface.\
  nDo not\ntilt.");
  delay(2000);
  runCalibrationTests(calData, AVERAGES, 500);

  Serial.println("\nStep 2: Turn X axis vertical.");
  bubbleLevel.printMessage("Step 2:\nVertical X");
  delay(2000);
  runCalibrationTests(calData, MIN_X, 250);

  Serial.println("\nStep 3: Turn X axis 180 degrees vertical.");
  bubbleLevel.printMessage("Step 3:\nVertical XFlip 180");
  delay(2000);
  runCalibrationTests(calData, MAX_X, 250);

  Serial.println("\nStep 4: Turn Y axis vertical.");
  bubbleLevel.printMessage("Step 4:\nVertical Y");
  delay(2000);
  runCalibrationTests(calData, MIN_Y, 250);

  Serial.println("\nStep 5: Turn Y axis 180 degrees vertical.");
  bubbleLevel.printMessage("Step 5:\nVertical YFlip 180");
  delay(2000);
  runCalibrationTests(calData, MAX_Y, 250);
  calData->calibrated = true;
```

```
    Serial.print("Bias X: ");
    Serial.print(calData->bias[X]);
    Serial.print(" Y: ");
    Serial.println(calData->bias[Y]);
    Serial.print("Range X: (");
    Serial.print(calData->min_values[X]);
    Serial.print(",");
    Serial.print(calData->max_values[X]);
    Serial.println(")");
    Serial.print("Range Y: (");
    Serial.print(calData->min_values[Y]);
    Serial.print(",");
    Serial.print(calData->max_values[Y]);
    Serial.println(")");

    Serial.println("\nCalibration complete.");
    delay(2000);
}
```

Now we can write the final portion of our project - the code for the class.

Class Code File

Click the tab named `BubbleLevel.cpp` to open the blank file. Here, we will add the code for the class. There are four functions and the constructor to write.

The constructor captures the sensitivity setting and the size of the center ring. We also create a variable with an instance of the Micro OLED class for use inside the class.

The `begin()` function initializes the Micro OLED and collects data to find the center position.

The setScale() function accepts the minimum and maximum values for the X and Y axes and uses them to calculate a mapping of the scale or range of X and Y values to the size of the screen. This is a nifty trick, so take some time to read through how the math works (it's a ratio).

The printMessage() function writes a string (message) to the Micro OLED and is a helper for the calibration sequence.

The drawBubble() function takes an X and Y value from the caller (sensor) and plots the pointer on the screen. It uses the ratio and mapping we described earlier to locate the pointer. However, there is a bit of a trick here too. Our Micro OLED is oriented where Y is forward to back and X is left to right, which is not the same orientation as our sensor. So we must gently alter the code to reverse the axes as shown in the following. Take some time to read through the code to see how it works. Aside from that, it simply draws the circles and the pointer at the calculated location:

```
// Plot the location on the OLED
// Note: we change the axis to orient the OLED correctly.
x = center_x - int(raw_y * oled_scale[X]);
y = center_y - int(raw_x * oled_scale[Y]);
```

Finally, the function writes the circles and pointer to the Micro OLED. Listing 10-7 shows the completed code for the class (documentation omitted for brevity). Since this is a much lower level than the code we're used to seeing, be sure to study it to understand how it works. While the protocol (how we talk to the module) is very simple, the functions we use are uncommon since we normally have libraries written for us.

Listing 10-7. Bubble Level Code File

```
#include "BubbleLevel.h"

BubbleLevel::BubbleLevel(float sens_ratio, int center_ring) {
  oled = new MicroOLED(PIN_RESET, DC_JUMPER);
  sensitivity = sens_ratio;
  inner_ring_size = center_ring;
}

bool BubbleLevel::begin() {
  if (!oled->begin()) {
    return false;
  }
  oled->begin();
  oled->clear(ALL);
  oled->clear(PAGE);

  center_y = oled->getLCDHeight() / 2;
  center_x = oled->getLCDWidth() / 2;

  oled->setFontType(0);
  oled->setCursor(0, 0);
  oled->print("Spirit");
  oled->setCursor(0, 10);
  oled->print("Level");
  oled->display();

  delay(2000);
  return true;
}
```

```
void BubbleLevel::setScale(float min_x, float min_y, float
max_x, float max_y) {
  oled_scale[X] = float(center_x) / max(abs(min_x), max_x) *
  sensitivity;
  oled_scale[Y] = float(center_y) / max(abs(min_y), max_y) *
  sensitivity;
  Serial.print("\nScales - X: ");
  Serial.print(oled_scale[X]);
  Serial.print("  Y: ");
  Serial.println(oled_scale[Y]);
  Serial.print("Center OLED - X: ");
  Serial.print(center_x);
  Serial.print("  Y: ");
  Serial.println(center_y);
  Serial.println();
}

void BubbleLevel::printMessage(const char *message) {
  oled->clear(PAGE);
  oled->setFontType(0);
  oled->setCursor(0, 0);
  oled->print(message);
  oled->display();
}

void BubbleLevel::drawBubble(int raw_x, int raw_y) {
  int x;  // Calculated X position
  int y;  // Calculated Y position

  // Plot the location on the OLED
  // Note: we change the axis to orient the OLED correctly.
  x = center_x - int(raw_y * oled_scale[X]);
  y = center_y - int(raw_x * oled_scale[Y]);
```

```
  oled->clear(PAGE);
  oled->pixel(x-1, y-1);
  oled->pixel(x+1, y-1);
  oled->pixel(x-1, y+1);
  oled->pixel(x+1, y+1);
  oled->pixel(x, y);
  oled->pixel(x+2, y);
  oled->pixel(x-2, y);
  oled->pixel(x, y+2);
  oled->pixel(x, y-2);
  // Outer guide ring
  oled->circle(center_x, center_y, center_y - 1);
  // Inner guide ring
  oled->circle(center_x, center_y, 6);
  oled->display();
}
```

As you can see, this class is another example of how to move code out of your main sketch and into helper modules. There is another possible class we could make for this project. We could move the calibration code to its own class. If you're curious, try it out yourself as an exercise.

Compile the Sketch

The last step is to compile the sketch before uploading it to your board. If you encounter any errors, be sure to fix them and recompile to ensure the sketch compiles without errors or serious warnings.

Once everything compiles, we're ready to start testing. But first, let's look at the code for the Raspberry Pi. You can skip to the "Sketch on the Arduino" section if you're curious to see how the project works. The code will run the same on both platforms.

Raspberry Pi

This section presents a walk-through of the Python code you will write to read values from the sensor and display them on the Micro OLED module. But first, we need to install a Python library on our Raspberry Pi.

Install a Software Library

We need only one more software library to use the Python code in this chapter. We need the accelerometer library from Adafruit as shown in the following:

```
$ pip3 install adafruit-circuitpython-lsm6ds
```

Tip If you haven't installed the SparkFun Qwiic Python libraries, you must install them to run this project (`pip3 install sparkfun_qwiic`).

Now we're ready to write the code.

Write the Code

The code for the Python version of this project is a bit shorter than the Arduino code. We will create a class for the bubble level, which is coding using similar techniques as the Arduino version. However, in the main script, instead of using a structure, we will use a Python dictionary for the calibration data. Let's get started!

Once again, we will not dive into every line of code. We will explore the code at a higher level and discuss the more complex or important parts in detail. You can read through the code and learn more about how it works at your leisure.

Like the Arduino example, we will use a class to contain the code for the bubble level. However, unlike the Arduino IDE, you can use any editor to create the class and main script. We will name the main script `spirit_level.py` and the class module `bubble_level.py`. Let's start with the class.

Bubble Level Class

Open the Thonny Python IDE under the *Main* ➤ *Programming* submenu. Create a new file `bubble_level.py`. We will name the class `BubbleLevel`, and we will need a few functions. In fact, we need the same functions as we used in the Arduino version as listed below for completion. While they perform the same operations as the Arduino version, the names of the functions differ. For a complete description of each of these functions, see the "Arduino" section.

The functions we will create include the following:

- *Constructor*: We need a constructor for the class to accept the sensitivity and size of the center ring.
- *begin()*: Allow the caller to initialize the class and the Micro OLED.
- *set_scale()*: Allow the caller to set the scale for the X and Y axes providing the minimum and maximum values for each.
- *print_message()*: Permit the caller to write a message (text) onto the screen. This is used in the calibration sequence.
- *draw_bubble()*: Draw the bubble level and position the pointer at the indicated X and Y coordinates.

Listing 10-8 shows the complete code for the class with documentation removed for brevity. Take a few moments to read through the code so that you understand all of the parts of the code. As you will see, it mimics the Arduino.

Listing 10-8. Bubble Level Class (Python)

```
import time
import qwiic

# Constants
X = 0
Y = 1

class BubbleLevel:
    oled = qwiic.QwiicMicroOled()
    center_x = 0
    center_y = 0
    inner_ring_size = 0
    oled_scale = [0.0, 0.0]
    sensitivity = 1.0

    def __init__(self, sens_ratio=1.0, center_ring=6):
        self.sensitivity = sens_ratio
        self.inner_ring_size = center_ring

    def begin(self):
        if self.oled.begin() is False:
            return False

        # Clear the screen and print greeting
        self.oled.clear(self.oled.PAGE)
        self.oled.clear(self.oled.ALL)
```

```
        # Learn size and find the center of the display
        # Note: we must invert the height and width to match
          orientation
        #       of the OLED so that the mount holes align.
        self.center_y = self.oled.get_lcd_height() / 2
        self.center_x = self.oled.get_lcd_width() / 2

        self.oled.set_font_type(0)
        self.oled.set_cursor(0, 0)
        self.oled.print("Spirit")
        self.oled.set_cursor(0, 10)
        self.oled.print("Level")
        self.oled.display()

        time.sleep(2)
        return True

    def set_scale(self, min_values, max_values):
        # Calculate OLED scale factors
        self.oled_scale[X] = float(self.center_x) / \
max(abs(min_values[X]), max_values[X]) * self.sensitivity
        self.oled_scale[Y] = float(self.center_y) / \
max(abs(min_values[Y]), max_values[Y]) * self.sensitivity
        print("\nScales - X: {0}  Y: {1}".format(
self.oled_scale[X], self.oled_scale[Y]))
        print("Center OLED - X: {0}  Y: {1}\n".format(
self.center_x, self.center_y))

    def print_message(self, message):
        self.oled.clear(self.oled.PAGE)
        self.oled.set_font_type(0)
        self.oled.set_cursor(0, 0)
        self.oled.print(message)
        self.oled.display()
```

```
def draw_bubble(self, raw_x, raw_y):
    # Plot the location on the OLED
    # Note: we change the axis to orient the OLED
      correctly.
    x_coordinate = self.center_x - int(raw_y * self.oled_
    scale[X])
    y_coordinate = self.center_y - int(raw_x * self.oled_
    scale[Y])

    self.oled.clear(self.oled.PAGE)
    self.oled.pixel(x_coordinate - 1, y_coordinate - 1)
    self.oled.pixel(x_coordinate + 1, y_coordinate - 1)
    self.oled.pixel(x_coordinate - 1, y_coordinate + 1)
    self.oled.pixel(x_coordinate + 1, y_coordinate + 1)
    self.oled.pixel(x_coordinate, y_coordinate)
    self.oled.pixel(x_coordinate + 2, y_coordinate)
    self.oled.pixel(x_coordinate - 2, y_coordinate)
    self.oled.pixel(x_coordinate, y_coordinate + 2)
    self.oled.pixel(x_coordinate, y_coordinate - 2)
    # Outer guide ring
    self.oled.circle(self.center_x, self.center_y, self.
    center_y - 1)
    # Inner guide ring
    self.oled.circle(self.center_x, self.center_y, self.
    inner_ring_size)
    self.oled.display()
```

Now we can write our main script.

Main Script (Python)

Open the Thonny Python IDE under the *Main* ➤ *Programming* submenu. Create a new file spirit_level.py. There is nothing new in this code as it follows the same flow as the Arduino example but much simplified. Listing 10-9 shows the complete code for the main script for this project. You can read through it to see how all of the code works.

Listing 10-9. Main Script (Python)

```
# Import libraries
import time
import board
import busio
import pprint
import sys

from adafruit_lsm6ds.lsm6ds33 import LSM6DS33
from bubble_level import BubbleLevel, X, Y

# Calibration operations enumeration
CAL_TESTS = ("AVERAGES", "MIN_X", "MAX_X", "MIN_Y", "MAX_Y")

# Global variables
i2c = busio.I2C(board.SCL, board.SDA)
lsm6ds33 = LSM6DS33(i2c)
bubble_level = BubbleLevel(1.25)

# Dictionary to save calibration data
calibration_data = {
    'bias': [999.0, 999.0],
    'min_values': [0.0, 0.0],
    'max_values': [0.0, 0.0],
    'calibrated': False
}
```

```
def run_calibration_tests(cal_data, operation, iterations):
    min_value = 32767.00
    max_value = -32767.00
    sum_values = [0.0, 0.0]

    # Loop reading sensor values recording data for specific
      test
    for i in range(0, iterations):
        x_coordinate = lsm6ds33.acceleration[0]
        y_coordinate = lsm6ds33.acceleration[1]
        if operation == "AVERAGES":
            sum_values[0] += x_coordinate
            sum_values[1] += y_coordinate
        elif operation == "MIN_X":
            min_value = min(x_coordinate, min_value)
        elif operation == "MAX_X":
            max_value = max(x_coordinate, max_value)
        elif operation == "MIN_Y":
            min_value = min(y_coordinate, min_value)
        elif operation == "MAX_Y":
            max_value = max(y_coordinate, max_value)
        print(".", end="")
        if (i % 100) == 99:
            print("")
        time.sleep(0.010)

    # Complete calculations
    if operation == "AVERAGES":
        cal_data["bias"][X] = sum_values[X] / float(iterations)
        cal_data["bias"][Y] = sum_values[Y] / float(iterations)
    elif operation == "MIN_X":
        cal_data["min_values"][X] = min_value
```

```
    elif operation == "MAX_X":
        cal_data["max_values"][X] = max_value
    elif operation == "MIN_Y":
        cal_data["min_values"][Y] = min_value
    elif operation == "MAX_Y":
        cal_data["max_values"][Y] = max_value
    print("done.")

def calibrate(cal_data):
    bubble_level.print_message("Begin     Calibrate sequence")
    print("\nStarting Calibration")
    time.sleep(2)
    # Calibrate at rest
    print("\nStep 1: Place on level surface. Do not tilt.")
    bubble_level.print_message("Step 1:  Place on  level
    surface."
"  Do not    tilt.")
    time.sleep(2)
    run_calibration_tests(cal_data, "AVERAGES", 500)

    print("\nStep 2: Turn X axis vertical.")
    bubble_level.print_message("Step 2:   Vertical X")
    time.sleep(2)
    run_calibration_tests(cal_data, "MIN_X", 250)

    print("\nStep 3: Turn X axis 180 degrees vertical.")
    bubble_level.print_message("Step 3:   Vertical XFlip 180")
    time.sleep(2)
    run_calibration_tests(cal_data, "MAX_X", 250)

    print("\nStep 4: Turn Y axis vertical.")
    bubble_level.print_message("Step 4:   Vertical Y")
    time.sleep(2)
    run_calibration_tests(cal_data, "MIN_Y", 250)
```

```
    print("\nStep 5: Turn Y axis 180 degrees vertical.")
    bubble_level.print_message("Step 5:   Vertical YFlip 180")
    time.sleep(2)
    run_calibration_tests(cal_data, "MAX_Y", 250)
    cal_data["calibrated"] = True

    print("\nCalibration Data:\n")
    pprint.pprint(cal_data, indent=4)
    print("\nCalibration complete.")
    time.sleep(2)

def main():
    print("\nSpirit Level")
    if not bubble_level.begin():
        print("ERROR: The OLED module is not found. "
  "Please check your connections!")
        sys.exit(1)

    # Calibrate the sensor.
    calibrate(calibration_data)

    # Calculate OLED scale factors
    bubble_level.set_scale(calibration_data["min_values"],
    calibration_data["max_values"])
    print("Readings:")
    while True:
        # Read sensor and adjust with bias from calibration
        x_coordinate = lsm6ds33.acceleration[0] - calibration_
        data["bias"][X]
        y_coordinate = lsm6ds33.acceleration[1] - calibration_
        data["bias"][Y]
```

```
        # Display the raw data for diagnostics
        print("X={0}\tY={1}".format(x_coordinate, y_coordinate))

        # Show the bubble level
        bubble_level.draw_bubble(x_coordinate, y_coordinate)
        time.sleep(0.5)

if __name__ == '__main__':
    try:
        main()
    except (KeyboardInterrupt, SystemExit) as exErr:
        print("\nbye!\n")
sys.exit(0)
```

Notice we have the same functions as we defined in the Arduino version for calibrating the sensor. In fact, while the code is organized differently – we use a `main()` function and we use a dictionary instead of a structure – the Python code is a port of the Arduino code.

OK, that's it! We've written the code. Unlike the Arduino, we do not need to compile the Python code. So we're now ready to execute the project!

Execute the Project

Now that we've spent many pages exploring the Qwiic modules and writing the code to interact with them, it is time to test the project by executing (running) it.

When the project runs (executes), you will see some diagnostic message written to the serial monitor (Arduino) or the terminal (Raspberry Pi). You will also see a welcome message appear on the Micro OLED followed by a short pause. Then the calibration process will begin.

Once the calibration is complete, you will see the bubble level displayed. Figure 10-9 shows an example of what you should see on the OLED. Notice the pointer is dead center of the inner ring. That puppy is level!

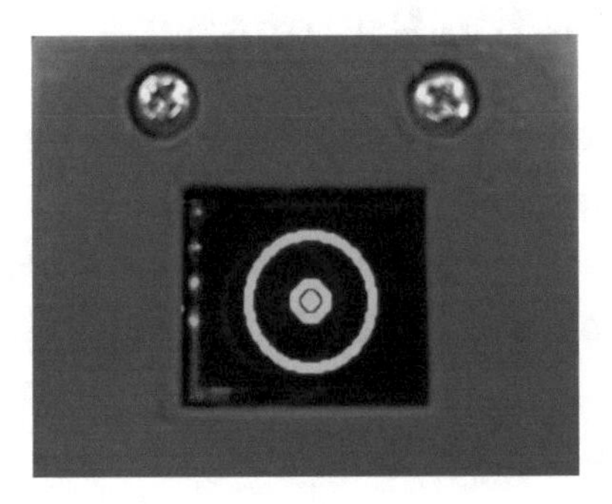

Figure 10-9. *Executing the digital level project*

Executing the code depends on which platform you're using. Let's look at the Arduino first.

Sketch on the Arduino

Executing the sketch on the Arduino requires connecting our board to our PC and then uploading the sketch to the Arduino. Recall the sketch will run so long as the USB cable is connected to our PC (and the Arduino).

Execute the Sketch

To execute the sketch, be sure your Arduino is connected and you've selected the correct board under the *Tools* ➤ *Board* menu. You also need to ensure you have the correct port selected under the *Tools* ➤ *Port* menu.

Once those items are set, you can click the *Upload* button or choose *Sketch* ➤ *Upload* from the menu. The Arduino IDE will compile the sketch and then upload it to your Arduino. Once you see the `Done uploading...` message, you can open the serial monitor. You should see the output begin momentarily that is the same as that on the OLED. Go ahead and try it out!

If something isn't working, check your connections or refer to Chapter 7 for troubleshooting tips.

Python Code on the Raspberry Pi

Executing the sketch on the Raspberry Pi requires running the Python code in a terminal after connecting your Qwiic daisy chain to your Raspberry Pi via a hat or the Qwiic female breakout cable. Recall the code will run until you stop it with *CTRL+C* on the keyboard.

Execute the Python Code

To run the Python code on the Raspberry Pi, you can issue the command `python3 ./spirit_level.py` from the same folder where the file was saved as shown in the following. If everything worked as executed, congratulations! You've just built your fourth Qwiic project.

If something isn't working, check your connections or refer to Chapter 7 for troubleshooting tips.

Going Further

While we didn't discuss them in this chapter, here are just a few suggestions you can try once we have learned how to take our projects to the cloud. Put your skills to work!

- *New class for the sensor*: Build a class to manage the sensor moving all of the code for reading the sensor to the new class.
- *New class for calibration*: Build a class to encapsulate the calibration functions. It could be a simple code module rather than a class.

- *Build a better enclosure*: Build an enclosure to contain your Arduino board and a battery supply to make the project portable. Use the 3D printer files from the book source code and expand the OpenSCAD code.
- *Calibrate button*: Add a button to trigger the calibration routine by either setting the calibration flag to false or running the calibration when the button is triggered. This could be helpful if you must move the sensor or want to use a specific plane for a base.
- *Orientation*: Add the ability to orient the sensor and enclosure in a different starting position. For example, instead of the horizontal, allow for a vertical orientation (or switch between them).
- *Robot level*: Use the project in this chapter to add the ability to sense when your robot is on level ground. While it may be a stretch, you could build a robot to move around your home checking for level floors recording the findings and displaying the results on a web page.

Summary

In this chapter, we got more hands-on experience making projects with Qwiic and STEMMA QT modules. We used an accelerometer sensor to detect movement in the X and Y axes. We then wrote a class library to represent a bubble level where the center position is the level position for the X and Y axes. We used a set of circles to help in using the project to find level positions for objects.

Along the way, we learned more about how to write class libraries to represent abstract constructs as well as how they can be used to model objects. Rather than write a class to manage the sensor, we wrote a class to mimic a bubble level represented on a Micro OLED module. This shows you how you can use classes not only to represent hardware (sensors) but also concepts (bubble level).

In the next chapter, we will learn how to use a magnetometer to create a digital compass. As you will see, the code for the project is more complex due to the need to use an alternative code library, trigonometric formulas, and more calibration.

CHAPTER 11

Digital Compass

To reinforce our experience with STEMMA QT and Qwiic modules, let's take a look at another project that has a bit more sophistication than the typical examples you find on the Internet.

In this chapter, we will see our last Qwiic and STEMMA QT project. We will learn how to use a magnetometer module to build a digital compass. In some respects, this project is similar to the project in the last chapter, but this one requires more mathematical formulas and thus is a bit more complex.

Part of the complexity results from the fact that the base code we need to use is not included in the vendor documentation. Specifically, there is no single example for this magnetometer we will use that shows how to use it as a compass. So we must develop that part on our own with some help from similar examples.

Of course, we will see how to implement this project on the Arduino and Raspberry Pi. Let's get started.

Caution This project is a rudimentary digital compass that is not guaranteed to be accurate under all conditions as outlined in the "Limitations" section.

C. Bell, *Beginning IoT Projects*, https://doi.org/10.1007/978-1-4842-7234-3_11

Project Overview

The project for this chapter is designed to demonstrate how to build a digital compass with a magnetometer sensor. We will use the sensor to read magnetic North and use that to calculate a heading (the direction we're facing). And, like the last chapter, there is a bit of calibration needed to ensure the sensor is ready for use.

For those with access to a 3D printer, we will also see a simple enclosure you can print to install the modules to protect the modules and make them easier to handle. In fact, any enclosure will make the calibration sequence easier.

What Will We Learn?

While we won't see anything new with the hardware other than a different sensor, by implementing this project, we will reinforce what we learned from the previous chapters.

Part of the complexity of this chapter is in learning how to calibrate and calculate a heading from a magnetometer to build a compass, which results in challenging programming because we will need to use trigonometry to calculate our heading.[1]

Let's look at what magnetometers are. Then we will discover calculations for determining the heading.

[1] See, you do need to pay attention in class!

What Is a Magnetometer?

A magnetometer measures magnetic fields and is capable of measuring the direction, strength, or relative change of a magnetic field. It can be used to measure the location of the magnetic field or as a compass to detect the earth's magnetic North pole. Most magnetometers are combined with accelerometers and other sensors to increase functionality.

However, a magnetometer is affected by two sources of interference known as hard and soft iron. Hard iron are objects that generate magnetic fields such as a speaker. Soft iron are distortions in the magnetic fields that deflect or alter existing magnetic fields. Sources include certain metals such as iron or nickel. Hard iron distortions affect the sensor more than soft iron, but both can be eliminated through calibration.

While there are a lot of different magnetometer modules available, not all are created equal. Some detect magnetic fields better (with more precision and sensitivity) than others. Further, some work better in the presence of metal than others. For example, if you have a lot of metal in the area, some magnetometers may not work correctly (or may need more calibration).

When choosing a magnetometer, be sure to choose one that meets your expectations for detecting magnetic North and can operate within your environment.

The magnetometers available for hobbyists and enthusiasts are not the same level of accuracy or sophistication as magnetometers used in aircraft, ships, etc. So, while we can build a nice compass, you should not use it in situations where one's life may be at risk such as navigation at sea, in space, or in your grandparents' basement or garage.

Mathematical Problems

This project has several mathematical problems we need to solve. We must calculate a heading from the data read from the magnetometer and calculate where to place the pointer on the OLED to represent our heading.

Let's begin with calculating a heading. We will use the magnetometer to read the X and Y values and convert those to a heading that we can plot on our display and show on the display in degrees.

Calculating a Heading

The magnetometer sensor will give us a set of values when read, but the two we will use are the X and Y values. These values change as the orientation of the sensor changes. That is, as you turn the sensor, the values reflect the orientation of the sensor to the reference point (magnetic North). We can use these values to determine our heading using trigonometry – the arctangent function. The basic formula is shown in the following:

```
heading = (atan2(y, x) * 180)/ PI;
```

However, this does not take into consideration our calibration values (more on that later). So we must subtract our calibration values (called bias) as follows. We will use an array to store the bias for X and Y:

```
heading = (atan2(y - bias[Y], x - bias[X]) * 180)/ PI;
```

There's one more thing to consider. The values for X and Y can result in a negative value for half the range. So we must add 360 to any heading that results in a negative value as follows:

```
heading = (atan2(y - bias[Y], x - bias[X]) * 180) / PI;
if (heading < 0) {
    heading += 360;
}
```

While this is the most basic calculation, it does not fully compensate for variations in the earth magnetic fields (called the declination angle). This value varies depending on your location. However, we can find the declination angle for our location by visiting `www.compassnetic-declination.com/`.

We must add this value to our equation. However, this value is in radians, so we must break out our preceding calculation to work with radians and then convert to degrees last. Let's look at the new calculations. Here, we use the declination angle for the Northeastern United States:

```
heading = atan2(y - bias[Y], x - bias[X]) + 0.174533;
// Fix signs when negative values are calculated
if (heading < 0) {
    heading += 2*PI;
}
// If the declination angle forces a wrap past 2 * PI,
// adjust by subtracting 2 * PI
if (heading > 2*PI) {
    heading -= 2*PI;
}
// Convert radians to degrees
heading = heading * 180 / PI;
```

OK, now we have a more accurate calculation for finding our heading. There is just one major limitation to consider.

Intentional Error

Since we are mounting the magnetometer sensor rotated 90 degrees from the orientation of the enclosure (90 degrees difference from the Micro OLED), we will need to compensate so our pointer is pointing correctly. Otherwise, it will be 90 degrees off. Figure 11-1 shows the magnetometer. Notice the label in the upper-right corner.

Figure 11-1. *Qwiic MLX90393 (courtesy of SparkFun)*

This is how you can tell the magnetometer is oriented on the board. Ideally, for North, we want the X coordinate pointing toward the polar North position, but we rotated the board 90 degrees to the left. Thus, to compensate, we will need to add 90 degrees to our heading calculation.

We have therefore introduced an intentional error. If you mount your board differently, be sure to alter the parts of the code where we make these adjustments. Hint: Look in the calibration animation, draw pointer, and heading functions.

Compass Calculations

We've examined the math for the heading calculation, but there is more math involved in the compass display, specifically the math for plotting the heading as a point on a circle and drawing a pointer to it.

We must translate a heading to a point on a circle. The display uses a line as the pointer, so we will be calculating two points: one for the point on the outer circle that is visible to the user and another that represents the starting point for the line. We'll call the outer circle compass radius and the inner circle mask radius. We call it a mask since we do not want to draw the line or pointer from the center since we will be displaying the heading in the center of the OLED. Figure 11-2 shows an example of what the circle with a pointer looks like when the heading is 45 degrees.

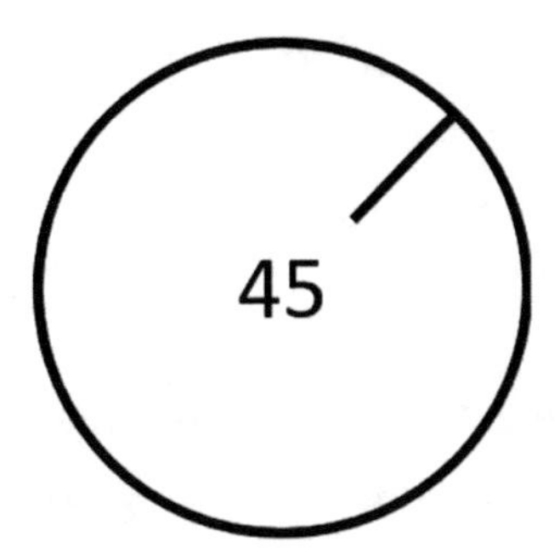

Figure 11-2. *Example pointer for compass display*

So how do we do that? Well, we must use a little of that math we learned in school but thought we'd never use again: trigonometry! Figure 11-3 shows a triangle placed over the pointer from the last image.

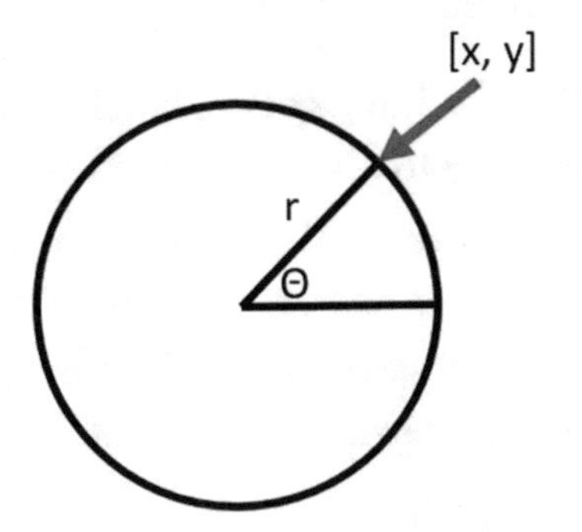

Figure 11-3. *Calculating the pointer location for compass display*

To plot a point using X and Y coordinates, we would use the angle Θ and use the following functions.

For X, we derive the following formula:

```
cos(Θ) = X / r
r * cos(Θ) = X
X = r * cos(Θ)
```

For Y, we derive the following formula:

```
sin(Θ) = Y / r
r * sin(Θ) = Y
Y = r * sin(Θ)
```

OK, that's great, but we don't know the value of Θ, do we? Wait. We do! While we have a heading in degrees, we can convert that back to an angle with the formula

```
Θ = PI * (heading - 90) / 180
```

Replacing Θ in the preceding formulas, we get

```
X = r * cos(PI * (heading - 90) / 180)
Y = r * sin(PI * (heading - 90) / 180)
```

Now, recall `r` is either the radius of the outer circle or the inner circle. Since we want both, we need to plot two points: one for the location of the line on the outer circle and one for the inner circle. With that in mind, we need to run the set of formulas twice as follows. Why subtract 90? Because we added it to the heading since our board is mounted 90 degrees offset from the enclosure. See, that "error" appears everywhere:

```
X1 = compassRadius * cos(PI * (heading - 90) / 180)
Y1 = compassRadius * sin(PI * (heading - 90) / 180)
X2 = MASK_RADIUS * cos(PI * (heading - 90) / 180)
Y2 = MASK_RADIUS * sin(PI * (heading - 90) / 180)
```

OK, we've got the line segment, so we draw that line, yes. Not so fast! Recall we must center the circles in the OLED, so we will need to add the center X and Y positions as follows:

```
oled->line(oledCenter[X] X1, oledCenter[Y] + Y1,
           oledCenter[X] + X2, oledCenter[Y] + Y2);
```

And that's it! We've taken the heading, converted it to an angle, and then plotted the pointer on the screen – all with the trickery of trigonometry!

Regarding the calibration animation, we will use four circles of decreasing size as an action sequence to tell the user to rotate the sensor. We will use the same preceding formulas to calculate the position of these circles on the outer circle. The circles will be spaced apart to form an action icon that travels around in a circle. You can explore the code that draws those circles as an exercise.

Note The Python library for the Micro OLED does not include a circle fill function. So the Python version will use circles without fill.

Limitations

Calculations for the heading have one major limitation to consider. The magnetometer calculates values in three dimensions, not two. Thus, there is a Z axis to consider. When the magnetometer is moved in three dimensions, the values for X and Y will be affected by the orientation in the Z axis.

This project is written to only consider values in the X and Y directions; thus, the magnetometer must be kept horizontal to get accurate readings. Any change in the Z axis while moving the magnetometer will affect the heading calculation.

An orientation limitation isn't so farfetched as you may think. Compasses used in the past were mounted on a gimbal designed to keep the compass face with the horizon no matter the orientation of the thing to which it was mounted (ship, plane, etc.).

If you are wondering if it is possible to use the Z axis values to ensure a correct heading is calculated, thereby removing this limitation, it is, but the solution adds complexity beyond the scope of this work. If you want to learn more about calibrating a magnetometer in three dimensions, see the following references:

- *Iron distortion*: `https://appelsiini.net/2018/calibrate-magnetometer`
- *Calibration concepts*: `https://teslabs.com/articles/magnetometer-calibration`
- *Simple calibration*: `https://github.com/kriswiner/MPU6050/wiki/Simple-and-Effective-Magnetometer-Calibration`
- *Calibration video*: `https://robotacademy.net.au/lesson/using-magnetometers`

There is one other limitation we should mention. The compiled code is too large for the older, smaller Arduino boards. You will need to use a board with more memory to run the sketch.

Let's see what hardware we will need.

Hardware Required

The hardware needed for this project is listed in Table 11-1. URLs for each component are included for ease of ordering including duplicate entries for alternative vendors.

Table 11-1. Hardware Needed for the Digital Compass Project

Component	URL	Qty	Cost
SparkFun Triple Axis Magnetometer Breakout - MLX90393	www.sparkfun.com/products/14571	1	$15.95
Micro OLED Breakout	www.sparkfun.com/products/14532	1	$16.95
Qwiic cable (any length can be used)	www.sparkfun.com/products/17259	2	$1.50
Qwiic cable kit (optional)	www.sparkfun.com/products/15081	1**	$7.95
SparkFun RedBoard Qwiic (Arduino Uno or compatible)	www.sparkfun.com/products/15123	1	$19.95
Raspberry Pi 3B or later	www.sparkfun.com/categories/233 www.adafruit.com/category/176	1	$35.00+
Qwiic pHAT for Raspberry Pi	www.sparkfun.com/products/15945	1	$5.95

Tip While there is only one Qwiic magnetometer from SparkFun, there are many I2C magnetometers from SparkFun and Adafruit (including an MLX90393 version). Most can easily be adapted to use in this project, so feel free to substitute a different module (if you just want to experiment).

About the Hardware

Let's discuss these components briefly. We will discover how to work with the hardware in more detail later in the chapter.

Sensor

The SparkFun Triple Axis Magnetometer Breakout - MLX90393 is a triple-axis magnetic sensor that can detect small magnetic fields as well as larger fields such as a magnetic source nearby. It can be used as a compass, for magnetic sensing for switches or similar nearness measurement, or as a position sensor such as an end stop for mechanical movements.

Perhaps the best part of this module is its low cost, so it makes purchasing the module for a permanent project a bit more economical. That is, if you decide to build the project in a permanent enclosure, you won't be spending a lot of money to build it.

OLED

The OLED module we will use is the same from the previous chapters. If you'd like to experiment with other output devices, you can, but it is recommended to use the Micro OLED so that the example code works without modification. That said, the code for the Arduino example is written so that it can be used with different OLED sizes. We'll see those details when we examine the code.

Assemble the Qwiic Modules

Recall from Chapter 7 we can use a single Qwiic cable to connect our MLX90393 sensor to the OLED module and then another to attach to the host adapter on our host board. Figure 11-4 shows an example of how you should connect your modules to form a Qwiic daisy chain.

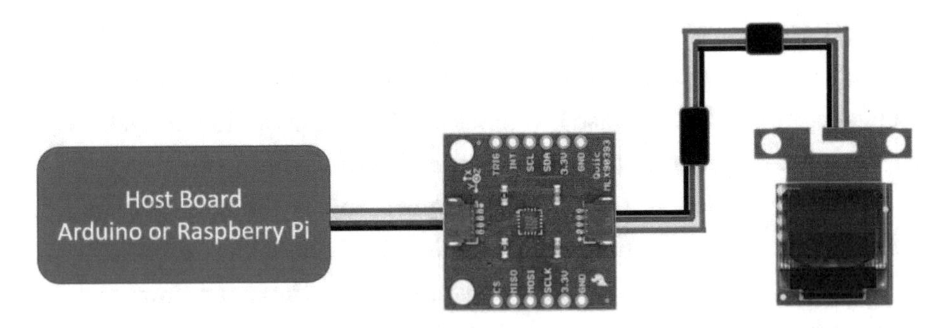

Figure 11-4. *Digital compass connections*

Using an Enclosure

Like the last project, the digital compass components we're using can be very cumbersome because we will need to move the module in a number of orientations during the calibration routine. We can fix this with an enclosure.

If you have your own or access to a 3D printer, you can print an enclosure. The source code for this chapter includes the 3D printing files you need to create a simple enclosure to house the magnetometer and OLED screen. Figure 11-5 shows the two parts of the enclosure with the bottom on the left.

Figure 11-5. *3D enclosure design for the digital compass project*

Notice we have a cutout in the bottom for the Qwiic cable as well as two sets of holes: one for mounting the Qwiic module and the other to attach the cover. The top also has holes for the OLED module as well as a cutout for the display. See Chapter 10 for a list of the hardware needed.

To print the files, you should use supports so that the recessed areas do not collapse. These are used to hide the heads of the bolts and screws.

To assemble the enclosure, begin by mounting the magnetometer module to the bottom as shown in Figure 11-6. Notice the sensor is mounted 90 degrees to the left. This is the intentional error mentioned earlier.

Figure 11-6. *Mounting the Qwiic board to the enclosure base*

Next, attach a short Qwiic cable to the OLED and mount it on the top using a ¾" (19mm) spacer from SparkFun as shown in Figure 11-7.

Figure 11-7. *Mounting the Micro OLED board to the enclosure top cover*

Finally, attach the Qwiic cable from the Micro OLED to the magnetometer on one side and another longer cable on the side with the cutout. Then, carefully press the bottom onto the top (or vice versa) and attach with two #40 screws. Figure 11-8 shows a completed example of the enclosure.

Figure 11-8. *3D printed enclosure for the digital compass project*

If you have experience creating 3D models for printing, feel free to experiment with creating your own enclosure - perhaps one that also includes a battery and a small form factor host board.

Calibrating the Sensor

The magnetometer must be calibrated to ensure we identify as much of the hard iron interference as possible. If you don't calibrate the sensor, the heading calculation will fail to determine the correct heading.

In fact, if your environment has a lot of metal in the area or you place a magnetic field-generating device nearby (your mobile phone), the calibration may not capture all of the anomalies, and it could make your heading calculations off by several degrees.

The calibration we run simply reads the sensor from as many different positions as possible in the horizontal plane (on a level surface). The hard iron (and some of the soft iron) distortions will appear as outliers or extreme values in the readings. We will capture those and calculate our bias to try and compensate.

Fortunately, the calibration sequence is easy. All we need to do is keep the sensor level and rotate it slowly several times to collect the minimum and maximum for the X and Y values. Then, we calculate an average of the range of values for each as the bias. The formula is shown in the following. Recall we use these values in our heading calculation:

```
bias[X]  = (maxValues[X] + minValues[X])/2;    // get average
                                                  x bias
bias[Y]  = (maxValues[Y] + minValues[Y])/2;    // get average
                                                  y bias
```

To execute the calibration, we simply ask the user to rotate the sensor a number of times. As you will see, the code for the project uses three rotations, but you may want to experiment with more rotations if your heading is a little off.

How would you know? Use your mobile phone compass application to check it. Keep in mind the sensor we're using is not going to be accurate, but it should be within about 5–10 degrees. If you find it off by more, try more calibration spins and spin the sensor more slowly to capture as many hard iron distortions as possible.

Now that we know more about the hardware for this chapter, let's write the code!

Write the Code

The code for this project follows the pattern of code layout we've learned. There are other ways to construct the code (mainly to do with order of operations and modularization), but the basics are still the same.

We begin by initializing the I2C bus and preparing the sensor and the compass face for use and then executing a loop that reads values from the sensor and then displays the values on the OLED. More specifically, we will read the X and Y axis values from the magnetometer.

Let's walk through how to prepare our computers to use the sensor and write code to read its values. We'll start with the Arduino.

Arduino

This section presents a walk-through of the sketch you will write to read values from the sensor and display a compass on the OLED module. But first, there are a couple of libraries we must install on our PCs.

Like we've done in previous chapters, we are not going to use the library that SparkFun lists for use with their MLX90393 module. Rather, we will use Adafruit's MLX90393 library. It has shown to be a more reliable library and has several examples for use. The SparkFun recommended library was created by a third party. It is old and provides only the bare minimum for working with the sensor. In this case, the Adafruit library is the better choice.

Install Software Libraries

We will need to install the Arduino libraries for the magnetometer and Micro OLED module separately. Fortunately, this is easy to do using the Library Manager. Simply open the Library Manager from the Arduino IDE menu (*Sketch* ➤ *Include Library* ➤ *Library Manager...*). Then search for `90393` and install the latest version of the Adafruit MLX90393 library as shown in Figure 11-9.

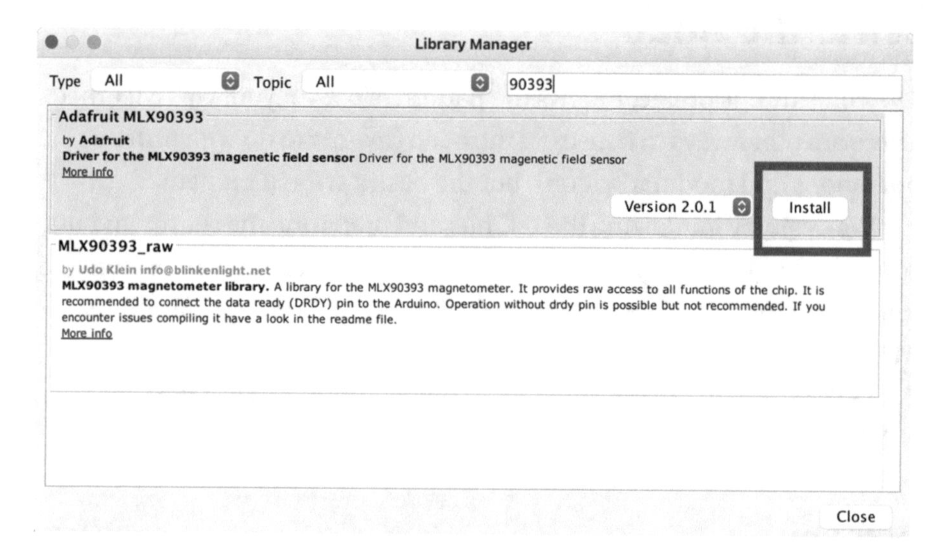

Figure 11-9. *Installing the Adafruit MLX90393 library (Arduino IDE)*

Similarly, we need to install the library for the OLED. Open the Library Manager and search for `micro OLED` and then install the latest version as shown in Figure 11-10.

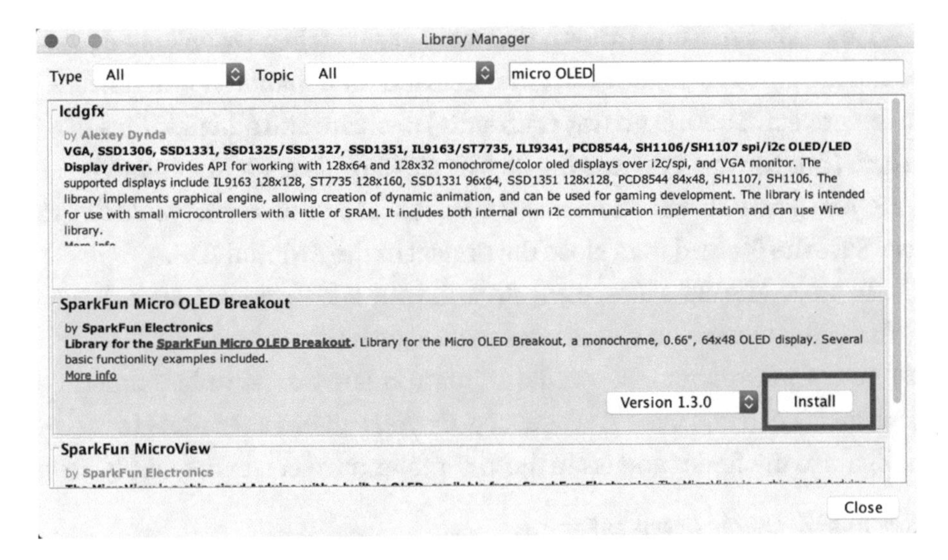

Figure 11-10. *Installing the Micro OLED library (Arduino IDE)*

Now that we have the software libraries installed, we can begin writing our sketch. Since this is not our first Arduino sketch, we will discuss the code at a high level and skip the line-by-line details focusing on the mechanics of how the code works. You can study the code at your leisure to ensure you understand the sketch in more detail.

Write the Sketch

Recall we are going to use a class to emulate the digital compass. Like we did in previous chapters, we will create a new project, add the class header and code file, write the class header and then the main sketch, and finally complete the class code.

The class will be named `CompassFace` and stored in two files: a header file named `CompassFace.h` and a source file named `CompassFace.cpp`. Effectively, we are moving the functions we would normally include in the main sketch to a class to make it easier to write, maintain, and understand.

However, since the Arduino IDE manages sketches, we will need to create the bare sketch file and folder first and then manually add the files to the project. There is no way (currently) to create and add new files to a sketch (but you can add existing files by clicking *Sketch* ➤ *Add File...*).

Open a new sketch and name it `compass.ino` or whatever you'd like to use. Save the file and then close the project in the Arduino IDE.

To create the class files, navigate with your File Explorer (Finder) to the folder where you stored your main sketch (`compass.ino`). Then, use your File Explorer or a text file editor to create two new files named `CompassFace.h` and `CompassFace.cpp`. Or you can use a terminal to navigate to the folder and issue these commands to create the empty files:

```
compass % touch CompassFace.h
compass % touch CompassFace.cpp
```

You can then open the project in the Arduino IDE and see all three files in the project. Let's see the code for each file starting with the header file.

Class Header File

Click the tab named `CompassFace.h` to open the blank file. Here, we will add the header or blueprint for the class. Recall the header file simply defines the class. We will use the `CompassFace.cpp` file to add the code for the functions in the class. Let's discuss those first.

This class is the same as the class from the last chapter. We will be using it to manage the digital compass. That is, we'll be putting all of the code that interacts with the Micro OLED to display the compass on the Micro OLED in this class. Reading the sensor will be done from the main sketch instead.

Thus, we will need a number of data that is used inside the class and a few functions for use outside the class. The functions we will create include the following. All are public unless indicated:

- *Constructor*: We need a constructor for the class to accept the sensitivity and size of the center ring.
- *begin()*: Allow the caller to initialize the class and the Micro OLED.
- *startCalibrationAnimation()*: Allow the caller to start the calibration animation. Recall we want to display a series of circles that spin around to tell the user to rotate the sensor. The animation continues until stopped.
- *stopCalibrationAnimation()*: Stop the animation for the calibration sequence.
- *isCalibrated()*: Returns true if the calibration has been run (animation has been run and then stopped).
- *drawCompass()*: Draw the compass and position the pointer at the indicated heading.
- *drawPointer()*: A private function to draw the pointer at a given heading.
- *drawCompassFace()*: Draw the compass circle and direction values.

We also need to store some data for use in the calculations as well as a few constants to make things easier. We leave the explanation for your own exploration. Let's look at the completed code for the header file. Listing 11-1 shows the file.

Listing 11-1. Compass Face Header File

```
#include <SFE_MicroOLED.h>

#define DC_JUMPER 1
#define PIN_RESET 9
#define MASK_RADIUS 12
#define X 0
#define Y 1
#define Z 2

class CompassFace {
public:
    CompassFace(int adjustment=0);
    void begin();
    void startCalibrateAnimation();
    void calibrateAnimation(int pass, int heading);
    void stopCalibrateAnimation() { calibrated = true; }
    bool isCalibrated() { return calibrated; }
    void drawCompass(int heading);

 private:
    MicroOLED *oled;

    int oledCenter[2] = {0,0};
    int compassRadius;
    int north[2] = {0, 0};
    int east[2] = {0, 0};
    int south[2] = {0, 0};
    int west[2] = {0, 0};
    int fontWidth;
    int fontHeight;
    bool calibrated = false;
```

```
    int orientation = 0;

    void drawPointer(int heading);
    void drawCompassFace();
};
```

The code should be easy to read, and there are no surprises other than the details described previously. Let's return to the main sketch to see how we can use this class.

Main Sketch

Now click the `compass.ino` tab to return to the main sketch. Let's begin with the preamble or top of the file. Recall here is where we include libraries we need, declare variables and constants, etc. We will also create functions to execute the calibration sequence, which is a bit more complex than what we are used to seeing, as well as a function to calculate the heading. We will walk through those functions separately.

In the preamble, we include the wire, MLX90393, and our new class header. There are a few variables we will need including variables for the MLX90393 and compass face class instances, values read from the sensor, and bias for storing the calibration data and a few constants to make things easier to read and adjust if needed. For example, we have a constant that defines the offset (intentional error) should we mount our sensor board in a different orientation than the OLED. Listing 11-2 shows the preamble code for the main sketch.

Listing 11-2. Main Sketch Preamble

```
#include <Wire.h>
#include <Adafruit_MLX90393.h>
#include "CompassFace.h"
```

```
// Pointer variables for class instances magnetometer and
// OLED compass display class.
Adafruit_MLX90393 mlx;
CompassFace *compass;

// Constants
#define CALIBRATION_SPINS 3  // Number of times to run
                                calibration
#define ORIENTATION 90       // Orientation difference for
                                mount location

// Global variables
float readValues[3] = {0.0, 0.0, 0.0}; // Values read from
                                          sensor
int heading = 0.0;                     // Current
float bias[2] = {0.0, 0.0};            // Bias values from
                                          calibration
```

Next, we will code the setup() function, which includes the sort of initialization code we've seen in other projects. Specifically, we will set up the magnetometer and the compass face class. Listing 11-3 shows the complete code for the setup() function. Read through it to ensure you understand all of the code included.

Listing 11-3. Main Sketch setup()

```
void setup()
{
  delay(100); // Give display time to power on
  Serial.begin(9600);
  while (!Serial);
  Wire.begin(); //Set up I2C bus
```

```
  if (!mlx.begin_I2C()) {
    Serial.println("No sensor found ... check your wiring?");
    while (1) { delay(100); }
  }

  mlx.setGain(MLX90393_GAIN_1X);

  Serial.println("Compass ready...");
  compass = new CompassFace(ORIENTATION);
  compass->begin();
}
```

Next, we have the `loop()` function. We simply read the values from the sensor and then call the `drawCompass()` function for our new class. We will also check to see if the sensor is calibrated. While we could have put this in the `setup()` function, placing it in the `loop()` function permits us to modify the code to run the calibration whenever needed. See the "Going Further" section for hints on how to do this.

Listing 11-4 shows the completed code for the `loop()` function. Go ahead and read through the code to ensure you understand how it works. There should not be any surprises or new techniques. In fact, most of the code is diagnostic statements.

Listing 11-4. Main Sketch loop()

```
void loop()
{
  float heading = 0.0;

  if (!compass->isCalibrated()) {
    calibrate(&bias[X], &bias[Y]);
  }
  mlx.readData(&readValues[X], &readValues[Y], &readValues[Z]);
  Serial.print("Bias: [");
```

```
  Serial.print(bias[X]);
  Serial.print(",");
  Serial.print(bias[Y]);
  Serial.print("]) Values: (");
  Serial.print(readValues[X]);
  Serial.print(",");
  Serial.print(readValues[Y]);
  Serial.print(") Heading (");
  heading = getHeading(readValues[X], readValues[Y], bias);
  Serial.print(heading);
  Serial.println(")");
  compass->drawCompass(heading);
  delay(250);
}
```

Now, let's look at the helper functions. We will use two functions: one to manage the calibration process named `calibrate()` and another to calculate the heading named `getHeading()`.

Let's look at the `calibrate()` function. Here, we simply initiate the calibration animation and execute a loop that reads the sensor capturing the minimal (smallest) value and maximum (largest) value for the X and Y coordinates. After we have executed the loop as specified in the `CALIBRATION_SPINS` constant, we will calculate the bias. Listing 11-5 shows the completed code for the function.

Listing 11-5. Main Sketch Calibration Function

```
void calibrate(float *xBias, float *yBias) {
  float readValues[3] = {0.0, 0.0, 0.0};
  float maxValues[2] = {-32767.0, -32767.0};
  float minValues[2] = {32767.0, 32767.0};
```

```
Serial.print("Calibrating..");
compass->startCalibrateAnimation();
for (int pass=CALIBRATION_SPINS; pass > 0; pass--) {
  for (int heading = 360; heading > 0; heading-= 10) {
    // Get latest values
    mlx.readData(&readValues[X], &readValues[Y],
    &readValues[Z]);
    // Get maximum values
    maxValues[X] = max(readValues[X], maxValues[X]);
    maxValues[Y] = max(readValues[Y], maxValues[Y]);
    // Get minimum values
    minValues[X] = min(readValues[X], minValues[X]);
    minValues[Y] = min(readValues[Y], minValues[Y]);
    // Update animation
    compass->calibrateAnimation(pass, heading);
    Serial.print("Calibration: (");
    Serial.print(readValues[X]);
    Serial.print(",");
    Serial.print(readValues[Y]);
    Serial.println(")");
  }
}
compass->stopCalibrateAnimation();
// Now, save the bias
bias[X]  = (maxValues[X] + minValues[X])/2;   // get average
                                                 x bias
bias[Y]  = (maxValues[Y] + minValues[Y])/2;   // get average
                                                 y bias
Serial.println("done.");
Serial.print("Bias (X, Y): (");
Serial.print(bias[X]);
```

```
  Serial.print(",");
  Serial.print(bias[Y]);
  Serial.println(")");
}
```

Once again, most of the code is diagnostic statements. Notice at the end we capture the calibration data, which is the difference between the minimum and maximum values for the X and Y coordinates.

Finally, we look at the getHeading() function. Since we have already discussed the math in this function, we leave the analysis of the code for an exercise. Listing 11-6 shows the completed code for the function.

Listing 11-6. Main Sketch Heading Function

```
float getHeading(float x, float y, float bias[2]) {
  float heading = atan2(y - bias[Y], x - bias[X]);

  // Compensate for declination angle. Be sure to look up your
  // declination angle and replace the value below with the
  // value for your geographic location.
  float declinationAngle = 0.174533;
  heading += declinationAngle;

  // Add the orientation shift if mounting the MLX90393 in the
     enclosure
  // discussed in the book. Recall, we are rotating the sensor
     90 degrees
  // to the left, so we must add 90 degrees to compensate.
     However, we are
  // working with radians here, so we convert degrees to
     radians for our
  // calculations.
  float sensorOrientation = (ORIENTATION * PI/180);
  heading += sensorOrientation;
```

```
  // If heading is negative, adjust for positive heading values
  if (heading < 0) {
    heading += 2*PI;
  }
  // If declination and orientation causes heading to overflow,
     adjust value
  if (heading > 2*PI) {
    heading -= 2*PI;
  }

  // Convert radians to degrees
  heading = heading * 180 / PI;
  return heading;
}
```

Now we can write the final portion of our project – the code for the class.

Class Code File

Click the tab named `CompassFace.cpp` to open the blank file. Here, we will add the code for the class. Since we have already discussed the math for the calibration animation as well as the function for drawing the compass, we leave examination of the code as an exercise. Listing 11-7 shows the completed code for the class (documentation omitted for brevity). Since this is a much lower level than the code we're used to seeing, be sure to study it to understand how it works. While the protocol (how we talk to the module) is very simple, the functions we use are uncommon since we normally have libraries written for us.

Listing 11-7. Compass Face Code File

```
#include "CompassFace.h"

CompassFace::CompassFace(int adjustment) {
  // Create an instance of the MicroOLED
  oled = new MicroOLED(PIN_RESET, DC_JUMPER);

  // Setup some variables for placing things on the screen
  fontWidth = oled->getFontWidth();
  fontHeight = oled->getFontHeight();
  oledCenter[X] = oled->getLCDWidth() / 2;
  oledCenter[Y] = oled->getLCDHeight() / 2;
  compassRadius = min(oledCenter[X], oledCenter[Y]) - 1;
  north[X] = oledCenter[X] - fontWidth / 2;
  north[Y] = oledCenter[Y] - compassRadius + 2;
  east[X] = oledCenter[X] + compassRadius - fontWidth - 1;
  east[Y] = oledCenter[Y] - fontHeight / 2;
  south[X] = north[X];
  south[Y] = oledCenter[Y] + compassRadius - fontHeight - 1;
  west[X] = oledCenter[X] - compassRadius + fontWidth - 2;
  west[Y] = east[Y];

  // Save the orientation for calculating points on the compass
     face
  orientation = adjustment;
}

void CompassFace::begin() {
  // Setup the OLED and initialize
  oled->begin();          // Start the display
  oled->clear(PAGE);      // Clear the display's internal memory
  oled->clear(ALL);       // Clear the library's display buffer
```

```
  oled->display();        // Display what's in the buffer
  oled->setFontType(0);  // Set the font type
}

void CompassFace::startCalibrateAnimation() {
  oled->clear(PAGE);
  oled->setCursor(0, 0);
  oled->print("Starting");
  oled->setCursor(0, 10);
  oled->print("Calibrate");
  oled->setCursor(0, 20);
  oled->print("Sequence");
  oled->display();
  delay(1500);
}

void CompassFace::calibrateAnimation(int pass, int heading) {
  int offset[2] = {fontWidth * 3.5, fontWidth / 2};
  int dot[2] = {0, 0};

  oled->clear(PAGE);
  int animationRadius = compassRadius - 2;
  dot[X] = animationRadius * cos(PI * ((float)heading -
  orientation) / 180);
  dot[Y] = animationRadius * sin(PI * ((float)heading -
  orientation) / 180);
  oled->circleFill(oledCenter[X] + dot[X],
                   oledCenter[Y] + dot[Y], 4);
  dot[X] = animationRadius *
        cos(PI * ((float)heading - (orientation - 20)) / 180);
  dot[Y] = animationRadius *
        sin(PI * ((float)heading - (orientation - 20)) / 180);
```

```
  oled->circleFill(oledCenter[X] + dot[X],
                   oledCenter[Y] + dot[Y], 3);
  dot[X] = animationRadius *
        cos(PI * ((float)heading - (orientation - 40)) / 180);
  dot[Y] = animationRadius *
        sin(PI * ((float)heading - (orientation - 40)) / 180);
  oled->circleFill(oledCenter[X] + dot[X],
                   oledCenter[Y] + dot[Y], 2);
  dot[X] = animationRadius *
        cos(PI * ((float)heading - (orientation - 55)) / 180);
  dot[Y] = animationRadius *
        sin(PI * ((float)heading - (orientation - 55)) / 180);
  oled->circleFill(oledCenter[X] + dot[X], oledCenter[Y] +
  dot[Y], 1);
  oled->setCursor(oledCenter[X] - offset[X], oledCenter[Y] -
  offset[Y]);
  oled->print("Rotate");
  oled->setCursor(0, 0);
  oled->print(pass);
  oled->display();
}

void CompassFace::drawCompass(int heading) {
  // Draw the COMPASS:
  oled->setFontType(0);
  oled->clear(PAGE);      // Clear page memory
  drawCompassFace();
  drawPointer(heading);
  oled->display();
}
```

```
void CompassFace::drawPointer(int heading)
{
  // Calculate the pointer start, end positions
  int pointer[4] = {
    compassRadius * cos(PI * ((float)heading - orientation) / 180),
    compassRadius * sin(PI * ((float)heading - orientation) / 180),
    MASK_RADIUS * cos(PI * ((float)heading - orientation) / 180),
    MASK_RADIUS * sin(PI * ((float)heading - orientation) / 180)
  };
  // Draw the pointer
  oled->line(
    oledCenter[X] + pointer[X], oledCenter[Y] + pointer[Y],
    oledCenter[X] + pointer[X+2], oledCenter[Y] + pointer[Y+2]
  );

  // Display the heading in the center of the screen
  oled->setCursor(oledCenter[X] - (fontWidth + (fontWidth/2)),
  east[Y]);
  if (heading < 10) {
    oled->print("00");
  } else if (heading < 100) {
    oled->print('0');
  }
  oled->print(heading);
}

void CompassFace::drawCompassFace()
{
  // Draw the compass outer border
  oled->circle(oledCenter[X], oledCenter[Y], compassRadius);
```

```
    // Draw the compass directions
    oled->setFontType(0);
    oled->setCursor(north[X], north[Y]);
    oled->print('N');
    oled->setCursor(east[X], east[Y]);
    oled->print('E');
    oled->setCursor(south[X], south[Y]);
    oled->print('S');
    oled->setCursor(west[X], west[Y]);
    oled->print('W');
}
```

As you can see, this class is another example of how to move code out of your main sketch and into helper modules. There is another possible class we could make for this project. We could move the calibration code to its own class. If you're curious, try it out yourself as an exercise.

Compile the Sketch

The last step is to compile the sketch before uploading it to your board. If you encounter any errors, be sure to fix them and recompile to ensure the sketch compiles without errors or serious warnings.

Once everything compiles, we're ready to start testing. But first, let's look at the code for the Raspberry Pi. You can skip to the "Sketch on the Arduino" section if you're curious to see how the project works. The code will run the same on both platforms.

Raspberry Pi

This section presents a walk-through of the Python code you will write to read values from the sensor and display them on the Micro OLED module. But first, we need to install a Python library on our Raspberry Pi.

Install a Software Library

We need only one more software library to use the Python code in this chapter. We need the magnetometer library from Adafruit as shown in the following:

```
$ pip3 install adafruit-circuitpython-mlx90393
```

Tip If you haven't installed the SparkFun Qwiic Python libraries, you must install them to run this project (`pip3 install sparkfun_qwiic`).

Now we're ready to write the code.

Write the Code

The code for the Python version of this project is about the same length and complexity as the Arduino code. Like the Arduino example, we will use a class to contain the code for the compass face. However, unlike the Arduino IDE, you can use any editor to create the class and main script. We will name the main script `compass.py` and the class module `compass_face.py`. Let's start with the class.

Compass Face Class

Open the Thonny Python IDE under the *Main* ➤ *Programming* submenu. Create a new file `compass_face.py`. We will name the class `CompassFace`, and we will need a few functions. In fact, we need the same functions as we used in the Arduino version as listed in the following for completion. While they perform the same operations as the Arduino version, the names of the functions differ. For a complete description of each of these functions, see the "Arduino" section.

The functions we will create include the following:

- *Constructor*: We need a constructor for the class to accept the sensitivity and size of the center ring.
- *begin()*: Allow the caller to initialize the class and the Micro OLED.
- *start_calibration_animation()*: Allow the caller to start the calibration animation. Recall we want to display a series of circles that spin around to tell the user to rotate the sensor. The animation continues until stopped.
- *stop_calibration_animation()*: Stop the animation for the calibration sequence.
- *is_calibrated()*: Returns true if the calibration has been run (animation has been run and then stopped).
- *draw_compass()*: Draw the compass and position the pointer at the indicated heading.
- *draw_pointer()*: A private function to draw the pointer at a given heading.
- *draw_compass_face()*: Draw the compass circle and direction values.

Listing 11-8 shows the complete code for the class with documentation removed for brevity. Take a few moments to read through the code so that you understand all of the parts of the code. As you will see, it mimics the Arduino.

Listing 11-8. Compass Face Class (Python)

```
import math
import time
import qwiic

MASK_RADIUS = 12
X = 0
Y = 1
Z = 2

class CompassFace:
    oled = None
    oled_center = [0, 0]
    compass_radius = 6
    north = [0, 0]
    east = [0, 0]
    south = [0, 0]
    west = [0, 0]
    font_width = 0
    font_height = 0
    calibrated = False

    def __init__(self, adjustment=0):
        self.oled = qwiic.QwiicMicroOled()
        self.oled.begin()

        self.font_width = self.oled.get_font_width()
        self.font_height = self.oled.get_font_height()
        self.oled_center[X] = self.oled.get_lcd_width() / 2
        self.oled_center[Y] = self.oled.get_lcd_height() / 2
        self.compass_radius = min(self.oled_center[X],
                                  self.oled_center[Y]) - 1
        self.north[X] = self.oled_center[X] - self.font_width / 2
```

```
        self.north[Y] = self.oled_center[Y] - self.compass_
        radius + 2
        self.east[X] = self.oled_center[X] +
            self.compass_radius - self.font_width - 1
        self.east[Y] = self.oled_center[Y] - self.font_height / 2
        self.south[X] = self.north[X]
        self.south[Y] = self.oled_center[Y] +
            self.compass_radius - self.font_height - 1
        self.west[X] = self.oled_center[X] -
            self.compass_radius + self.font_width - 2
        self.west[Y] = self.east[Y]
        self.orientation = adjustment

    def begin(self):
        self.oled.clear(self.oled.PAGE) # Clear the display's
                                          internal memory
        self.oled.clear(self.oled.ALL)  # Clear the library's
                                          display buffer
        self.oled.display()             # Display what's in the
                                          buffer
        self.oled.set_font_type(0)      # Set the font type

    def start_calibrate_animation(self):
        self.oled.clear(self.oled.PAGE)
        self.oled.set_cursor(0, 0)
        self.oled.print("Starting")
        self.oled.set_cursor(0, 10)
        self.oled.print("Calibrate")
        self.oled.set_cursor(0, 20)
        self.oled.print("Sequence")
        self.oled.display()
        time.sleep(1.5)
```

```
def calibrate_animation(self, pass_number, heading):
    offset = [self.font_width * 3.5, self.font_width / 2]
    dot = [0, 0]

    self.oled.clear(self.oled.PAGE)
    animation_radius = self.compass_radius - 2
    dot[X] = animation_radius *
        math.cos(math.pi * (heading - self.orientation) / 180)
    dot[Y] = animation_radius *
        math.sin(math.pi * (heading - self.orientation) / 180)
    self.oled.circle(
        self.oled_center[X] + dot[X],
        self.oled_center[Y] + dot[Y],
        4
    )
    dot[X] = animation_radius *
        math.cos(math.pi * (heading -
        (self.orientation - 20)) / 180)
    dot[Y] = animation_radius *
        math.sin(math.pi * (heading -
        (self.orientation - 20)) / 180)
    self.oled.circle(
        self.oled_center[X] + dot[X],
        self.oled_center[Y] + dot[Y],
        3
    )
    dot[X] = animation_radius *
        math.cos(math.pi * (heading -
        (self.orientation - 40)) / 180)
    dot[Y] = animation_radius *
        math.sin(math.pi * (heading -
        (self.orientation - 40)) / 180)
```

```
        self.oled.circle(
            self.oled_center[X] + dot[X],
            self.oled_center[Y] + dot[Y],
            2
        )
        dot[X] = animation_radius *
            math.cos(math.pi * (heading - (self.orientation -
            55)) / 180)
        dot[Y] = animation_radius *
            math.sin(math.pi * (heading - (self.orientation -
            55)) / 180)
        self.oled.circle(
            self.oled_center[X] + dot[X],
            self.oled_center[Y] + dot[Y],
            1
        )
        self.oled.set_cursor(self.oled_center[X] - offset[X],
                             self.oled_center[Y] - offset[Y])
        self.oled.print("Rotate")
        self.oled.set_cursor(0, 0)
        self.oled.print(pass_number)
        self.oled.display()

    def stop_calibrate_animation(self):
        self.calibrated = True

    def draw_compass(self, heading):
        # Draw the COMPASS:
        self.oled.set_font_type(0)
        self.oled.clear(self.oled.PAGE)       # Clear page
                                                memory
```

```
        self.draw_compass_face()
        self.draw_pointer(heading)
        self.oled.display()

    def draw_pointer(self, heading):
        # Calculate the pointer start, end positions
        pointer = [
            self.compass_radius *
                math.cos(math.pi * (heading - self.orientation)
               / 180),
            self.compass_radius *
                math.sin(math.pi * (heading - self.orientation)
                / 180),
            MASK_RADIUS *
                math.cos(math.pi * (heading - self.orientation)
                / 180),
            MASK_RADIUS *
                math.sin(math.pi * (heading - self.orientation)
                / 180)
        ]
        # Draw the pointer
        self.oled.line(self.oled_center[X] + pointer[X],
            self.oled_center[Y] + pointer[Y],
            self.oled_center[X] + pointer[X+2],
            self.oled_center[Y] + pointer[Y+2]
        )

        # Display the heading in the center of the screen
        self.oled.set_cursor(self.oled_center[X] -
            (self.font_width + (self.font_width/2)), self.
            east[Y])
```

```
        if heading < 10:
            self.oled.print("00")
        elif heading < 100:
            self.oled.print('0')
        self.oled.print("{0:.0f}".format(heading))

    def draw_compass_face(self):
        # Draw the compass outer border
        self.oled.circle(self.oled_center[X], self.oled_
        center[Y], self.compass_radius)

        # Draw the compass directions
        self.oled.set_font_type(0)
        self.oled.set_cursor(self.north[X], self.north[Y])
        self.oled.print('N')
        self.oled.set_cursor(self.east[X], self.east[Y])
        self.oled.print('E')
        self.oled.set_cursor(self.south[X], self.south[Y])
        self.oled.print('S')
        self.oled.set_cursor(self.west[X], self.west[Y])
        self.oled.print('W')

    def is_calibrated(self):
        return self.calibrated
```

Now we can write our main script.

Main Script (Python)

Open the Thonny Python IDE under the *Main* ➤ *Programming* submenu. Create a new file `compass.py`. There is nothing new in this code as it follows the same flow as the Arduino example but much simplified. Listing 11-9 shows the complete code for the main script for this project. You can read through it to see how all of the code works.

Listing 11-9. Main Script (Python)

```
import math
import time
import sys
import board
import busio
import adafruit_mlx90393
from compass_face import CompassFace, X, Y, Z

CALIBRATION_SPINS = 3  # Number of calibration spins
ORIENTATION = 90       # Orientation difference for mount
                         location

i2c = busio.I2C(board.SCL, board.SDA)
mlx = adafruit_mlx90393.MLX90393(i2c, gain=adafruit_mlx90393.
      GAIN_1X)
compass_face = CompassFace(ORIENTATION)
read_values = [0.0, 0.0, 0.0]
bias = [0.0, 0.0]

def calibrate():
    maxValues = [-32767.0, -32767.0]
    minValues = [32767.0, 32767.0]

    print("Calibrating..")
    compass_face.start_calibrate_animation()
    for pass_num in range(CALIBRATION_SPINS, 0, -1):
        for heading in range(360, 0, -10):
            # Get latest values
            read_values[X], read_values[Y], read_values[Z] =
            mlx.magnetic
            # Get maximum values
            maxValues[X] = max(read_values[X], maxValues[X])
```

```
            maxValues[Y] = max(read_values[Y], maxValues[Y])
            # Get minimum values
            minValues[X] = min(read_values[X], minValues[X])
            minValues[Y] = min(read_values[Y], minValues[Y])
            # Update animation
            compass_face.calibrate_animation(pass_num, heading)
            print("Calibration: ({0},{1})".format(read_
            values[X], read_values[Y]))

    compass_face.stop_calibrate_animation()
    # Now, save the bias
    bias[X] = (maxValues[X] + minValues[X])/2   # get average x
                                                  bias
    bias[Y] = (maxValues[Y] + minValues[Y])/2   # get average y
                                                  bias
    print("done.")
    print("Bias (X, Y): ({0},{1})".format(bias[X], bias[Y]))
    return bias

def get_heading(x, y, bias_values):
    heading = math.atan2(y - bias_values[Y], x - bias_values[X])

    declinationAngle = 0.174533
    heading += declinationAngle

    sensor_orientation = (ORIENTATION * math.pi/180)
    heading += sensor_orientation

    # If heading is negative, adjust for positive heading values
    if heading < 0:
        heading += 2*math.pi
```

```
    # If declination and orientation causes heading to
      overflow, adjust value
    if heading > 2*math.pi:
        heading -= 2*math.pi

    # Convert radians to degrees
    heading = heading * 180 / math.pi
    return heading

def main():
    print("\nDigital Compass")
    compass_face.begin()
    while True:
        # Calibrate the sensor if not already calibrated
        if not compass_face.is_calibrated():
            calibrate()
            print("Readings:")
        # Read sensor and adjust with bias from calibration
        read_values[X], read_values[Y], read_values[Z] = mlx.
        magnetic
        # Check for errors
        if mlx.last_status > adafruit_mlx90393.STATUS_OK:
            mlx.display_status()
        heading = get_heading(read_values[X], read_values[Y],
        bias)
        # Show diagnostics
        print("Values: ({0:5.2f},{1:5.2f}) Heading ({2:.0f})"
              "".format(read_values[X], read_values[Y],
              heading))
        compass_face.draw_compass(heading)
        time.sleep(0.25)
```

```
if __name__ == '__main__':
    try:
        main()
    except (KeyboardInterrupt, SystemExit) as exErr:
        print("\nbye!\n")
sys.exit(0)
```

OK, that's it! We've written the code. Unlike the Arduino, we do not need to compile the Python code. So we're now ready to execute the project!

Execute the Project

Now that we've spent many pages exploring the Qwiic modules and writing the code to interact with them, it is time to test the project by executing (running) it.

When the project runs (executes), you will see some diagnostic message written to the serial monitor (Arduino) or the terminal (Raspberry Pi). You will also see a welcome message appear on the Micro OLED followed by a short pause. Then the calibration process will begin.

Once the calibration is complete, you will see the digital compass displayed. Figure 11-11 shows an example of what you should see on the OLED. Notice how the pointer is drawn along with the heading in the center. Here we see our heading is 296 degrees, and the pointer is indicating a west-northwest (WNW) heading.

Figure 11-11. *Executing the digital compass project*

Executing the code depends on which platform you're using. Let's look at the Arduino first.

Sketch on the Arduino

Executing the sketch on the Arduino requires connecting our board to our PC and then uploading the sketch to the Arduino. Recall the sketch will run so long as the USB cable is connected to our PC (and the Arduino).

Execute the Sketch

To execute the sketch, be sure your Arduino is connected and you've selected the correct board under the *Tools* ➤ *Board* menu. You also need to ensure you have the correct port selected under the *Tools* ➤ *Port* menu.

Once those items are set, you can click the *Upload* button or choose *Sketch* ➤ *Upload* from the menu. The Arduino IDE will compile the sketch and then upload it to your Arduino. Once you see the `Done uploading...` message, you can open the serial monitor. You should see the output begin momentarily that is the same as that on the OLED. Go ahead and try it out!

If something isn't working, check your connections or refer to Chapter 7 for troubleshooting tips.

Python Code on the Raspberry Pi

Executing the sketch on the Raspberry Pi requires running the Python code in a terminal after connecting your Qwiic daisy chain to your Raspberry Pi via a hat or the Qwiic female breakout cable. Recall the code will run until you stop it with *CTRL+C* on the keyboard.

Execute the Python Code

To run the Python code on the Raspberry Pi, you can issue the command `python3 ./compass.py` from the same folder where the file was saved as shown in the following. If everything worked as executed, congratulations! You've just built your fifth Qwiic project.

If something isn't working, check your connections or refer to Chapter 7 for troubleshooting tips.

Going Further

This project isn't one you're likely to use for an IoT project by itself; rather, you will likely want to incorporate it as part of a larger project. And while we didn't discuss them in this chapter, here are just a few suggestions you can try once we have learned how to take our projects to the cloud. Put your skills to work!

- *New class for the sensor*: Build a class to manage the sensor moving all of the code for reading the sensor to the new class.
- *Build a better enclosure*: Build an enclosure to contain your Arduino board and a battery supply to make the project portable. Use the 3D printer files from the book source code and expand the OpenSCAD code.

- *Build a gimbal*: Build a gimbal to keep the compass level to compensate for changes in the Z axis.
- *Calibrate button*: Add a button to trigger the calibration routine by either setting the calibration flag to false or running the calibration when the button is triggered. This could be helpful if you must move the sensor or want to use a specific plane for a base.
- *Robot compass*: Use the project in this chapter to add the ability to determine the direction of movement for a robot.
- *General direction*: Some robotics projects use compass directions (16, 32, or greater point compass) rather than precise headings (in degrees). You could convert this project to make such a compass by replacing the code in the compass face class with a class that presents the compass directions based on the heading. For more information about compass directions, see `https://en.wikipedia.org/wiki/Points_of_the_compass`.

Summary

In this chapter, we got more hands-on experience making projects with Qwiic and STEMMA QT modules. We used a magnetometer sensor to detect magnetic fields to determine the direction (heading). We used some basic trigonometry to calculate the location and covert radians to degrees for the heading.

We then wrote a class library to represent the compass face to draw a compass and display the heading in the center, the compass directions around a circle, and a pointer to represent a graphic presentation of the heading.

Along the way, we learned more about how to write class libraries to represent abstract constructs as well as how they can be used to model objects. Rather than write a class to manage the sensor, we wrote a class to represent a compass represented on a Micro OLED module.

This concludes our journey to learn how to build IoT projects using Qwiic and STEMMA QT modules. While we did not complete the IoT portions, we learned quite a bit about the hardware and had some fun along the way. We will circle back to some of these projects after we take a tour of another component system named Grove from Seeed Studio.

PART III

The Grove Component System

This part introduces the Grove component system including a series of chapters containing example projects that detail the steps needed to implement the system with the Arduino and Raspberry Pi. While the example projects are not complete IoT solutions in that they are not integrated with the cloud, they are a good starting point to learn how to program IoT projects for the Arduino and Raspberry Pi.

CHAPTER 12

Introducing Grove

Thus far in the book, we have learned how to use two versatile component systems to build electronic projects quickly and easily without soldering or complicated connections. We saw many examples and five projects that demonstrated how to use the Qwiic and STEMMA QT systems.

Both systems provide a vast array of modules, and given they are compatible (can be interchanged with the same cables), these systems are very powerful and an excellent base to build IoT projects. However, these systems restrict you to a single protocol, and while the list of modules supported is long, there are other protocols and other ways to connect components.

Recall from Chapter 6 we were introduced to three component systems and Chapters 6–11 covered the Qwiic and STEMMA QT component systems. In this chapter and the next three chapters, we will explore the third component system named Grove from Seeed Studio (`https://wiki.seeedstudio.com/Grove/`).

Overview

In this section, we will discover the Grove component system. We will learn about the capabilities and limitations of the system as well as examples of the components available. The chapter also includes details on how to start using the components in projects.

C. Bell, *Beginning IoT Projects*, https://doi.org/10.1007/978-1-4842-7234-3_12

Grove is designed to make building projects faster using pluggable modules containing sensors, input, output, and other functions. Unlike the Qwiic and STEMMA QT component systems, the Grove component system supports a variety of protocols[1] that operate over the same set of wires!

Grove supports the analog, digital, and universal asynchronous receiver-transmitter (UART[2]) protocols. Furthermore, Grove supports all of these protocols using the same wiring and connectors, so there's no need to remember what cables go with what protocols. Cool!

Now that we know what protocols Grove supports and how the cables are wired, let's see how easy the Grove component system makes using the modules.

The Grove Component System

Grove was created and released in 2010 by Seeed Studio (`seeedstudio.com`). They wanted to create an open source, modular component system to simplify rapid prototyping. But they didn't stop there. They continued to refine and develop more modules to include an impressive array of modules that contain small circuits that include sensors, input devices, output devices, and more. They also produce host adapters for many platforms.

Note Seeed Studio also uses the term breakout board for host adapters.

---

[1] You can call them "interfaces" or "connections" if it helps keep them sorted.

[2] A form of serial communication (`https://en.wikipedia.org/wiki/Universal_asynchronous_receiver-transmitter`).

Each host adapter supports many Grove connectors that match the capabilities of the host board. If the host board supports all of the protocols that Grove supports, the host adapter will have several connectors for each of the protocols.

These host adapters simply connect to your host board enabling the use of Grove modules without the need for additional electronics such as breadboards and discrete components and without soldering.

Unlike the Qwiic and STEMMA QT, the Grove cabling system is not designed for daisy chaining. Rather, Grove modules are connected to the host adapter directly. The "Y" cables are a rare deviation from this configuration. That's why most host adapters have so many Grove connectors. We will see some examples of host adapters in a later section.

Each Grove module is self-contained; all of the supporting electrical components are on the module mounted on a small PCB (most come in a pretty blue color in fact) of various sizes. All you need to do is connect the modules to your host adapter using a Grove cable, and your hardware is done.

Capabilities

The capabilities of the Grove system include the following:

- Modularized cabling supporting four protocols (I2C, digital, analog, and UART)
- Easy, polarized connectors (no incorrect or reversed connections[3])
- No soldering required!

[3] Perhaps the greatest bane of anyone working with I2C is inadvertently reversing the data and clock connections. Grove eliminates that guesswork entirely.

How Does It Work?

Grove wiring is very similar to the Qwiic and STEMMA QT wires - you can only connect the cable to the device one way, so you always know the connections are correct. Grove uses a four-wire cable of various lengths with a larger keyed connector. Like Qwiic and STEMMA QT, you can't misconnect a Grove cable. Nice. Figure 12-1 shows a typical Grove cable and connectors.

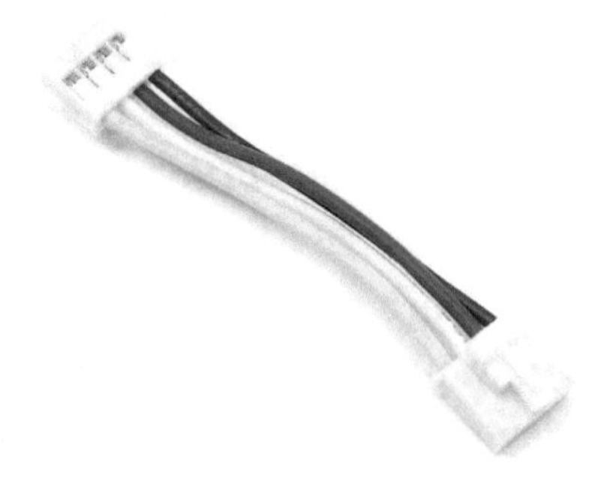

Figure 12-1. *Grove connectors (courtesy of seeedstudio.com)*

Figure 12-2 shows the Qwiic connector on the left and the Grove connector on the right. Notice the size difference. Clearly, this is one difference between the systems. Another is the size of the modules, which we will discover in a later section. And, yes, there is a cable you can buy that has a Grove connector on one side and a Qwiic connector on the other.

Figure 12-2. *Comparing Qwiic and Grove connectors (courtesy of sparkfun.com)*

We will discuss how the cables are used for each of the protocols in more detail.

I2C

Recall from Chapter 6 I2C is a fast digital protocol that uses two wires (plus power and ground) to read data from circuits (or devices). It is the protocol of choice (the only protocol) used in Qwiic and STEMMA QT. The Grove system also supports it making it possible to use the modules together. I2C over the Grove cabling system uses all four wires as shown in Table 12-1.

Table 12-1. *Grove Cable (I2C)*

Pin	Color	Description
1	Yellow	SCL
2	White	SDA
3	Red	VCC (power)
4	Black	GND (ground)

If this looks familiar, it is because it is the same layout as the Qwiic and STEMMA QT cables albeit with different colored wires.

Digital

The digital protocol is used for modules that produce a digital value, typically a positive integer in the range 0–1024 or larger. Digital wiring uses three wires: ground (GND), power (VCC of 3.3V or 5V), and signal. The digital protocol for Grove allows for up to two signal lines (named D0 and D1) using two of the four wires as shown in Table 12-2. Some modules may be labeled in such a way to indicate three signal lines (D0/D1 and D1/D2), but the interface supports only two signal lines. Signal lines can be used for input or output.

Table 12-2. *Grove Cable (Digital)*

Pin	Color	Description
1	Yellow	D0 – primary signal line
2	White	D1 – secondary signal line
3	Red	VCC (power)
4	Black	GND (ground)

Analog

The analog protocol supports communicating with modules using voltage. Like the digital protocol, the analog protocol supports up to two analog lines as well as the ground (GND) and power (VCC). The first analog line is named A0 and the second A1. Similar to the digital protocol, some modules may label the analog lines A0/A1 and A1/A2. Table 12-3 shows the layout of the analog protocol over the Grove cabling.

Table 12-3. *Grove Cable (Analog)*

Pin	Color	Description
1	Yellow	A0 – primary analog line
2	White	A1 – secondary analog line
3	Red	VCC (power)
4	Black	GND (ground)

UART

The UART protocol is a special serial protocol that uses two lines for transmit (TX) and receive (RX). Pins 1 and 2 are used for these lines, and the other two are the common ground and power lines as shown in Table 12-4.

Table 12-4. *Grove Cable (UART)*

Pin	Color	Description
1	Yellow	RX – serial receive
2	White	TX – serial transmit
3	Red	VCC (power)
4	Black	GND (ground)

Having all of the cables wired the same means you don't need to have any special cables for each of the four protocols, but there are some cases where we may need a slightly different cable. We will discuss the available Grove cables in a later section.

Grove modules come in a variety of sizes, and most have only a single Grove connector but may host a number of other connectors depending on the features supported. Grove modules are designed to support a single function using a dedicated circuit.

Like the connectors, Grove modules are a bit larger than Qwiic and STEMMA QT modules. While Grove modules are not uniform in size, they do conform to one of several formats as shown in Table 12-5. Most of the formats support a Grove connector in either a vertical (cable plugs in at a right angle to the board) or horizontal orientation.

Table 12-5. *Grove Module Sizes (courtesy of seeedstudio.com)*

Format	Size	Example
1×1	20×20mm	
1×2	20×40mm	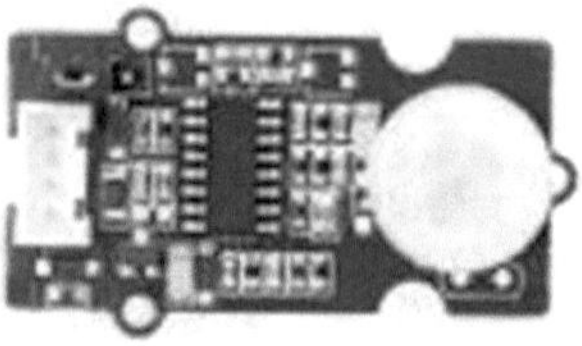
1×3	20×60mm	

(continued)

Table 12-5. (*continued*)

Format	Size	Example
2×2	40×40mm	
2×3	40×60mm	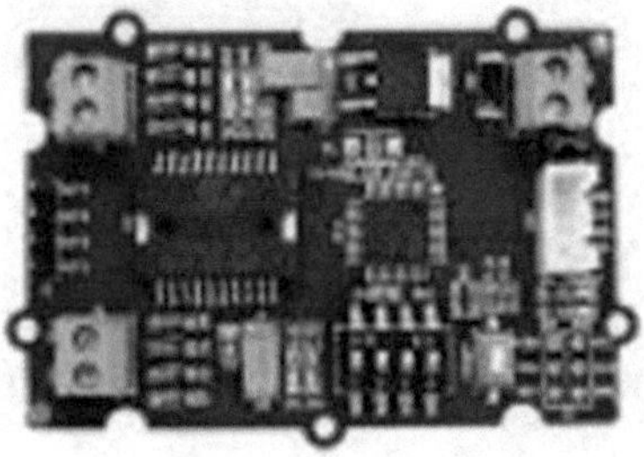

The host adapter has multiple Grove connectors that you can use to connect modules (depending on the protocol as there are dedicated connectors for each protocol). There are a variety of host adapters available for a growing list of host boards. This includes several for the Arduino, NodeMCU, Raspberry Pi, and many more. You can discover the latest offerings by visiting `https://wiki.seeedstudio.com/Grove_System/#how-to-connect-grove-to-your-board`.

The host adapter most will want to use for the Raspberry Pi is named simply GrovePi+ (`www.seeedstudio.com/GrovePi.html`) that provides 15 Grove connectors (3 I2C, 7 digital, 3 analog, and 2 UART connectors - one for the Raspberry Pi and one for Grove) as well as some most used GPIO pins broken out. It is designed to mount on the Raspberry Pi GPIO header so that the hat extends beyond the Raspberry Pi board making it possible to use them in addition to the Grove connectors. Figure 12-3 shows a GrovePi+ for the Raspberry Pi.

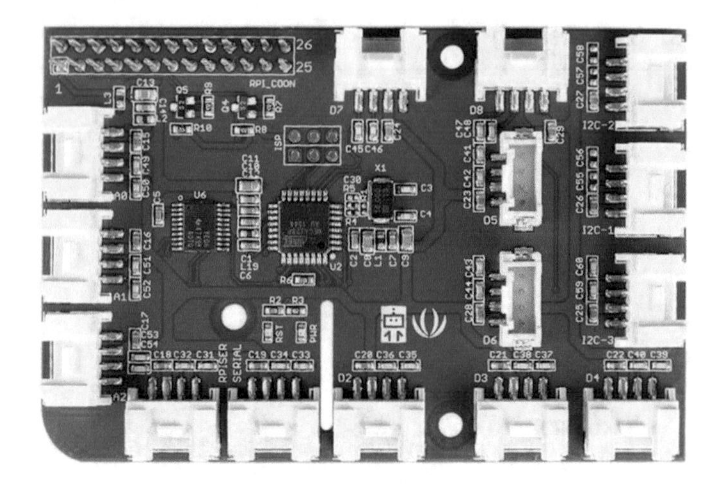

Figure 12-3. *GrovePi+ host adapter for Raspberry Pi (courtesy of seeedstudio.com)*

Notice the most commonly used GPIO pins are exposed in the upper-left corner. This allows you to use the header for additional connections. Another cool feature!

There are six host adapters for the Arduino. The one most will use for the Uno platform is the Grove Base Shield V2.0 (`www.seeedstudio.com/Base-Shield-V2.html`) that provides 16 Grove connectors (4 I2C, 7 digital, 4 analog, and 1 UART connector). Figure 12-4 shows a Grove Base Shield V2.0 for the Arduino.

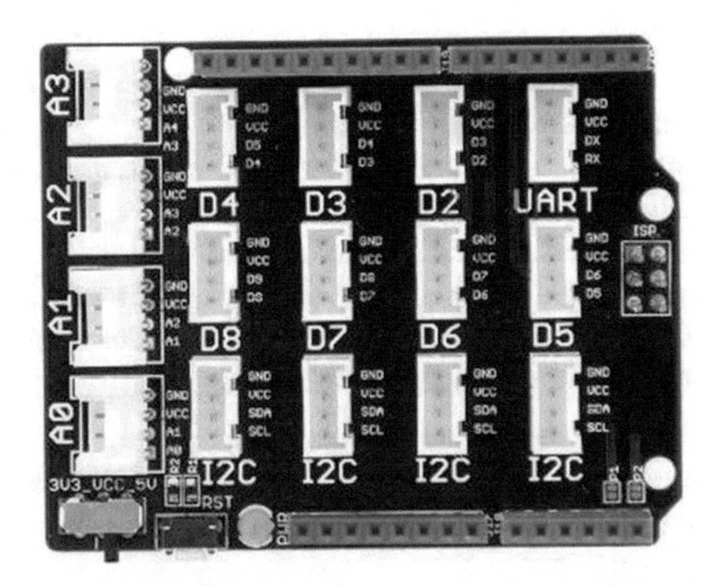

Figure 12-4. *Grove Base Shield V2.0 host adapter for Arduino (courtesy of seeedstudio.com)*

Now that we know what the Grove system is and how it works, let's examine some of the limitations.

Caution There is a small switch in the lower-right corner of the board. This allows you to switch the voltage from 3.3V to 5V. Use this switch to match the voltage output on your Arduino board. For example, if your board outputs 3.3V, you must set the switch to 3.3V.

Limitations

Like most systems, there are some limitations. Fortunately, there are few, and only the largest or most complex projects may need to heed. The limitation you may encounter for larger projects is the maximum number of modules that can be supported is limited to the number of connections available on the host adapter, which is often limited by the host device or by the size of the host adapter.

For example, if you want to use the Raspberry Pi Zero, the GrovePi Zero host adapter has only a single Grove I2C connector and thus can use only one I2C module. Similarly, most Grove host adapters have limited numbers of digital and analog connectors. However, there are some things you can do to mitigate some of these limitations. Seeed Studio offers a number of modules that can help out (called interfaces).

Tip You can discover the latest interface boards available for a variety of uses at `www.seeedstudio.com/interfaces-c-946.html`.

For example, if you want to use more I2C connections than what are available on the host adapter, you can use the Grove 8 Channel I2C Hub (`www.seeedstudio.com/Grove-8-Channel-I2C-Hub-TCA9548A-p-4398.html`) to extend the number of I2C connections. With this module, you can use one I2C connector on your host adapter and connect up to eight I2C modules to the hub. Figure 12-5 shows the Grove 8 Channel I2C Hub.

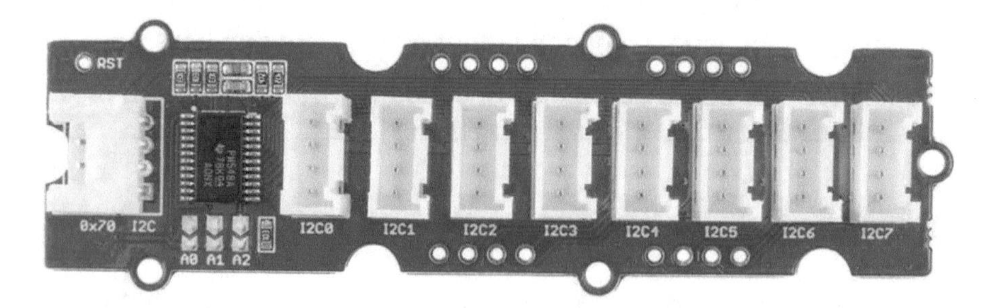

***Figure 12-5.** Grove 8 Channel I2C Hub (courtesy of seeedstudio.com)*

To increase the number of connections for analog sensors, you can take a different route and use an analog-to-digital (ADC) module. The Grove 4 Channel 16-bit ADC (ADS1115) module (`www.seeedstudio.com/Grove-ADS1115-16-bit-ADC-p-4599.html`) allows you to connect up to four analog sensors connected via the onboard screw terminals. Figure 12-6 shows the Grove 4 Channel 16-bit ADC (ADS1115) module.

Figure 12-6. *Grove 4 Channel 16-bit ADC (ADS1115) module (courtesy of seeedstudio.com)*

Another limitation is the length of the Grove cables. Currently, the longest Grove cable from Seeed Studio is 50cm. If you need to use a longer cable, you can use two Grove Screw Terminal modules and a set of twisted pair wires (such as an Ethernet cable) to create your own longer cable. Figure 12-7 shows the Grove Screw Terminal module.

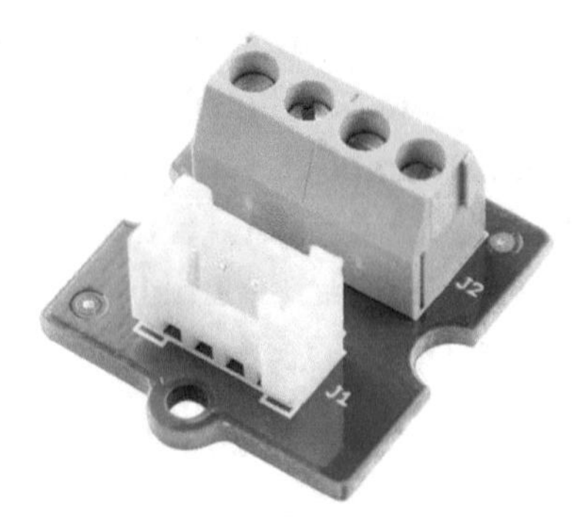

Figure 12-7. *Screw Terminal module (courtesy of seeedstudio.com)*

Tip See `www.seeedstudio.com/cables-c-949.html` for the list of Grove cables from Seed Studio.

Now that we know more about Grove, let's see what components (host adapters and modules) are available.

Components Available

There are a lot of components available for the Grove system. This section highlights some of the categories of modules available. We won't see everything that is available because the catalog is quite large. Since the product has been around for some time, there are several versions of some of the modules. Rather than attempt to view all of the latest modules, we will see the more popular host adapters and modules as well as those we will use in upcoming chapters. Figure 12-8 shows a snapshot of the top-level index from the Seeed Studio Grove online store. As you can see, there are a lot of categories!

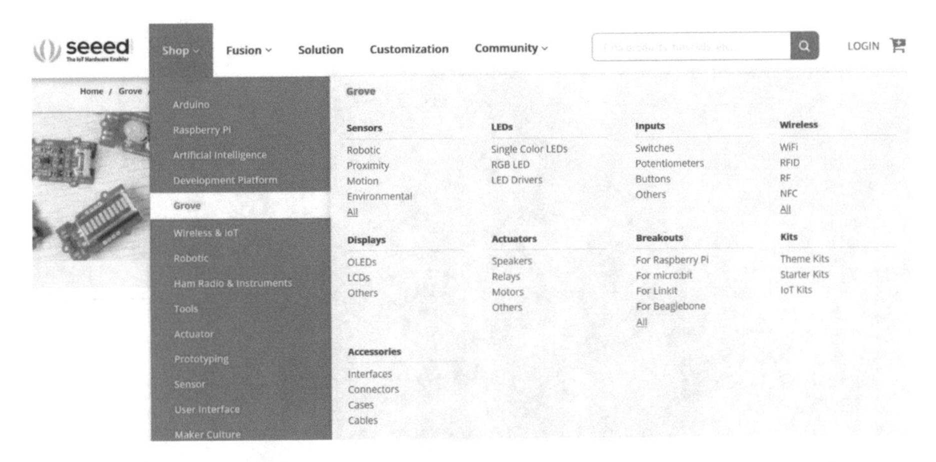

Figure 12-8. *Seeed Studio Grove online store index*

Note While you may encounter older versions of some Grove components, the older versions are still usable and can sometimes be found used for a discount.

Host Adapters

Aside from the impressive list of modules, the list of host adapters available from Seeed Studio is very impressive. Since we are working with Arduino and Raspberry Pi in this book, let's look at versions for these platforms.

There are six Grove shields available for the Arduino including the Nano, Uno, and Mega. The Grove Base Shield V2.0 shown in Figure 12-9 is the best one to use for the projects in this book assuming you will be using an Arduino with the Uno shield headers. Recall we saw this host adapter in an earlier section.

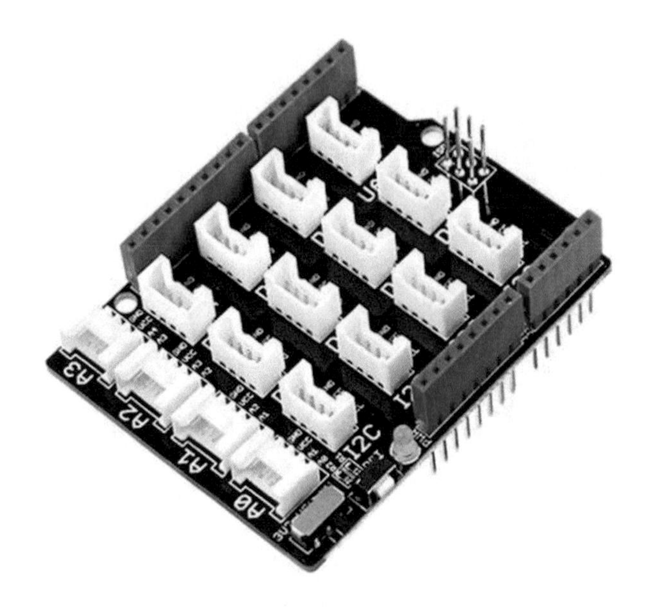

Figure 12-9. *Grove Base Shield V2.0 for Arduino (courtesy of seeedstudio.com)*

Notice the number of Grove connectors on this board. There are enough for most projects including all of the projects in this book. Notice also the shield has a set of headers that allow you to access the Arduino GPIO for more advanced projects.

Tip See `www.seeedstudio.com/breakouts-c-933.html?cat=939` for the complete list of Grove shields available for the Arduino and other platforms.

There are also two hats available for the Raspberry Pi: one for the Raspberry Pi Zero discussed earlier and the GrovePi+. The GrovePi+ is the host adapter (hat) we will use in this chapter and is also the same one we saw earlier. Figure 12-10 shows the GrovePi+ host adapter.

Figure 12-10. *GrovePi+ for Raspberry Pi (courtesy of seeedstudio.com)*

This host adapter is one that has an older version named the GrovePi. If you find the GrovePi, you can use that instead of the GrovePi+ for the projects in this book.

Modules

Seeed Studio offers a wide variety of modules that contain sensors, input, output, and display capabilities similar to those available for Qwiic. However, the categories and number of modules available are several times that of the other systems. So many that it is not possible to list them all here. Table 12-6 list the categories of modules available with a link to each category for further reading. You will find most have subcategories that you can explore to find more about the modules in the category. All URLs (links) begin with `www.seeedstudio.com/category/`.

Table 12-6. Categories of Grove Modules

Category	Description	Category Link
Sensors	Modules that allow you to sample the world around us	`Sensor-for-Grove-c-24.html`
LEDs	Modules that contain various forms of LEDs	`leds-c-891.html`
Input	Modules that contain devices that permit input of data or input actions like buttons	`Input-c-21.html`
Wireless	Modules that support wireless technologies	`wireless-c-899.html`
Displays	Modules with output devices	`displays-c-929.html`
Actuators	Modules with devices that produce movement, drive motors, or produce sound	`actuators-c-940.html`
Accessories	Grove accessories such as cables, headers, and more	`accessories-c-945.html`

So what are the modules available in these categories? We will use the same list of subcategories we used for the other component systems. You will find the Seed Studio website organized a bit different, but all of these subcategories are present:

- *Sensors*: Typically contain a single sensor that produces output (readings or values) on the I2C bus. Examples include temperature, humidity, pressure, distance, magnetometer, light, and environmental (gases) sensors.
- *Displays*: Modules that contain an output device for displaying data. Examples include OLED and LED displays.

- *Relays*: Modules that contain relays that permit you to switch higher-power devices on or off.
- *Motors*: Modules that permit you to control small electric motors.
- *Input*: Modules that contain one or more buttons, potentiometers, keypads, or switches.
- *ADC/DAC*: Modules that provide analog-to-digital conversion (ADC) or digital-to-analog conversion (DAC) that permit incorporation of other circuits into your project.
- *Accessories*: Various modules that provide handy operations such as data loggers, cryptographic operations, and more.

Now, let's look at a sample of the Grove modules we will be using in the upcoming chapters as we explore how to write the code for IoT projects using the Grove system beginning with an output device.

We will make use of several Grove LED modules. These modules contain one LED of a particular color. Figure 12-11 shows a Grove LED module. You can discover all of the Grove LED modules at `www.seeedstudio.com/single-color-leds-c-914.html`.

Figure 12-11. *Grove Red LED module (courtesy of seeedstudio.com)*

We will also use an LCD screen in some of the projects. The Grove OLED Display 0.96″ (`www.seeedstudio.com/Grove-OLED-Display-0-96-SSD1315-p-4294.html`) is a nifty, small OLED similar to the one we used in the Qwiic projects. Figure 12-12 shows the OLED display.

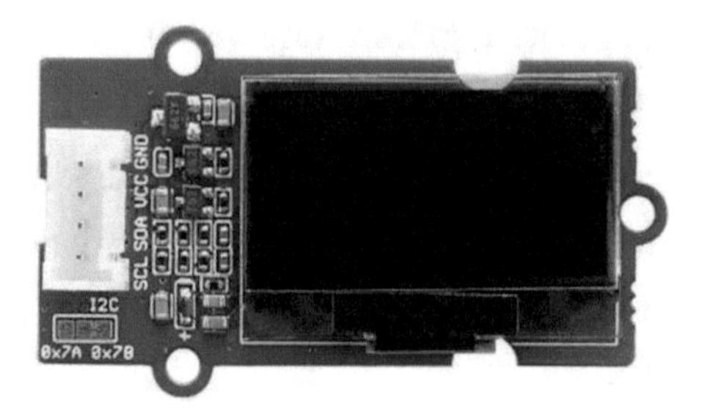

Figure 12-12. *Grove OLED Display 0.96″ (SSD1315) (courtesy of seeedstudio.com)*

The Grove Sound Sensor module (`www.seeedstudio.com/Grove-Sound-Sensor-Based-on-LM386-amplifier-Arduino-Compatible.html`) provides the ability to detect sound or noise as an analog value similar to a microphone. Figure 12-13 shows the module with the sensor facing side.

Figure 12-13. *Grove Sound Sensor (courtesy of seeedstudio.com)*

We will also use input devices such as the Grove Dual Button module that has two buttons mounted (`www.seeedstudio.com/Grove-Dual-Button-p-4529.html`). The module comes with a variety of colored caps permitting you to match the color of the cap to your project features. Figure 12-14 shows the module with caps on the buttons. You can also use them without the caps. Note that the Grove connector is located on the bottom of the board.

Figure 12-14. *Grove Dual Button module (courtesy of seeedstudio.com)*

The Grove AHT20 I2C Temperature and Humidity Sensor module can measure both temperature and humidity (`www.seeedstudio.com/Grove-AHT20-I2C-Industrial-grade-temperature-and-humidity-sensor-p-4497.html`). Figure 12-15 shows the module with the sensor facing side.

Figure 12-15. *Grove AHT20 I2C Temperature and Humidity Sensor (courtesy of seeedstudio.com)*

Notice this module has additional pins for advanced users. In this case, we see the I2C pins broken out on the right side of the module. Modules with these features typically come without headers mounted.

Once again, there are many modules available. These are just a sampling of the modules available from Seed Studio. See a compact list of all Grove devices and modules at `www.seeedstudio.com/category/Grove-c-1003.html`.

Cabling and Connectors

The Grove system includes cables of various lengths including 5, 20, 30, 40, and 50cm. Most modules produced by Seeed Studio include a 20cm cable or longer. The 5cm cables are great for projects that include modules mounted in close proximity or inside an enclosure.

There are also special cables available for a variety of uses such as a branch or "Y" cable that lets you connect two modules to a single source, a Grove-to-servo cable for using servos connected directly to a host, and even cables for connecting directly to your host board with a Grove connector on one side and the other side broken out with individual male or female pins.

See `www.seeedstudio.com/cables-c-949.html` for a list of cables and connectors to support the Grove system.

Developer Kits

Since the Grove system has been around a lot longer, Seeed Studio has had time to develop a number of developer kits that include a host adapter and a set of modules. There are over 50 kits available designed for a particular purpose (theme), starter kits, and even kits for IoT development. Let's look at two starter kits that you may find interesting and that are great for beginners.

If you use an Arduino, the Grove Beginner Kit for Arduino comes as a single PCB that has 10 sensors and supports 12 different projects (`www.seeedstudio.com/Grove-Beginner-Kit-for-Arduino-p-4549.html`). In the center is an Arduino Uno–compatible board, and all of the modules are connected through the PCB. You can either run the projects without breaking off the modules or, when you've run all of the projects, break them off so you can use them in other projects. Figure 12-16 shows the Grove Beginner Kit for Arduino. It comes in a sturdy reusable case and standoffs to make using the large board easier.

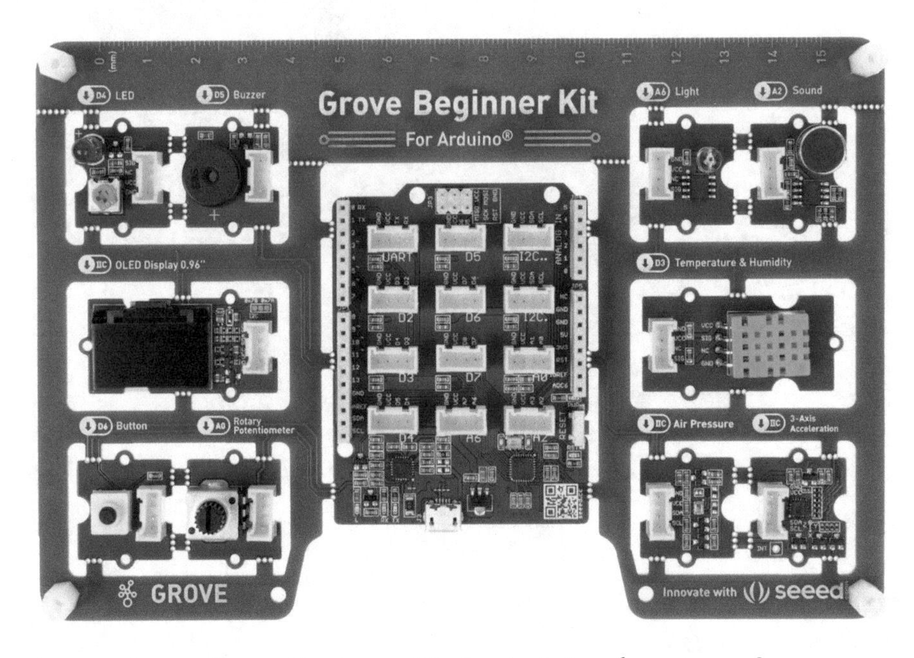

Figure 12-16. *Grove Beginner Kit for Arduino (courtesy of seeedstudio.com)*

Another great kit for those learning the Raspberry Pi is the GrovePi+ Starter Kit for Raspberry Pi (`www.seeedstudio.com/GrovePi-Starter-Kit-for-Raspberry-Pi-p-2240.html`). This kit comes with the GrovePi+ host adapter and ten modules including sensors, LED, button, LCD, and more. It is an excellent choice as a starting point in collecting Grove modules. Figure 12-17 shows the Raspberry Pi starter kit.

Figure 12-17. *GrovePi+ Starter Kit for Raspberry Pi (courtesy of seeedstudio.com)*

Tip See `www.seeedstudio.com/category/Kits-for-Grove-c-28.html` for all of the Grove kits available from Seeed Studio.

Where to Buy Grove Components

You can purchase Grove components directly from Seeed Studio (seeedstudio.com), who are based in China. They often ship products quickly, but shipping may take longer than expected. Fortunately, you can often find Grove modules on popular online retail sites such as amazon.com and online auction sites. In fact, I have seen select starter kits in brick-and-mortar stores that sell electronic components. If you live in the United States, check out the online retail stores first or buy in bulk to save on shipping from Seeed Studio.

Now, let's discuss how to use these systems in your projects.

Using the Components in your Projects

Plugging your choice of Grove host adapter onto your host board and plugging the modules together with the cables is pretty easy. Recall the connectors only go one way so you can't cross-connect anything.

However, like the Qwiic and STEMMA QT modules, Grove modules are not designed to be hot pluggable. You should not connect and disconnect modules while your board is powered on. This could lead to damaging the module(s) or your host board.

Caution Do not plug or unplug Grove modules while your board is powered on.

Once the hardware is plugged together, the next step is to start working on the code to enable your modules and complete your project. To do so, you are likely required to load one or more software libraries.

Like the vast array of modules, the software libraries required for the Grove modules vary and depend on the module itself. Fortunately, Seeed Studio is very good about providing samples for use of each of their hundreds of modules.

The following summarizes the steps necessary for the Arduino and Raspberry Pi. The following does not include all of the steps needed for all of the projects in the book; rather, the section is an overview of what you can expect to configure your PC to implement the projects. Specific details for each example are included in each chapter.

Fortunately, most Grove modules have examples on how to use them that include, at a minimum, sample code for the Arduino. For example, there is a Wiki page for the OLED Display 0.96" module that shows you how to get started using it (`https://wiki.seeedstudio.com/Grove-OLED_Display_0.96inch/`).

Loading Grove Libraries for the Arduino

Recall from Chapter 2 we can install software libraries using the Library Manager in the Arduino IDE. Simply open the Library Manager and search for Grove or the name of the module.

For example, in a later chapter, we will use the OLED Display 0.96" module. According to the hookup guide, all we need to do is search the Library Manager for "Grove OLED" (no quotes). Figure 12-18 shows how that would appear in the Arduino IDE.

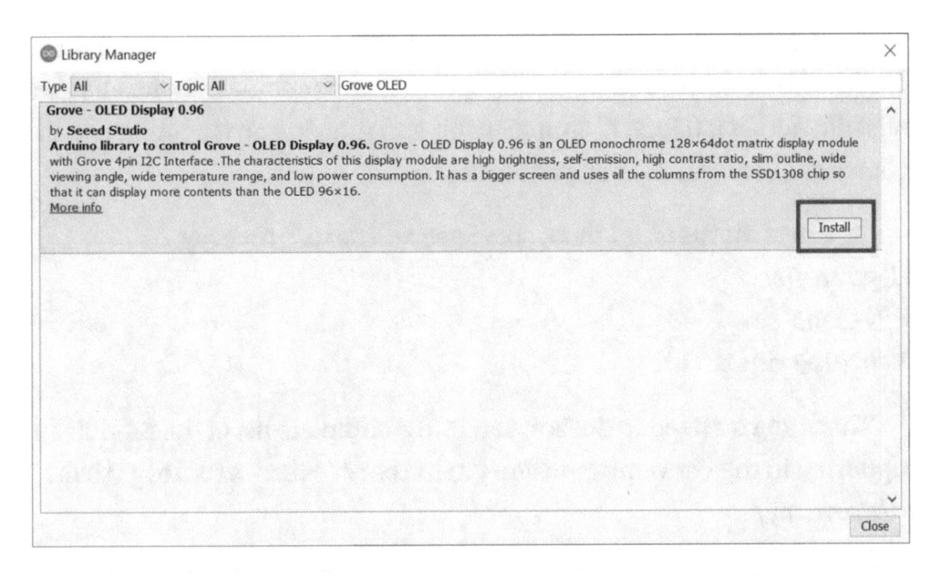

Figure 12-18. *Searching the Library Manager for Grove libraries*

Seeed Studio sometimes uses third-party libraries of which you can also download the source code from GitHub. For example, the OLED library can be found at `https://github.com/olikraus/u8g2`, which happens to be a more generic library than the one in the Arduino Library Manager. It is best to try and use the libraries from the Library Manager first and seek the source when you need to modify the code or just want to see how it works.

If you do decide to download and use the library from GitHub (or elsewhere), follow the instructions in Chapter 6 to make it available in your Arduino IDE.

Loading Grove Libraries for the Raspberry Pi

Software libraries on the Raspberry Pi are a little different. Recall Python libraries are installed differently. We would install them either using the `pip` command or, in some rare cases, by downloading the library and copying it to our project folder.

For example, Seeed Studio has created a Python library that contains all of the libraries needed for their popular Grove modules. To install it, we must download it (clone it) from GitHub and then install it as shown in the following. Once you do that, you're all set:

```
$ git clone https://github.com/Seeed-Studio/grove.py
cd grove.py
$ Python3
sudo pip3 install .
```

That's all you need to do! You can find a complete list of the modules supported in the Grove.py repository at `https://seeed-studio.github.io/grove.py/`.

Tip Seeed Studio has kindly provided a getting started page for the Raspberry Pi. To learn more about how to set up your environment to work with Grove modules in Python, see the Wiki page for the GrovePi hat for Raspberry Pi at `https://wiki.seeedstudio.com/Grove_Base_Hat_for_Raspberry_Pi/#installation`.

If you encounter a Grove module where there isn't a Python library, do not despair. Again, most Grove modules have a Wiki page that will show you how to get started. But if there isn't a specific Python library, you most likely can find a similar one from the Internet that you can use. All it takes is a bit of exploring, and you can find Python libraries for what you need.

Summary

Like the Qwiic and STEMMA QT component systems, Grove provides a simple, no-error connector that you can use to connect a variety of components together using several protocols – all from the same board.

Now that the hardware challenges have been nearly eliminated, we can turn our attention back to learning how to write the code for our projects. As you saw in this chapter, this may require installing software libraries to support the modules you are using or adapting existing libraries to suit your needs.

The next chapter begins a series of projects that use Grove components to teach you how to work with the system for both the Arduino and Raspberry Pi. As you will see, except for the hardware itself, the pattern of building the projects is the same as the previous project chapters.

CHAPTER 13

Example: Knock-Knock!

Thus far in the book, we've discovered how to write code for Arduino and Python (on the Raspberry Pi) for projects that use the Qwiic and STEMMA QT modules. We have also discovered a third component system named Grove. While the Grove system is also a rapid prototyping system, it supports more than a single protocol (interface), which makes wiring a bit more complicated only so far as you must know which protocols your modules require and select the correct connector on the host adapter.

In this chapter, we will keep the project simple with respect to the hardware so that we can get some practice working with Grove modules. In fact, we will use two LEDs, a button, and a sound sensor module to build a secret knock detector. We will only focus on the knock detection, but you will see where in the code you could add an actuator such as a solenoid-powered lock or similar security device to restrict access to something.

Like the previous project chapters, we will see how to implement this project on the Arduino and Raspberry Pi but with an interesting twist. Let's get started.

C. Bell, *Beginning IoT Projects*, https://doi.org/10.1007/978-1-4842-7234-3_13

Project Overview

The project for this chapter is designed to demonstrate how to use analog and digital Grove modules. We will use them to build a secret knock detector that permits you to record a secret knock and requires users enter the correct knock that matches. We won't be recording sound levels with the sound sensor (but we could); rather, we will keep it interesting and use the frequency or timing of the knocks. After all, the best secret knocks are melodic in nature.

To enter the secret knock, the user presses a button, and we check the knock against the recorded secret knock. If the timing of the knocks plus/minus a percentage to allow some variance (such as knocking a bit faster or slower but the same cadence) matches, the knock is accepted and we illuminate a green LED, or if the knock is rejected, we illuminate a red LED. We will also allow a long press on the button to enter a recording mode so that you can store a new secret knock where will illuminate both LEDs.

Once again, we won't complete the project with a locking mechanism because they are rather expensive, but the knock detection and recording are by far the more complex components of any secret knock application. Fortunately, incorporating a locking mechanism is no more difficult than determining the triggering of the lock.

For example, some mechanical locks can be activated by using a servo, which can be activated using the Grove Servo module (`https://www.seeedstudio.com/Grove-Servo.html`). You would write your code to move the servo so that it turns the lock into a closed or open position.

Or you could use a solenoid lock that requires a 12V signal to trigger, which can be accomplished using a relay and the `analogWrite()` method in Arduino to trigger the relay, which causes the lock to engage or disengage.

Tip See `https://create.arduino.cc/projecthub/projects/tags/lock` for several examples of locks you can use in Arduino sketches. See `https://makezine.com/projects/remote-camera-doorbell-and-smart-lock-with-raspberry-pi/` for an intriguing example of a Raspberry Pi locking mechanism.

What Will We Learn?

By implementing this project, we will learn how to connect Grove modules to our host boards and how to connect Grove modules to the various protocol connectors on the host adapter. We will also pick up a few tips on working with the hardware along the way. Thus, the project itself is very simple and is not likely to impress, but it is well suited for learning all of the nuances of building Grove projects that use analog and digital modules.

The programming tasks will reveal how to read values from the sound sensor to detect a knock sequence and how to interpret the value to make a decision on the validity of the knock sequence. We will also learn how to use LED modules to communicate the project operations.

Let's see what hardware we will need.

Hardware Required

The hardware needed for this project is listed in Table 13-1. URLs for each component are included for ease of ordering including duplicate entries for alternative vendors.

Table 13-1. *Hardware Needed for the Secret Knock Project*

Component	URL	Qty	Cost
Grove Sound Sensor	www.seeedstudio.com/Grove-Sound-Sensor-Based-on-LM386-amplifier-Arduino-Compatible.html	1	$4.90
Grove Red LED	www.seeedstudio.com/Grove-Red-LED.html	1	$1.90
Grove Green LED	www.sparkfun.com/products/14532	1	$1.90
Grove Button	www.seeedstudio.com/buttons-c-928/Grove-Button.html	1	$1.90
Grove cables(any length can be used)	Included with each preceding module	5	
Arduino MKR 1010 WiFi	www.sparkfun.com/products/15251	1	$35.95
Raspberry Pi 3B or later	www.sparkfun.com/categories/233 www.adafruit.com/category/176	1	$35.00+
Grove Base Shield V2.0 for Arduino	www.seeedstudio.com/Base-Shield-V2.html	1	$4.45
GrovePi+	www.sparkfun.com/products/15945	1	$5.95

Note If you use the MKR series Arduino, you will need to purchase the MKR2UNO adapter (`https://store.arduino.cc/usa/mkr2uno-adapter`) in order to use the Grove shield. You will also need to purchase a set of stacking headers (`https://www.sparkfun.com/products/10007` or `https://www.adafruit.com/product/85`) for the Arduino to allow for the Grove shield to sit high enough to clear the MKR series components.

About the Hardware

Let's discuss these components briefly. We will discover how to work with the hardware in more detail later in the chapter. Once again, some of the hardware is used in only one version.

Sound Sensor

The Grove Sound Sensor is an analog module that incorporates a microphone and a small amplifier. It can be used to detect sound in the area and even the intensity of the sound. We will use both features in this project. Figure 13-1 shows the Grove secret knock sensor.

Figure 13-1. *Grove Sound Sensor (courtesy of seeedstudio.com)*

Grove LED

The two LED modules we will use are single LEDs mounted on a board with a small circuit to ensure they are powered correctly. If you've built your own electronic circuits with LEDs in the past, you will recall each LED requires a specific resistor that matches the LED's power requirements. These modules eliminate those concerns and also use a digital interface, which makes it easy to turn them on and off. Figure 13-2 shows the Grove Red LED (all single LED modules look similar but include a different color LED).

Figure 13-2. *Grove Red LED module (courtesy of seeedstudio.com)*

One interesting facet of the Grove single LED modules is that they sometimes ship without the LED installed, so you may need to plug it into the socket. Fortunately, they come with a small card that explains how to do so (the LEDs must be oriented correctly). In short, the longest leg on the LED is positive, so it should be inserted so that the long leg goes in the socket next to the + symbol on the board.

The boards also have a small potentiometer on them that allows more advanced uses such as changing the color of the LED and tuning the board to match the new LED's power requirements. Cool!

Grove Button

There are several button modules with the Grove connector mounted either vertically or horizontally and some that have more than one button. For this project, we need only a single-button module. Figure 13-3 shows the Grove Button module.

Figure 13-3. *Grove Button module (courtesy of seeedstudio.com)*

Grove Base Shield V2.0 for the Arduino

The base shield is the host adapter we will use for our Arduino version of the project. It is a full-sized Uno-compatible shield that has all of the Grove connectors you will need for this and other projects. Figure 13-4 shows the Grove Base Shield V2.0 for Arduino.

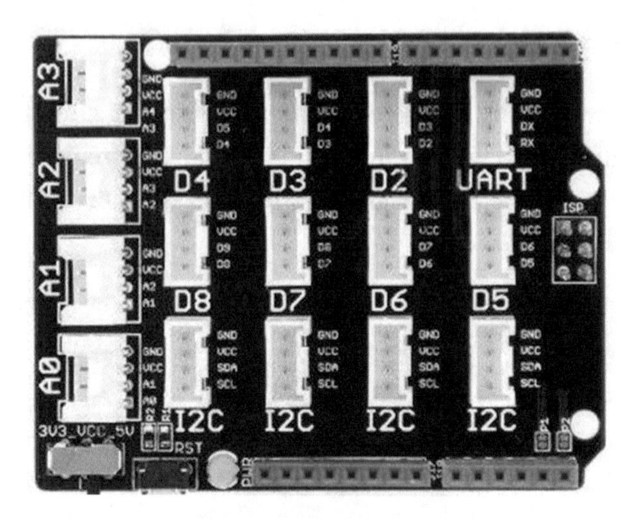

***Figure 13-4.** Grove Base Shield V2.0 for Arduino (courtesy of seeedstudio.com)*

GrovePi+ for the Raspberry Pi

The GrovePi+ (or the older GrovePi) is the host adapter we will use for our Python version of the project. It is a full-sized Raspberry Pi hat that has all of the Grove connectors you will need for this and other projects.

Note You may encounter older Grove host adapters, namely, the Grove Base Hat. This board is not compatible with the latest software libraries and should not be used.

Interestingly, this board is manufactured by Dexter Industries (`www.dexterindustries.com`), and the software we will install to use it is provided by them. The documentation for the GrovePi on the Seeed Studio wiki refers to the Dexter Industries instructions. Thus, we will be using the Dexter Industries instructions (`https://www.dexterindustries.com/grovepi/`) to configure our board a bit later in the chapter.

Figure 13-5 shows the GrovePi+.

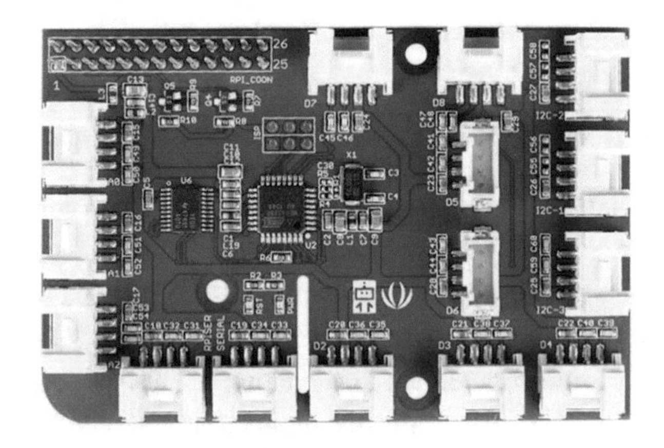

Figure 13-5. *GrovePi+ for the Raspberry Pi (courtesy of seeedstudio.com)*

You may notice that the GPIO header on the bottom of the GrovePi has 26 pins and is smaller than the GPIO header on the Raspberry Pi. This is by design, and it will still work with the newer Raspberry Pi boards. To connect the GrovePi to your Raspberry Pi, insert the GrovePi female header into the leftmost portion of the Raspberry Pi GPIO header as shown in Figure 13-6.

Figure 13-6. *Connecting the GrovePi to the Raspberry Pi GPIO*

Notice the arrows that indicate the portion of the Raspberry Pi GPIO you will need to use to connect the GrovePi. Be sure to double-check your connections before powering on your Raspberry Pi.

Connect the Grove Modules

Recall from Chapter 12 we can use a single Grove cable to connect each Grove module separately to a specific connector on the host adapter on our host board. The host adapter for the Raspberry Pi has a different layout but has the same connectors we will use. Table 13-2 includes the details of each connection on the host adapter to help you make the right connections. Simply use the table to connect a Grove cable from the module to the Grove connector on the host adapter as marked in the table.

Table 13-2. *Grove Connections*

Module	Protocol	Grove Connector on the Host Adapter
Button	Digital	D6
Red LED	Digital	D5
Green LED	Digital	D4
Sound Sensor	Analog	A2

Figure 13-7 shows an example of how you should connect your modules for this project. Notice the figure shows the Arduino host adapter.

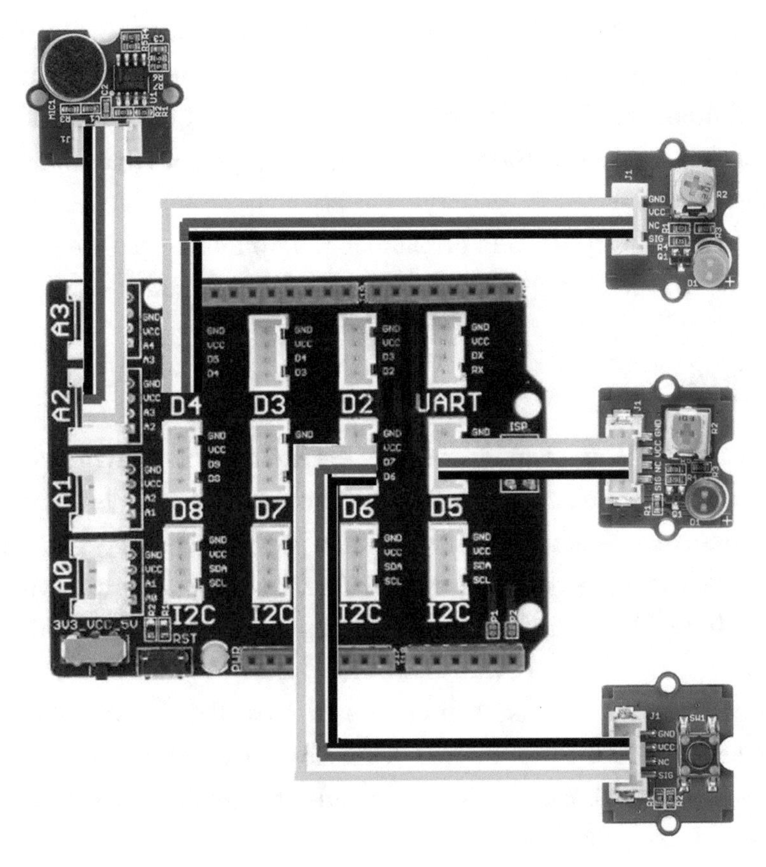

Figure 13-7. *Secret knock project Grove connections*

Now that we know more about the hardware for this chapter, let's write the code!

Write the Code

The code for this project follows what should be a familiar pattern by now if you've worked the previous project chapters. Specifically, we will create a class to contain the code for the sound sensor, and our main sketch/script will contain the code that makes it all work (the user interface).

We will be using analog and digital modules rather than I2C modules, so there won't be a need to install any software libraries, but there are some differences in how you write your code for using the modules between Arduino and Python. We'll cover those specifics in each language-specific section.

Let's walk through how to prepare our computers to use the sensor and write code to read its values. We'll start with the Arduino.

Arduino

This section presents a walk-through of the sketch you will write to read values from the sound sensor and detect the secret knock. We will skip the software installation section we used in previous chapters because the modules do not require any special libraries. As you will see, we will use the built-in methods from the Arduino platform to interact with the modules.

Since this is not our first Arduino sketch, we will discuss the code at a high level and skip the line-by-line details focusing on the mechanics of how the code works. You can study the code at your leisure to ensure you understand the sketch in more detail.

Write the Sketch

Recall we are going to use a class to read the sound sensor, which will be in the form of a separate file. Rather than write the main sketch file first and then add the class, we will write the class header first and then the main sketch and finally complete the class code. This is typically how programmers develop code with class modules (but not always). By creating the header for the class first, we can understand how to use the class making writing the main sketch easier.

The class will be named KnockSensor and stored in two files: a header file named KnockSensor.h and a source file named KnockSensor.cpp. Recall we must first create a blank sketch and then add the empty files. Open a new sketch and name it secret_knock.ino or whatever you'd like to use. Save the file, then close the project in the Arduino IDE, and then create the class files. For example, use a terminal to navigate to the folder and issue these commands to create the empty files:

```
secret_knock % touch KnockSensor.h
secret_knock % touch KnockSensor.cpp
```

Now, let's see the code for each file starting with the header file.

Class Header File

Click the tab named KnockSensor.h to open the blank file. Here, we will add the header or blueprint for the class. Recall the header file simply defines the class. We will use the KnockSensor.cpp file to add the code for the functions in the class. Let's discuss those first.

We will write a new class that encapsulates the behavior of our knock sensor. Specifically, we will need to read a secret knock from the user and validate the knock against a stored sequence, which is stored in memory, and we will need a way to store (record) a new secret knock.

Aside from those functions, we will also create a constructor so we can pass in some tuning parameters. We want to allow the caller to change the analog pin for the sound sensor and a percent value we can use to compare the knock read to the stored secret knock. Let's see how we can write the header file for the class.

Let's begin with one constant that we can use to define the maximum number of knocks and a delay value we can use to ensure we read different knocks, more specifically a length of time we wait before reading the next knock. We use constants like this so that we can tune the code by changing the value in only one place:

```
#define MAX_KNOCKS 10
#define KNOCK_DELAY 200
```

Next, we will define the class to include the three methods plus the constructor as follows. We make them public since we will be calling them from our main sketch:

```
class KnockSensor {
public:
  KnockSensor(int analogPin=A2, float sensitivity=0.25) {
    soundSensor = analogPin;
    rejectPercent = sensitivity;
  }
  bool readSecret();
  void recordSecret();
  boolean validateSecret();
...
}
```

Notice the constructor. Here, we define two parameters: one for the analog pin that we want to use to connect the sound sensor and a sensitivity parameter we can use to make the secret knock comparison more (lower value) or less (higher value) sensitive. We set the default to A2 since that is the connector shown in the connection diagram. Since the code for the constructor is only two lines, we define them in the header file, but constructors with more than a couple of lines of code should be implemented in the code file.

Next, we will need a few variables for use in reading and comparing the knock sequences. These include variables we will use to configure, tune, and store data. All of these will be placed in the private section.

The configuration section contains two variables: one for the sensitivity and another to store the pin to use for the sound sensor. Recall we accept both of these via the constructor:

```
// Configuration variables
float rejectPercent = 0.25;  // Percent timing difference
int soundSensor = A2;        // Sound Sensor
```

The tuning section contains two more variables for setting the sound threshold, which is the lowest value read from the sound sensor to establish a knock, and the maximum delay to wait for a knock. The timeout variable is how we detect the end of the knock sequence. You may need to experiment to determine the threshold you want to use for the sound detection. Higher means you will have to knock louder/harder:

```
// Tuning variables
int soundThreshold = 300;    // Minimum value to detect a knock
int knockTimeout = 2000;     // Wait timeout for knock
                                detection
```

The storage section contains variables to store the knock sequence read and the known secret knock sequence. We'll initialize the sequence to a classic (perhaps overused) tune:

```
// Secret Knock Storage: "Shave and a haircut - two bits"
int secretKnock[MAX_KNOCKS] = {100,53,20,21,44,87,54,0,0,0};
int numSecretKnocks = 7;
int valuesRead[MAX_KNOCKS];  // Values read from sound sensor
```

Let's look at the completed code for the header file. Listing 13-1 shows the file.

Listing 13-1. Grove Secret Knock Header File

```
#include <Arduino.h>

#define MAX_KNOCKS 10
#define KNOCK_DELAY 200

class KnockSensor {
public:
  KnockSensor(int analogPin=A2, float sensitivity=0.25) {
    soundSensor = analogPin;
    rejectPercent = sensitivity;
  }
  bool readSecret();
  void recordSecret();
  boolean validateSecret();

private:
  // Configuration variables
  float rejectPercent = 0.25;  // Percent timing difference
  int soundSensor = A2;        // Sound Sensor

  // Tuning variables
  int soundThreshold = 300;    // Minimum value to detect a
                                  knock
  int knockTimeout = 2000;     // Wait timeout for knock
                                  detection

  // Secret Knock Storage: "Shave and a haircut - two bits"
  int secretKnock[MAX_KNOCKS] = {100,53,20,21,44,87,54,0,0,0};
  int numSecretKnocks = 7;
  int valuesRead[MAX_KNOCKS];  // Values read from sound sensor
};
```

OK, let's return to the main sketch to see how we can use this class.

Main Sketch

Now click the secret_knock.ino tab to return to the main sketch. Let's begin with the preamble or top of the file. Recall here is where we include libraries we need, declare variables and constants, etc. Listing 13-2 shows the code for the main sketch preamble. As you can see, we only need to include our new class header.

Listing 13-2. Main Sketch Preamble

```
#include "KnockSensor.h"

// Global Variables
KnockSensor *knockSensor;

// Constants
#define TRIGGER_PROGRAM_SECONDS 4.0
#define BUTTON_PIN 6
#define RED_LED 5
#define GREEN_LED 4
#define PROGRAM_PRESS 2
#define KNOCK_PRESS 1
#define NO_PRESS 0
```

Notice we define a global variable for our knock sensor class. We also define a number of constants for the LEDs and button and several constants we will use to determine if the button is pressed, if it was held for at least 4 seconds (using the TRIGGER_PROGRAM_SECONDS constant), or if it isn't pressed. As you've guessed, we will write a function to return these values.

Next, we will code the setup() function, which includes the sort of initialization code we've seen in other projects. Specifically, we will set up the LEDs using the pinMode() Arduino function as output pins and the button as an input pin. We also set up the serial monitor so we can use diagnostic statements. Finally, we initialize our new knock sensor class. Listing 13-3 shows the complete code for the `setup()` function. Read through it to ensure you understand all of the code included.

Listing 13-3. Main Sketch setup()

```
void setup() {
  pinMode(BUTTON_PIN, INPUT);
  pinMode(RED_LED, OUTPUT);
  pinMode(GREEN_LED, OUTPUT);
  Serial.begin(115200);
  while(!Serial);
  knockSensor = new KnockSensor(A2, 0.33);
  Serial.println("Welcome to the Secret Knock program.");
}
```

Finally, we have the `loop()` function. At the highest level, we want to loop until the button is pressed. If it is a normal press (momentary), we use the knock sensor class to read the knock sequence from the user. If it is a long press, we record a new secret knock. That's it!

Listing 13-4 shows the completed code for the `loop()` function. Go ahead and read through the code to ensure you understand how it works. There should not be any surprises or new techniques.

Listing 13-4. Main Sketch loop()

```
void loop() {
  int buttonStatus = readButtonStatus();
  // If this is a program trigger, turn on both LEDs
  if (buttonStatus == PROGRAM_PRESS) {
    digitalWrite(RED_LED, HIGH);
    digitalWrite(GREEN_LED, HIGH);
    Serial.println("Program Mode. Enter new secret knock.");
    delay(100);
```

```
    if (knockSensor->readSecret()) {
      knockSensor->recordSecret();
    }
    digitalWrite(RED_LED, LOW);
    digitalWrite(GREEN_LED, LOW);
  } else if (buttonStatus == KNOCK_PRESS) {
    Serial.println("What's the secret knock?");
    knockSensor->readSecret();
    // If knock accepted, turn on green LED, or red LED if
       failure
    if (knockSensor->validateSecret()) {
      Serial.println("Secret knock accepted.");
      digitalWrite(GREEN_LED, HIGH);
      delay(3000);
      digitalWrite(GREEN_LED, LOW);
    } else {
      Serial.println("ERROR: Secret knock rejected. Go away!");
      digitalWrite(RED_LED, HIGH);
      delay(3000);
      digitalWrite(RED_LED, LOW);
    }
  }
  delay(100);
}
```

Notice how we use the LEDs to signal the user for a successful knock sequence detection, a rejection, or storing a new knock sensor. Cool, eh?

Now, let's look at the function we will use to read the button. We move this code to a new function not because it is complicated, but because it makes the code easier to read. Listing 13-5 shows the code for the `readButtonStatus()` function. There isn't any complicated code here, but be sure to read through it to see how we return one of three values: button not pressed, button pressed, and program press (long press). The only tricky part is measuring the time expired for the long press.

Listing 13-5. Main Sketch readButtonStatus()

```
int readButtonStatus() {
  int ticsPressed = 0;

  if (digitalRead(BUTTON_PIN) == HIGH) {
    while (digitalRead(BUTTON_PIN)) {
      ticsPressed++;
      delay(100);
    }
    if ((ticsPressed/10.0) >= TRIGGER_PROGRAM_SECONDS) {
      return PROGRAM_PRESS;
    }
    return KNOCK_PRESS;
  }
  return NO_PRESS;
}
```

Now we can write the final portion of our project – the code for the class.

Class Code File

Click the tab named `KnockSensor.cpp` to open the blank file. Here, we will add the code for the class. There are only three functions to write. Diagnostic messages are printed to the serial monitor in each of the functions. You may want to add more messages if you encounter problems.

The `readSecret()` function implements a loop to read values from the sound sensor. It loops until an error occurs or time expires on the knock timeout (`knockTimeout`). It ignores any sounds (knocks) with a value less than the minimum knock threshold (soundThreshold). Finally, it returns a boolean value where true means a successful read and false indicates an error occurred.

The recordSecret() function writes the last read knock sequence to the internal variables in the class. To use this method, you must call readSecret() first. We will use a special function named map() to change the values read ranging from 0 to the largest value read to a range of 0–100. In other words, we will "normalize" the values to a scale of 0:100. This will allow us to store a secret knock and compare it to a knock sequence that may vary slightly in timing (faster, slower) but still have the same cadence (relative timing between knocks). We will use the map() function again in the validateSecret() function when we do the compare.

The validateSecret() function implements an algorithm to compare the last entered knock sequence (so you must call readSecret() first) to the stored secret knock. The function returns a boolean value where true means a successful comparison and false indicates the knock sequence doesn't match the stored secret. We do a simple check first to ensure the number of knocks entered matches the number of knocks in the stored secret knock. If it doesn't, we reject and stop the compare (return false).

However, the code isn't as simple as that because we also incorporate a variable to check the sensitivity (rejectPercent) that helps to match the knock timings within a percentage threshold. For example, if the sensitivity is set to 25%, a knock timing is considered a match if the value is +/- 25% of the stored secret. So, if the secret timing for a given knock is 50, acceptable values with a 25% sensitivity range from 37.5 to 62.5. Recall these values are the millisecond measurement since the last knock.

Listing 13-6 shows the completed code for the class (documentation omitted for brevity). Since this is complicated code than we're used to seeing, be sure to study it to understand how it works. Spend some time in the validation function so you can understand how we are doing the compare. Notice how the map function is used for normalization of the timing differences.

Listing 13-6. Grove Secret Knock Code File

```
#include "KnockSensor.h"

bool KnockSensor::readSecret() {
  int startTime = millis();
  int now = millis();
  int value = 0;
  int num = 0;

  // Clear the read values array
  for (int i = 0; i < MAX_KNOCKS; i++) {
    valuesRead[i] = 0;
  }
  // Wait for knocks that exceed threshold
  do {
    value = analogRead(soundSensor);
    if (value >= soundThreshold) {
      now = millis();
      valuesRead[num] = now - startTime;
      startTime = now;
      num++;
      delay(KNOCK_DELAY);
      Serial.println("Knock heard.");
    }
    now = millis();
  } while (((now - startTime) < knockTimeout) && (num < MAX_
  KNOCKS));
  // Check for errors
  if (num == 0) {
    Serial.println("ERROR: no knocks detected.");
    return false;
  }
```

```
  if (num >= MAX_KNOCKS) {
    Serial.println("ERROR: maximum number of knocks exceeded.");
    return false;
  }
  return true;
}

void KnockSensor::recordSecret() {
  int maxKnockInterval = 0;
  // Normalize the knock timings
  numSecretKnocks = 0;
  for (int i = 0; i < MAX_KNOCKS; i++) {
    // Collect normalization data while we're looping.
    if (valuesRead[i] > maxKnockInterval) {
      maxKnockInterval = valuesRead[i];
    }
    secretKnock[i] = map(valuesRead[i], 0, maxKnockInterval,
    0, 100);
    // Save number of secret knocks
    if (secretKnock[i] > 0) {
      numSecretKnocks++;
    }
  }
  Serial.println("Secret knock saved.");
}

boolean KnockSensor::validateSecret() {
  boolean knockMatches = true;
  int num = 0;
  int knockInterval = 0;      // We use this later to normalize
                                 the times.
  float timeDiff = 0.0;
```

```
// Do the simple check first - do the number of knocks match?
// Loop through the number of knocks counting the knocks in
   the secret
// and in the values read.
for (int i = 0; i < MAX_KNOCKS; i++) {
  if (valuesRead[i] > 0) {
    num++;
  }
  // Capture max knock interval from values read for
     normalization
  if (valuesRead[i] > knockInterval) {
    knockInterval = valuesRead[i];
  }
}
if (num != numSecretKnocks) {
  return false;
}

// Diagnostic Messages:
Serial.print("Diagnostics - Reject% = ");
Serial.println(rejectPercent);
Serial.println("Secret\tRead\tDiff");
Serial.println("----\t----\t----");
for (int i = 0; i < MAX_KNOCKS; i++) {
  timeDiff = 0.0;
  // Normalize the values to remove timing issues.
  valuesRead[i] = map(valuesRead[i], 0, knockInterval, 0, 100);
  // Check for differences and compensate with error percentage.
  if (valuesRead[i] != secretKnock[i]) {
    if (secretKnock[i] > valuesRead[i]) {
      timeDiff = 1.0 - (float(valuesRead[i]) /
      float(secretKnock[i]));
```

```
      } else {
        timeDiff = 1.0 - (float(secretKnock[i]) /
        float(valuesRead[i]));
      }
      if (timeDiff > rejectPercent) { // Value too far out of
                                         range
        knockMatches = false;
      }
    }
    Serial.print(secretKnock[i]);
    Serial.print("\t");
    Serial.print(valuesRead[i]);
    Serial.print("\t");
    Serial.println(timeDiff);
  }
  return knockMatches;
}
```

As you can see, while the code to read from the sound sensor is simple, the code becomes more complicated when we attempt to compare a knock sequence read to the secret stored. That's why we use the sensitivity setting. When testing the project for the first time, you may want to set the sensitivity higher to 40% or 50% until you're confident you can reproduce the default knock sequence (you get a green light).

Compile the Sketch

The last step is to compile the sketch before uploading it to your board. If you encounter any errors, be sure to fix them and recompile to ensure the sketch compiles without errors or serious warnings.

Once everything compiles, we're ready to start testing. But first, let's look at the code for the Raspberry Pi. You can skip to the "Sketch on the Arduino" section if you're curious to see how the project works. While the code will execute the same on both platforms, the values differ due to the differences in how the sensors are read (the range of values differs).

Raspberry Pi

This section presents a walk-through of the Python code you will write to create a Python version of our secret knock project. As you will see, it is a bit different in places because of the GrovePi libraries. But first, we must configure and install software for the GrovePi or GrovePi+ host adapter on our Raspberry Pi. Fortunately, there is a single command we can use to download the software we need and install it. You only need to run this command once.

Note If you are using a Raspberry Pi 4B or 400, you must also make a minor change to your operating system to enable the GrovePi and GrovePi+.

Let's start with configuring the GrovePi host adapter.

Install Software Libraries

The GrovePi and GrovePi+ boards require a number of Python libraries to work. Fortunately, you can install them all at once with a single command as follows. Note that you must be connected to the Internet to use this command:

```
$ curl -kL dexterindustries.com/update_grovepi | bash
```

This command will download everything you need and install the Python libraries. It takes a while to run, and it is recommended that you restart your Raspberry Pi after the process completes. Listing 13-7 shows an excerpt of the command executing.

Listing 13-7. Installing the GrovePi and GrovePi+ Software

```
$ curl -kL dexterindustries.com/update_grovepi | bash
  % Total    % Received % Xferd  Average Speed Time Time Time Current
  Dload  Upload   Total   Spent    Left  Speed
100   251  100   251  0  0  143  0  0:00:01  0:00:01 --:--:--  143
  0  0     0    0     0  0  0   0 --:--:--  0:00:02 --:--:--  0
100 11903  100 11903  0  0  3325  0  0:00:03  0:00:03 --:--:-- 16812

  _____            _
 |  __ \          | |
 | |  | | _____  _| |_ ___ _ __
 | |  | |/ _ \ \/ / __/ _ \ '__|
 | |__| |  __/>  <| ||  __/ |
 |_____/ \___/_/\_\\__\___|_|                 _
 |_   _|         | |         | |     (_)
   | |  _ __   __| |_   _ ___| |_ _ __ _  ___  ___
   | | | '_ \ / _` | | | / __| __| '__| |/ _ \/ __|
  _| |_| | | | (_| | |_| \__ \ |_| |  | |  __/\__ \
 |_____|_| |_|\__,_|\__,_|___/\__|_|  |_|\___||___/
```

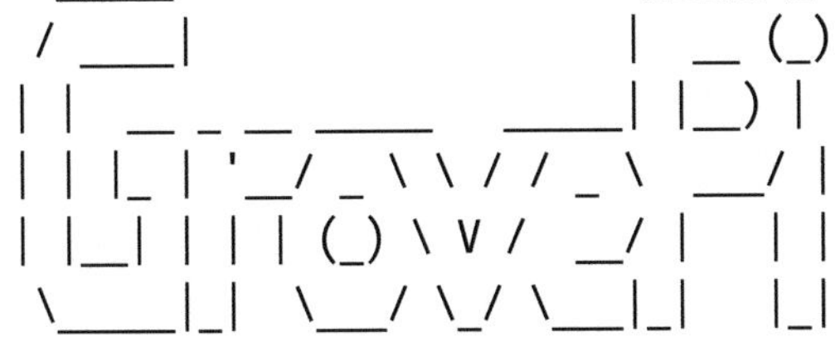

```
Welcome to GrovePi Installer.
...
```

The software installs all of the libraries available as well as documentation. To see the documentation, use your browser and open `file:///home/pi/Dexter/GrovePi/docs/index.html`.

That's it. We're done and ready to use our GrovePi. That is, unless you're using a Raspberry Pi 4B or 400.

Configure the Raspberry Pi 4 or 400

If you are using a Raspberry Pi 3B or older board, you do not need to make the following changes. This applies only to the Raspberry Pi 4B and 400. The software libraries provided by Dexter Industries do not work well with the newer Raspberry Pi boards. The problem is in the timing or speed of the processor. Fortunately, we can slow our Raspberry Pi down just a bit to make it work. It's not an ideal solution, and hopefully it will be fixed in a future release.

The default speed for the Raspberry Pi 4B and 400 is 700Mhz, but we need to lower that to 600Mhz. To change the speed of the ARM processor, run the following command from a terminal. This will allow you to edit the configuration file read at boot time and adjust the speed to a lower value. You may encounter this process if you want to overclock your Raspberry Pi where we set the value > 700Mhz[1]:

```
$ sudo nano /boot/config.txt
```

Scroll down through the file until you find the `arm_freq` value and uncomment it (remove the # sign) and change the value as follows:

```
arm_freq=600
```

[1] https://www.raspberrypi.org/documentation/configuration/config-txt/overclocking.md

When you're ready to save the file, press *CTRL+X* and *Y* to save the file. Then, reboot your computer for the changes to take effect. Now you are ready to start using the GrovePi and GrovePi+ host adapters.

GrovePi/GrovePi+ Troubleshooting Tips

The GrovePi host adapter works well under most conditions. If you encounter problems installing the software or using the board, there are several things you can try.

First, use the following command to see if the board is detected by the Raspberry Pi. If you do not see the output as shown – specifically, the I2C address 0x04 is present (shown in bold) – you may need to power off the Raspberry Pi to ensure the GrovePi is plugged into the GPIO header correctly:

```
$ sudo i2cdetect -y 1
     0  1  2  3  4  5  6  7  8  9  a  b  c  d  e  f
00:          -- 04 -- -- -- -- -- -- -- -- -- -- --
10: -- -- -- -- -- -- -- -- -- -- -- -- -- -- -- --
20: -- -- -- -- -- -- -- -- -- -- -- -- -- -- -- --
30: -- -- -- -- -- -- -- -- -- -- -- -- -- -- -- --
40: -- -- -- -- -- -- -- -- -- -- -- -- -- -- -- --
50: -- -- -- -- -- -- -- -- -- -- -- -- -- -- -- --
60: -- -- -- -- -- -- -- -- -- -- -- -- -- -- -- --
70: -- -- -- -- -- -- -- --
```

If the GrovePi is detected on the I2C bus, but you cannot get any of the sample code to work, you can update the firmware. This has been known to fix such problems. Should you want to update the firmware of your GrovePi board, you can do so by using the following command. You should not need to do this more than once:

```
$ cd ~/Dexter/GrovePi/Firmware
$ ./firmware_update.sh
Updating the GrovePi firmware
=============================
 http://www.dexterindustries.com/grovepi
 Run this program:
 sudo ./firmware_update.sh

=============================
Do you want to update the firmware? [y,n]y
Make sure that GrovePi is connected to Raspberry Pi
Firmware found
Press any key to start firmware update
. . .
avrdude done.  Thank you.
```

Finally, if the board stops working or your projects hang, you can try resetting the board with the following command. Or try rebooting your Raspberry Pi:

```
$ avrdude -c gpio -p m328p
```

Now we're ready to write the code.

Write the Code

The code for the Python version of this project is very similar to the Arduino version except for the usual differences such as not needing a header file and the naming scheme being different. Recall we use underscores and mostly lowercase names in Python and lowercase first word and initial capitals for internal words in Arduino, for example, `knock_timeout` for Python and `knockTimeout` for Arduino.

Once again, we will not dive into every line of code. We will explore the code at a higher level and discuss the more complex or important parts in detail. You can read through the code and learn more about how it works at your leisure.

Like the Arduino example, we will use a class to contain the code to read the sensor. However, unlike the Arduino IDE, you can use any editor to create the class and main script. We will name the main script `secret_knock.py` and the class file `knock_sensor.py`. Let's start with the class.

Secret Knock Class

Open the Thonny Python IDE under the *Main* ➤ *Programming* submenu. Create a new file `knock_sensor.py`. We will name the class `KnockSensor`, and we will need the same three functions we used in the Arduino version.

The constructor accepts two values that specify the pin number of the sound sensor and the sensitivity of the sensor. The remaining functions operate in the same manner as the Arduino version with some major differences.

Recall in the Arduino version, we used the function `millis()` to read the milliseconds expired, but that method does not exist in Python. In fact, we must change the code to use the `datetime` library to keep track of time. The changes are clear when you read through the code. We must also use a decimal value for milliseconds for the `time.sleep()` methods since they accept values in seconds.

The other major difference is the use of the map() function. Once again, there is no function in Python to map one range of values to another. So we must write our own. The following shows the replacement function named map_range():

```
def map_range(value, istart, istop, ostart, ostop):
    if value == istart:
        return value
    return int(ostart + (ostop - ostart) * ((value - istart) /
    (istop - istart)))
```

The rest of the differences are due to the difference in writing in Python vs. Arduino C++ code. Listing 13-8 shows the complete code for the class with documentation removed for brevity. Take a few moments to read through the code so that you understand all of the parts of the code.

Listing 13-8. Secret Knock Class (Python)

```
import time
from grovepi import pinMode, analogRead

MAX_KNOCKS = 10
KNOCK_DELAY = 0.200

def map_range(value, istart, istop, ostart, ostop):
    if value == istart:
        return value
    return int(ostart + (ostop - ostart) * ((value - istart) /
    (istop - istart)))

class KnockSensor:
    """Knock Sensor Class"""

    # Configuration variables
    reject_percent = 0.25
    sound_sensor = 2  # Sound Sensor
```

```
# Tuning variables
sound_threshold = 500        # Minimum value to detect a
                               knock
knock_timeout = 2.0          # Wait timeout for knock
                               detection

# Secret Knock Storage: "Shave and a haircut - two bits"
secret_knock = [100, 53, 20, 21, 44, 87, 54, 0, 0, 0]
num_secret_knocks = 7
values_read = [0, 0, 0, 0, 0, 0, 0, 0, 0, 0]  # Values from
                                                 sound sensor

def __init__(self, analogPin=2, sensitivity=0.25):
    self.sound_sensor = analogPin
    self.reject_percent = sensitivity
    pinMode(self.sound_sensor, "INPUT")

def read_secret(self):
    start_time = time.time()
    now = time.time()
    value = 0
    num = 0

    # Clear the read values array
    for i in range(0, MAX_KNOCKS):
        self.values_read[i] = 0
    # Listen for knocks that exceed threshold
    while (((now - start_time) < self.knock_timeout) and
           (num < MAX_KNOCKS)):
        value = analogRead(self.sound_sensor)
        if value >= self.sound_threshold:
            now = time.time()
```

```
                self.values_read[num] = now - start_time
                start_time = now
                num = num + 1
                time.sleep(KNOCK_DELAY)
                print("Knock heard.")
            now = time.time()
        # Check for errors
        if num == 0:
            print("ERROR: no knocks detected.")
            return False
        if num >= MAX_KNOCKS:
            print("ERROR: maximum number of knocks exceeded.")
            return False
        return True

    def record_secret(self):
        max_knock_interval = 0
        # Normalize the knock timings
        self.num_secret_knocks = 0
        for i in range(0, MAX_KNOCKS):
            # Collect normalization data while we're looping.
            if self.values_read[i] > max_knock_interval:
                max_knock_interval = self.values_read[i]
            self.secret_knock[i] = map_range(self.values_
            read[i], 0,
                                             max_knock_
                                             interval, 0, 100)
            # Save number of secret knocks
            if self.secret_knock[i] > 0:
                self.num_secret_knocks = self.num_secret_
                knocks + 1
        print("Secret knock saved.\n")
```

```
def validate_secret(self):
    # Check to see the number of knocks match
    num = 0
    knock_interval = 0     # We use this later to normalize
                             the times.
    for i in range(0, MAX_KNOCKS):
        if self.values_read[i] > 0:
            num = num + 1
        # Capture max knock interval from values read for
          normalization
        if self.values_read[i] > knock_interval:
            knock_interval = self.values_read[i]
    if num != self.num_secret_knocks:
        return False

    time_difference = 0
    for i in range(0, MAX_KNOCKS):  # Normalize the times
        # Normalize the values to remove timing issues.
        self.values_read[i] = map_range(self.values_
        read[i], 0, knock_interval, 0, 100)
        # Check for differences and compensate with error
          percentage.
        if self.values_read[i] != self.secret_knock[i]:
            if self.secret_knock[i] >= self.values_read[i]:
                time_difference = 1.0 -
                    (self.values_read[i] / self.secret_
                    knock[i])
            else:
                time_difference = 1.0 -
                    (self.secret_knock[i] / self.values_
                    read[i])
```

```
                if time_difference > self.reject_percent:
                    return False
        return True
```

Now we can write our main script.

Main Script (Python)

Open the Thonny Python IDE under the *Main* ➤ *Programming* submenu. Create a new file secret_knock.py. There is nothing new in this code as it follows the same flow as the Arduino example. Listing 13-9 shows the complete code for the main script for this project. You can read through it to see how all of the code works.

Listing 13-9. Main Script (Python)

```
# Import libraries
import sys
import time

from grovepi import pinMode, digitalRead, digitalWrite
from knock_sensor import KnockSensor

# Global variables
sensor = KnockSensor(sensitivity=0.30)

# Constants
TRIGGER_PROGRAM_SECONDS = 4.0
BUTTON_PIN = 6
RED_LED = 5
GREEN_LED = 4
PROGRAM_PRESS = 2
KNOCK_PRESS = 1
NO_PRESS = 0
HIGH = 1
LOW = 0
```

```
def read_button_status():
    tics_pressed = 0

    button_pressed = digitalRead(BUTTON_PIN)
    if button_pressed:
        while button_pressed:
            tics_pressed = tics_pressed + 1
            time.sleep(0.100)
            if (tics_pressed/10.0) >= TRIGGER_PROGRAM_SECONDS:
                return PROGRAM_PRESS
            button_pressed = digitalRead(BUTTON_PIN)
        return KNOCK_PRESS
    return NO_PRESS

def main():
    print("\nWelcome to the Secret Knock program.")
    pinMode(RED_LED, "OUTPUT")
    pinMode(GREEN_LED, "OUTPUT")
    pinMode(BUTTON_PIN, "INPUT")
    while True:
        button_status = read_button_status()
        # If this is a program trigger, turn on both LEDs
        if button_status == PROGRAM_PRESS:
            digitalWrite(RED_LED, HIGH)
            digitalWrite(GREEN_LED, HIGH)
            print("Program Mode. Enter new secret knock.")
            time.sleep(0.100)
            if sensor.read_secret():
                sensor.record_secret()
            digitalWrite(RED_LED, LOW)
            digitalWrite(GREEN_LED, LOW)
```

```
        elif button_status == KNOCK_PRESS:
            print("What's the secret knock?")
            sensor.read_secret()
            # If knock accepted, turn on green LED, or red LED
              if failure
            if sensor.validate_secret():
                print("Secret knock accepted.")
                digitalWrite(GREEN_LED, HIGH)
                time.sleep(3)
                digitalWrite(GREEN_LED, LOW)
            else:
                print("ERROR: Secret knock rejected. Go away!")
                digitalWrite(RED_LED, HIGH)
                time.sleep(3)
                digitalWrite(RED_LED, LOW)
        time.sleep(0.100)

if __name__ == '__main__':
    try:
        main()
    except (KeyboardInterrupt, SystemExit) as err:
        print("\nbye!\n")
sys.exit(0)
```

OK, that's it! We've written the code. Unlike the Arduino, we do not need to compile the Python code. So we're now ready to execute the project!

Execute the Project

Now that we've spent many pages exploring the Grove modules and writing the code to interact with them, it is time to test the project by executing (running) it.

When the project runs (executes), you will see some diagnostic message written to the serial monitor (Arduino) or the terminal (Raspberry Pi). You will also see the LEDs illuminate as follows:

- The red LED is turned on if a knock doesn't match the stored secret knock.
- The green LED is turned on when a knock matches the stored secret knock.
- The green and red LEDs are turned on when we store a new secret knock.

Also, recall the button behavior is as follows:

- A single press is used to initiate a knock sequence to match.
- A long press (of 4 seconds or more) initiates a recording mode where we can record a new knock sequence.

Executing the code depends on which platform you're using. Let's look at the Arduino first.

Sketch on the Arduino

Executing the sketch on the Arduino requires connecting our board to our PC and then uploading the sketch to the Arduino. Recall the sketch will run so long as the USB cable is connected to our PC (and the Arduino).

Execute the Sketch

To execute the sketch, be sure your Arduino is connected and you've selected the correct board under the *Tools* ➤ *Board* menu. You also need to ensure you have the correct port selected under the *Tools* ➤ *Port* menu.

Once those items are set, you can click the *Upload* button or choose *Sketch* ➤ *Upload* from the menu. The Arduino IDE will compile the sketch and then upload it to your Arduino. Once you see the `Done uploading...` message, you can open the serial monitor. You should see the diagnostic output begin momentarily. Go ahead, and try it out! You should see values similar to Listing 13-10. Notice the several attempts to reproduce the secret knock and one successful attempt.

Listing 13-10. Sample Execution Transcript (Arduino)

```
Welcome to the Secret Knock program.
What's the secret knock?
Knock heard.
Knock heard.
Knock heard.
Knock heard.
Knock heard.
Knock heard.
Knock heard.
Diagnostics - Reject% = 0.33
Secret     Read     Diff
----       ----     ----
100        100      0.00
53         29       0.45
20         15       0.25
21         14       0.33
44         24       0.45
87         48       0.45
```

```
54      26      0.52
0       0       0.00
0       0       0.00
0       0       0.00
ERROR: Secret knock rejected. Go away!
What's the secret knock?
ERROR: no knocks detected.
ERROR: Secret knock rejected. Go away!
What's the secret knock?
Knock heard.
Knock heard.
Knock heard.
Knock heard.
Knock heard.
Knock heard.
Knock heard.
Diagnostics - Reject% = 0.33
Secret     Read     Diff
----       ----     ----
100     100     0.00
53      43      0.19
20      20      0.00
21      22      0.05
44      40      0.09
87      64      0.26
54      45      0.17
0       0       0.00
0       0       0.00
0       0       0.00
Secret knock accepted.
```

Notice also the diagnostic statements printed to show the knock sequence read vs. the stored secret knock and the percent difference in timing. You can use this when you are practicing the default secret knock. It's not as easy as it may appear and takes some practice to get it right (or you can increase the sensitivity).

If something isn't working, check your connections or refer to Chapter 7 for troubleshooting tips.

Python Code on the Raspberry Pi

Executing the sketch on the Raspberry Pi requires running the Python code in a terminal after connecting your Grove modules to your Raspberry Pi via a hat or the Grove female breakout cable. Recall the code will run until you stop it with *CTRL+C* on the keyboard.

Execute the Python Code

To run the Python code on the Raspberry Pi, you can issue the command `python3 ./secret_knock.py` from the same folder where the file was saved as shown in Listing 13-11. You will see output similar to the Arduino version but without the diagnostic statements.

Listing 13-11. Sample Execution Transcript (Python)

```
$ python3 ./secret_knock.py
Welcome to the Secret Knock program.
What's the secret knock?
Knock heard.
Knock heard.
Knock heard.
Knock heard.
Knock heard.
Knock heard.
```

```
Knock heard.
Secret knock accepted.
Program Mode. Enter new secret knock.
Knock heard.
Knock heard.
Knock heard.
Secret knock saved.

What's the secret knock?
Knock heard.
Knock heard.
Knock heard.
Knock heard.
ERROR: Secret knock rejected. Go away!
What's the secret knock?
Knock heard.
Knock heard.
Knock heard.
Secret knock accepted.
```

Notice the transcript also includes a sequence where we change the secret knock and test it. Try it yourself! If everything worked as executed, congratulations! You've just built your first Grove project! If something isn't working, check your connections or refer to Chapter 7 for troubleshooting tips.

Going Further

While we didn't discuss them in this chapter, there are some ideas where you could make this project into an IoT project. Here are just a few suggestions you can try once we have learned how to take our projects to the cloud. Put your skills to work!

- *Add a lock*: You can add a locking mechanism as described earlier to create a smart lock for a box, door, window, etc. When doing so, it is a good idea to remove the secret knock programming feature or, better, add a second button inside the box that triggers the programming mode.
- *Add a logger*: It would be easy to add a short bit of code to write the date and time a secret knock is accepted. This is easy to do in Python but requires adding a storage mechanism for the Arduino such as an SD card reader or a logging module.
- *IoT smart lock*: Store the access log that you can view on the Internet.
- *Remote programming*: Add the ability to send a new secret knock sequence over the Internet (or local network). This is a more advanced enhancement that will require using networking programming to make it work. Fortunately, there are many examples you can follow including a client/server protocol, message queue, and more.

Summary

In this chapter, we got hands-on experience making projects with Grove modules. We used a sound sensor to detect a series of knocks to simulate a secret knock locking mechanism. We used two LEDs to provide feedback to the user as well as diagnostic statements written to the serial monitor.

Along the way, we learned more about how to work with Grove modules including how to write our own class for managing the sensor. We also learned how to install the Python software for the Raspberry Pi as well as how to update the firmware for the GrovePi host adapter. Finally, we saw some potential to make this project better as well as some ideas for how to adapt the project for practical uses.

In the next chapter, we will see another project that demonstrates how to use a light sensor, a temperature sensor, and a red-green-blue LED to create a mood light. Groovy, eh?

CHAPTER 14

Mood Lighting

At one time, mood lighting was (perhaps still is) an artistic mechanism to add some ambiance to a room. We see mood lighting in cars, TVs, and more. Mood lighting in that form is typically a series of LEDs placed behind an object or in dark corners to provide a relaxing glow.

There are also various forms of jewelry that use a thermochromic mechanism that changes color depending on the temperature of the skin or finger. The most popular form is a mood ring.[1]

In this chapter, we will see how to create a mood lamp that works like a mood ring to read the temperature of your finger and display the color that represents your mood. We'll see more analog and digital modules as well as our first I2C Grove modules.

Like the last chapter, we will see how to implement this project on the Arduino and Raspberry Pi but with an interesting twist. Let's get started.

Project Overview

The project for this chapter is designed to demonstrate how to use analog, digital, and I2C devices on the same Grove host adapter to build a mood lamp. It works like a mood ring by using surface (skin) temperature of your finger to calculate a mood. The RGB will change depending on the temperature read, and your mood will be displayed on the LCD.

---

[1] `https://en.wikipedia.org/wiki/Mood_ring`

C. Bell, *Beginning IoT Projects*, https://doi.org/10.1007/978-1-4842-7234-3_14

We will use a light sensor as a way to wake up the mood detector and a temperature sensor to read the surface temperature of your skin. Thus, you will need to use two fingers: one to cover the temperature sensor and another to cover the light sensor.

When the light sensor reading reaches the threshold, the code will read the temperature and display the mood. When not reading and presenting your mood, the RGB will default to a pleasant shade of light blue. In essence, we're building a mood detector.

What Will We Learn?

By implementing this project, we will get more practice in using analog and digital Grove modules and how to connect Grove modules to the various protocol connectors on the host adapter. We will also pick up a few tips on working with the I2C protocol on the Grove platform as well as discovering how to use other I2C devices along the way. Rather than a typical educational project, this project is fun to use as well as suited for learning all of the nuances of building Grove projects that use a mixture of analog, digital, and I2C modules.

WHAT PROTOCOL DOES MY GROVE MODULE USE?

You may be thinking, *Since all Grove modules use the same cables and the connectors are all the same, how can I tell which protocol is required for a given module?*

You can visit the Wiki page for the module, and it will be clearly described there. For example, visit `wiki.seeedstudio.com` and click the Grove menu to the left to read about all of the Grove modules.

You can also determine the protocol by looking at the connectors on the module. The first two pins are always labeled GND, VCC. It is the last two pins that determine the protocol. Analog and digital modules use NC, SIG, some digital modules use DIN, CIN (sometimes only DI, CI), I2C modules use SDA, SCL, and UART modules use RX, TX. So the modules you need to pay attention to when connecting are the analog and digital because the printed connector labels may not indicate which protocol they use. Always check the documentation for the module before use.

The programming tasks will reveal how to read values from the temperature and light sensors to detect the initiation of a mood reading and how to interpret the value to make a decision on the color to display.

Let's see what hardware we will need.

Hardware Required

The hardware needed for this project is listed in Table 14-1. URLs for each component are included for ease of ordering including duplicate entries for alternative vendors. We will use the Grove Light Sensor, Grove Chainable RGB LED, Grove LCD RGB Backlight, Grove Qwiic Hub, and a Qwiic TMP102 temperature sensor. Yes, we will be combining Grove and Qwiic in the same project!

Table 14-1. Hardware Needed for the Mood Detector Project

Component	URL	Qty	Cost
Grove Light Sensor	wiki.seeedstudio.com/Grove-Light_Sensor	1	$2.90
Grove Chainable RGB LED	www.seeedstudio.com/Grove-Chainable-RGB-Led-V2-0.html	1	$5.99
Grove LCD RGB Backlight	www.seeedstudio.com/Grove-LCD-RGB-Backlight.html	1	$11.90
Grove Qwiic Hub	www.seeedstudio.com/Grove-Qwiic-Hub-p-4531.html	1	$1.90
SparkFun Qwiic TMP102	www.sparkfun.com/products/16304	1	$6.50
Grove cables(any length can be used)	Included with each preceding module	4	
Qwiic cable	Included with the Qwiic Hub	1	
Arduino MKR 1010 WiFi	www.sparkfun.com/products/15251	1	$35.95
Raspberry Pi 3B or later	www.sparkfun.com/categories/233 www.adafruit.com/category/176	1	$35.00+
Grove Base Shield V2.0 for Arduino	www.seeedstudio.com/Base-Shield-V2.html	1	$4.45
GrovePi+	www.sparkfun.com/products/15945	1	$5.95

About the Hardware

Let's discuss these components briefly. We will discover how to work with the hardware in more detail later in the chapter. If you implement the Python version, be sure to note the changes in the libraries used.

Grove Light Sensor

If you want to detect the amount of light in a room (or inside a container), you can use a light sensor that will return a value that you can use to determine brightness. Or you can use it like a switch to turn on or off external lighting much like your backlit laptop keyboard.

The Grove Light Sensor is an analog module that, when read, produces a value in a range of 0–255 or higher depending on the ADC used on your host board. It uses a photodiode to detect the intensity of light. Note that there are many different Grove light sensors, so if you cannot find this exact module, some of the older or newer versions will work as well.

Figure 14-1 shows the Grove Light Sensor.

Figure 14-1. *Grove Light Sensor (courtesy of seeedstudio.com)*

Note If you do not have a Grove Light Sensor, you can substitute a button or touch sensor to achieve the same goal.

Grove Chainable RGB LED

The lamp used in this project is a bright red, green, blue (RGB) LED that can be used to produce a vast array of colors by specifying a value of 0–255 for each color. The higher the value, the brighter (intensity) that color is shown. By mixing the intensity, we can see a wide range of colors.

For example, values of (255, 0, 0) are for red or (127, 0, 127) for purple. To see what this might look like, an RGB chooser (`www.w3schools.com/colors/colors_rgb.asp`) can help you visualize the color. Navigate there now and try it out yourself.

The Grove Chainable RGB LED module allows you to produce just about any color you want. Figure 14-2 shows the Grove Chainable RGB LED.

***Figure 14-2.** Grove Chainable RGB LED (courtesy of seeedstudio.com)*

So what does the chainable in the name mean? It means if you want to use more than one RGB LED, you can "chain" the modules together. In fact, on the bottom of the module, you will see two Grove connectors: one marked "IN" and another "OUT." Figure 14-3 shows what the connectors look like. Notice the labels for each.

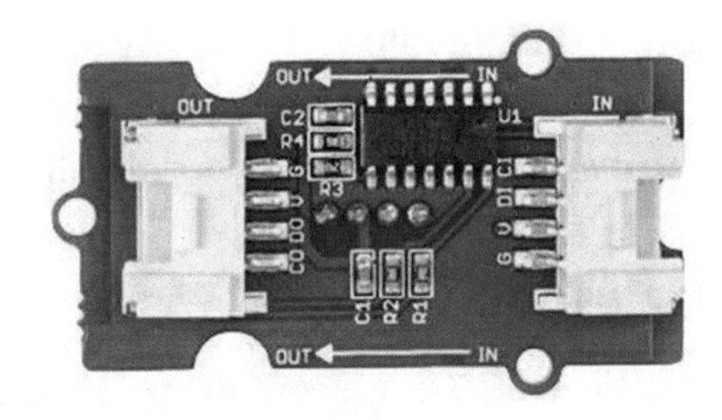

Figure 14-3. *Grove Chainable RGB LED connectors on the bottom (courtesy of seeedstudio.com)*

To chain multiple modules together, simply connect the first Grove cable from your host adapter to the "IN" connector, then another Grove cable to the "OUT," and then the "IN" to the next module and so on. You can connect up to 1024 RGBs together.

Before we move on to the next module, let's discuss the colors we will use for this project. We will use a variety of colors that map to temperature ranges. Table 14-2 shows the moods as well as the temperature and color for each.

Table 14-2. *Available Moods*

Mood	Temperature Range (F)	Color
Off/resting	< 72	Light blue
Troubled	72–73	Orange
Alert	74–77	Purple
Calm	78–81	Green
Happy	82–85	Yellow
Romantic	86–89	Red
Nervous	90–93	Violet
Stressed	> 94	White

You may need to adjust these values if you'd like to tune the readings a bit. There is no magic to selecting the temperature ranges (yes, mood rings are largely hokum), so feel free to adjust them to, er, suit your mood.

Grove LCD RGB Backlight

If you've used monochrome LCD displays in the past, you may appreciate the interesting option on the Grove LCD RGB Backlight. While the text color remains dark gray, you can change the background using an RGB color similar to the Chainable RGB LED. Figure 14-4 shows the Grove LCD RGB Backlight.

Figure 14-4. *Grove LCD RGB Backlight (courtesy of seeedstudio.com)*

While this module does not offer the option, some Grove I2C modules support address changes like the Qwiic/STEMMA QT modules by opening or closing jumpers on the bottom of the board. For example, you can change the address from `0x76` to `0x77` via a jumper on the Grove barometric pressure (BMP280) module (`https://wiki.seeedstudio.com/Grove-Barometer_Sensor-BMP280/`).

Grove Qwiic Hub

The Grove Qwiic Hub allows you to do just that – use Qwiic modules with a Grove host adapter. All we need to do is use this module to provide two Qwiic connectors, plug in our Qwiic module(s), and start programming. Figure 14-5 shows the Grove Qwiic Hub.

Figure 14-5. *Grove Qwiic Hub (courtesy of seeedstudio.com)*

The hub can also be used with a Qwiic host adapter to allow the use of Grove modules. Cool! This is one module you will want to add to your stores if you plan to continue to experiment with Grove and Qwiic/STEMMA QT modules. And, best of all, it comes with a Qwiic cable so you don't need to buy one if you have not already invested in the Qwiic Component System.

Qwiic TMP102

Our choice for a temperature sensor is a Qwiic TMP102 sensor from SparkFun. The TMP102 can read temperatures to a resolution of 0.0625°C and is accurate up to 0.5°C.

Figure 14-6. *Sparkfun Qwiic Temperature Sensor (courtesy of sparkfun.com)*

Tip See Chapter 13 for more details on the Grove host adapters.

Connect the Grove Modules

Recall from Chapter 12 we can use a single Grove cable to connect each Grove module separately to a specific connector on the host adapter on our host board. The host adapter for the Raspberry Pi has a different layout but has the same connectors we will use. Table 14-3 includes the details of each connection on the host adapter to help you make the right connections. Simply use the table to connect a Grove cable from the module to the Grove connector on the host adapter as marked in the table.

Table 14-3. *Grove Connections*

Module	Protocol	Grove Connector on the Host Adapter
Light Sensor	Analog	A0
RGB LED	Digital	D7
LCD RGB Backlight	I2C	I2C1
Qwiic Hub + TMP102	I2C	I2C2

Figure 14-7 shows an example of how you should connect your modules for this project. Notice the figure shows the Arduino host adapter.

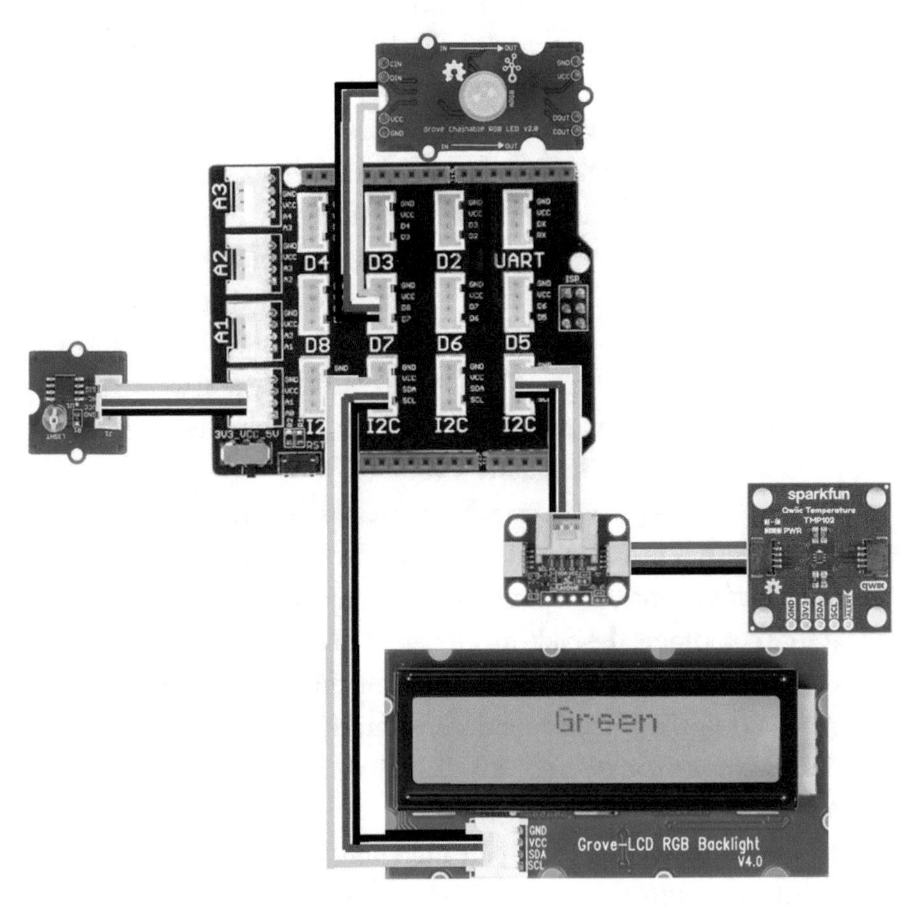

Figure 14-7. *Mood detector project Grove connections*

Now that we know more about the hardware for this chapter, let's write the code!

Write the Code

The code for this project involves following the usual pattern. For this project, that means using analog and digital modules as well as two I2C devices. The light sensor is an analog sensor, the chainable RGB LED is a digital module, and the TMP102 and LCD are both I2C devices.

As you will see, the code isn't overly complicated for the Arduino version, but we will have some more work to do for the Python version. This is because some of the Grove modules used are not directly supported by the GrovePi library and even some that are supported use libraries that are not built using classes. Rather, they are implemented as a set of modules using normal functions, which makes reading and working with them a bit more challenging. Never fear, though. The code we will see is fully functional albeit not "standard" Python.

Like the previous projects, we will use classes to wrap our functionality. In the past, this has been focused on a single sensor or module. This time, we'll focus on making the mood lamp concept its own class. This will require combining the code for all four of the modules (light and temperature sensors, LCD, and RGB LED) into the class. It is an excellent example of the power of classes (software libraries) to contain not only small but also more complex objects. This also makes our main code much smaller.

Let's walk through how to prepare our computers to use the components and write the code. We'll start with the Arduino.

Arduino

This section presents a walk-through of the sketch and classes you will write to read values from the light and temperature sensors and display the mood value on the LCD module and change the RGB to a specific color. But first, there are a couple of libraries we must install on our PCs.

Install Software Libraries

We will need to install the Arduino libraries for the RGB LED, LCD, and TMP102 modules separately. Open the Library Manager from the Arduino IDE menu (*Sketch* ➤ *Include Library* ➤ *Library Manager...*). Then search for `chainable` and install the latest version of the Grove Chainable RGB LED library as shown in Figure 14-8.

Figure 14-8. *Installing the Grove Chainable RGB LED library (Arduino IDE)*

Similarly, we need to install the library for the Grove LCD RGB Backlight. Open the Library Manager and search for `LCD RGB Backlight` and then install the latest version as shown in Figure 14-9.

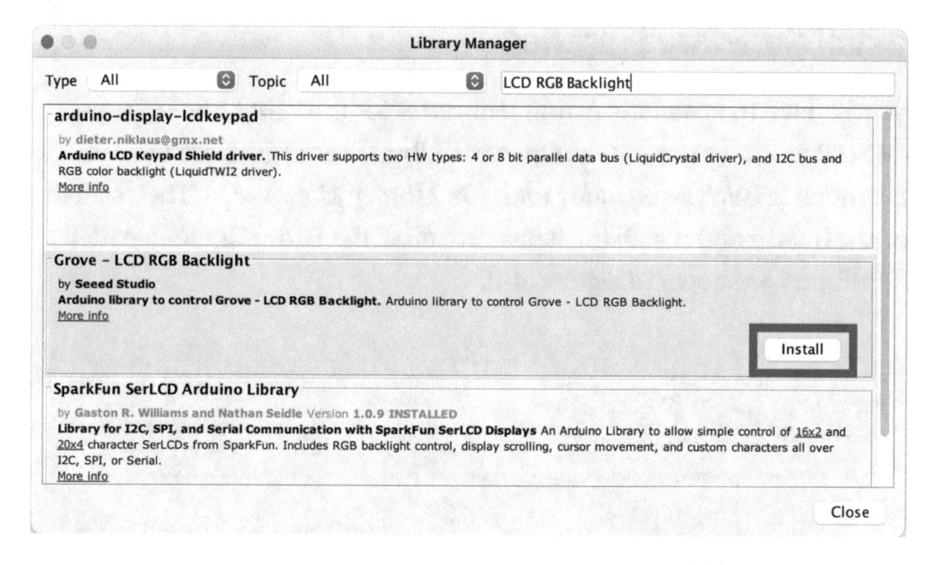

Figure 14-9. *Installing the Grove LCD RGB Backlight library (Arduino IDE)*

Similarly, we need to install the library for the TMP102 sensor. Open the Library Manager and search for `TMP102` and then install the latest version as shown in Figure 14-10.

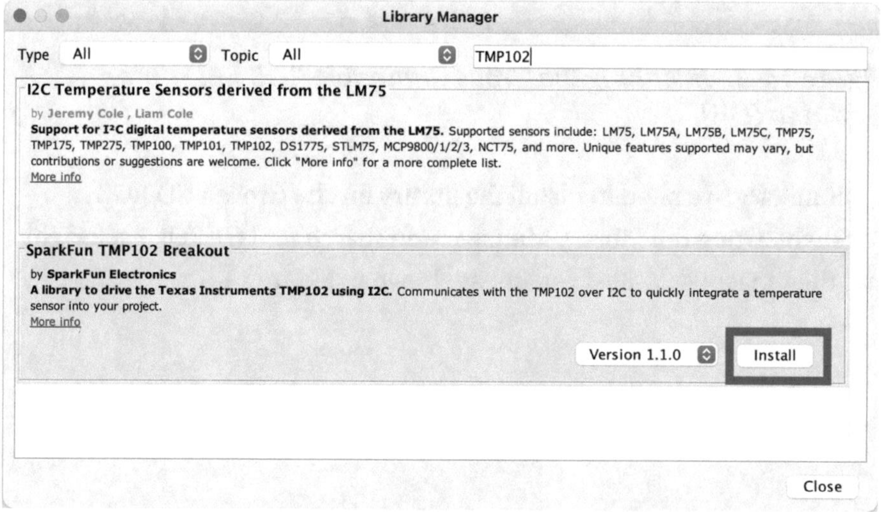

Figure 14-10. *Installing the SparkFun TMP102 library (Arduino IDE)*

Now that we have the software libraries installed, we can begin writing our sketch. Since this is not our first Arduino sketch, we will discuss the code at a high level and skip the line-by-line details focusing on the mechanics of how the code works. You can study the code at your leisure to ensure you understand the sketch in more detail.

Write the Sketch

Recall we are going to use a class to encapsulate the mood lamp. In fact, we will create two classes: one to define the moods so we can change the temperature ranges and colors without affecting the rest of the code and the mood lamp itself.

We will write the class headers first and then the main sketch and finally complete the code for the classes. This is typically how programmers develop code with class modules (but not always). By creating the headers first, we can understand how to use the classes making writing the main sketch easier. The mood lamp class will be named `MoodLamp`, and the moods class will be named `Moods`.

Recall there is no way (currently) to create and add new files to a sketch (but you can add existing files by clicking *Sketch* ➤ *Add File...*). So we will once again create the main sketch and add the code header and source files manually.

Open a new sketch and name it `mood_detector.ino` or whatever you'd like to use. Save the file and then close the project in the Arduino IDE.

To create the class files, navigate with your File Explorer (Finder) to the folder where you stored your main sketch (`mood_detector.ino`). Then, use your File Explorer or a text file editor to create four new files named `Moods.h`, `Moods.cpp`, `MoodLamp.cpp`, and `MoodLamp.h`. Or you can use a terminal to navigate to the folder and issue these commands to create the empty files:

```
mood_detector % touch Moods.h
mood_detector % touch Moods.cpp
```

```
mood_detector % touch MoodLamp.h
mood_detector % touch MoodLamp.cpp
```

Now, let's see the code for each file starting with the header files.

Class Header File: Moods

Click the tab named `Moods.h` to open the blank file. Here, we will add the header or blueprint for the class. The module provides a mechanism to store and retrieve the mood values for the temperature ranges and color values.

Thus, we will need a way to get the default color (the color used when a mood is not being read or displayed) and the color values for a given temperature. To make things easier, we will store the color values and mood name (string) in a structure. Thus, we need only two functions: `getDefaultMood()` and `getMood()`.

Aside from those functions, we will also create a constructor so we can define the default color and store it for later retrieval.

Since the header code is not difficult to read, let's look at the completed code for the header file. Listing 14-1 shows the file.

Listing 14-1. Moods Header File

```
#include <Arduino.h>

// Mood definitions by temperature
#define MOOD_TROUBLED 72
#define MOOD_ALERT 74
#define MOOD_CALM 78
#define MOOD_HAPPY 82
#define MOOD_ROMANTIC 86
#define MOOD_NERVOUS 90
#define MOOD_STRESSED 94
```

```
typedef struct{
    int red;
    int green;
    int blue;
    String mood;
} MoodValue;

class Moods {
public:
  Moods() { defaultMood = {102, 178, 255, "Resting"}; }
  MoodValue getMood(int temperature);
  MoodValue getDefaultMood() { return defaultMood; }
private:
  MoodValue defaultMood;
};
```

Now, let's look at the header file for the mood lamp.

Class Header File: MoodLamp

Click the tab named `MoodLamp.h` to open the blank file. Here, we will add the header or blueprint for the class. Recall this class will manage all four (we don't count the Qwiic Hub as it is a passive device) of our modules. So there is a lot of code devoted to set up each of the modules.

We will also need to use the Moods header file and define a number of constants such as which port the analog light sensor is using, the lowest value to read from the light sensor to indicate a read event, the minimum temperature to permit a mood determination, and the number of LEDs we're using.

Aside from that, we will also need a number of methods including a `begin()` method for initialization, `detectMood()` to read the temperature and determine if it is a successful read, and `clearMood()` to reset the RGB LED to return it to the default color.

We will also define a couple of private variables and methods such as variables for the other hardware libraries (and the moods library) and the previous mood (for a fading effect) and a method to read the temperature (`readTemperature()`), `reset()` to reset the LCD, `off()` to turn off the RGB LED and LCD, and a method set the mood (`setMood()`). What these functions do will become clearer once you see the code for them.

Since the header code is not difficult to read, let's look at the completed code for the header file. Listing 14-2 shows the file.

Listing 14-2. MoodLamp Header File

```
#include <Arduino.h>
#include <rgb_lcd.h>
#include <SparkFunTMP102.h>
#include <ChainableLED.h>
#include "Moods.h"

// Default value read from light sensor to initiate a mood
   detection.
#define LIGHT_TRIGGER_LEVEL 300
#define LIGHT_SENSOR A0
#define NUM_LEDS 1
#define MINIMUM_TEMPERATURE 72 // Degrees Fahrenheit

class MoodLamp {
public:
  void begin();
  bool detectMood();
  void clearMood();

private:
  Moods *moods;
  rgb_lcd lcd;                        // Grove LCD
  TMP102 temperatureSensor;           // Qwiic TMP102
```

```
  ChainableLED *rgbLed;                    // Grove RGB LED
  MoodValue prevMood;

  void reset();
  void off();
  int readTemperature();
  void setMood(MoodValue moodValue, boolean show_message=true);
};
```

OK, let's return to the main sketch.

Main Sketch

Now click the `mood_detector.ino` tab to return to the main sketch. Since we are placing all of the hardware work in the mood lamp class, all we need to do here is instantiate a new class instance (stored in a variable named `moodLamp`), set up the serial class, initialize the mood lamp, and print the greeting. All of these are done in the `setup()` function.

The `loop()` function simply calls the `detectMood()` method, and if it returns true, we wait for a certain amount of time as defined in the constant `MOOD_TIMEOUT` and then clear the mood lamp with the `clearMood()` function. That's it!

Now you can see why building a robust class module that encapsulates all of the hardware makes it easy to write our main code. Listing 14-3 shows the code for the main sketch. Take a few moments and read through the code.

Listing 14-3. Main Sketch

```
#include <Wire.h>
#include "MoodLamp.h"

// Constants
#define MOOD_TIMEOUT 10
```

```
// Global variables
MoodLamp *moodLamp;

void setup() {
  Serial.begin(115200);
  while(!Serial);
  // Setup mood lamp
  moodLamp = new MoodLamp();
  Serial.println("Welcome to the mood detector!");
  moodLamp->begin();
  Serial.println("Mood detector is ready.");
  Serial.println("Place fingers over the light and temperature
  sensor.");
  Serial.println("Reading takes about 15 seconds.");
}

void loop() {
  // If the mood detector was initiated, clear the value after
     a timeout
  if (moodLamp->detectMood()) {
    delay(MOOD_TIMEOUT * 1000);
    moodLamp->clearMood();
  }
  delay(2000);
}
```

Now we can write the final portion of our project – the code for the classes.

Class Code File: Moods

Click the tab named `Moods.cpp` to open the blank file. Here, we will add the code for the class. There is only one function to write: `getMood()`. Note that we defined the `getDefaultMood()` and constructor in the class header file.

The getMood() function takes as a parameter the temperature value (as an integer) and then uses the constants defined in the header file to set up a series of if statements that return a mood represented by the MoodValue structure based on the temperature. The code is not overly complicated, and you can read it for yourself. Listing 14-4 shows the completed code for the class (documentation omitted for brevity).

Once again, this code was moved to its own class for ease of maintenance and modification. Otherwise, we would have complicated the mood lamp code and made it harder to change the temperature ranges and colors.

Listing 14-4. Moods Code File

```
#include "Moods.h"

MoodValue Moods::getMood(int temperature) {
  if (temperature == 0) {
    return {0, 0, 0, ""};
  } else if (temperature < MOOD_TROUBLED) {
    return {defaultMood.red, defaultMood.green, defaultMood.
    blue, ""};
  } else if ((temperature >= MOOD_TROUBLED) && (temperature <
  MOOD_ALERT)) {
    return {255, 128, 0, "Troubled"};
  } else if ((temperature >= MOOD_ALERT) &&
(temperature < MOOD_CALM)) {
    return {153, 0, 153, "Alert"};
  } else if ((temperature >= MOOD_CALM) &&
(temperature < MOOD_HAPPY)) {
    return {0, 102, 0, "Calm"};
  } else if ((temperature >= MOOD_HAPPY) &&
(temperature < MOOD_ROMANTIC)) {
```

```
    return {255, 255, 0, "Happy"};
  } else if ((temperature >= MOOD_ROMANTIC) &&
(temperature < MOOD_NERVOUS)) {
    return {204, 0, 0, "Romantic"};
  } else if ((temperature >= MOOD_NERVOUS) &&
(temperature <= MOOD_STRESSED)) {
    return {255, 0, 255, "Nervous"};
  }
  return {255, 255, 255, "Stressed!"};
}
```

Now let's look at the code file for the mood lamp class. As you will see, this class contains a lot of code, but most is rather easy to read and understand.

Class Code File: MoodLamp

Click the tab named `MoodLamp.cpp` to open the blank file. Here, we will add the code for the class. Since there are a lot of functions in the class, we will first list the functions and then highlight one of the more complex ones in a detailed walk-through. You can discover how the other functions work as an exercise, but the code is similar to what we've seen in previous projects.

The functions defined in this class include the following. Included with each are details of what the function is used for and how it works:

- `void begin()`: Set up the hardware in the module. Initializes the pin for the light sensor, initializes the temperature sensor, defines the LCD size and initializes it, instantiates the moods class variable, and clears the mood (sets the default mood color).
- `bool detectMood()`: This is the main function used to initiate the mood detection. It first reads the light sensor and, if it is covered, reads the temperature.

If the temperature is above the minimum value, it uses the `Moods` class to get the color and text for the mood. It then sets the RGB LED to the mood color and displays the mood text on the LCD.

- `void clearMood()`: Returns the RGB LED to the default color and clears the LCD text. This uses the `Moods` class to get the default color values and the reset to display the normal message (instructions) on the LCD.
- `void reset()`: Displays the main screen on the LCD, which is the instructions for use (however terse).
- `void off()`: Turns off the LCD and RGB LED.
- `int readTemperature()`: Reads the temperature from the TMP102. We also use the TMP102 library to turn the sensor on before reading and off after to save power. Most advanced sensors have similar capabilities, and since we could be waiting a long time between mood detections, it makes the sensor last longer.
- `void setMood(MoodValue moodValue, boolean show_message)`: Used to set the current mood using the values read from the `Moods` class and the previous mood to fade the RGB LED from one color to another. The `show_message` parameter is used to toggle displaying of the mood name on the LCD. False means don't display the text.

This last function, `setMood()`, is the most complex code-wise because it uses an algorithm to fade the RGB LED. Why fade? Fading the RGB LED makes the color changes more interesting and slower. Rather than blinking from one color to another, it eases the color changes slowly. It's a neat effect when you see it in operation.

However, the fade is not as simple as using a `for` loop because we not only must change from a low value to a higher value but also from a high value to a lower value. Thus, we must calculate an increment to use inside a `while` loop that loops until all three values (red, green, and blue) become equal. That is essentially the most difficult part of this function, and that code is shown in Listing 14-5.

Listing 14-5. Excerpt from setMood() – Fade Effect

```
int red;
int green;
int blue;
int red_increment = 0;
int green_increment = 0;
int blue_increment = 0;
...
// Prepare starting position and increment values for R,G,B
red = prevMood.red;
green = prevMood.green;
blue = prevMood.blue;
if (red < moodValue.red) red_increment = 1;
if (red > moodValue.red) red_increment = -1;
if (green < moodValue.green) green_increment = 1;
if (green > moodValue.green) green_increment = -1;
if (blue < moodValue.blue) blue_increment = 1;
if (blue > moodValue.blue) blue_increment = -1;

// If the colors are the same, just reset the display
if ((red == moodValue.red) && (green == moodValue.green) &&
(blue == moodValue.blue)) {
  rgbLed->setColorRGB(0, red, green, blue);
  delay(10);
}
```

```
// Implement a smooth transition to the next values
while ((red != moodValue.red) || (green != moodValue.green) ||
(blue != moodValue.blue)) {
  // Check for increment stops
  if (red == moodValue.red) red_increment = 0;
  if (green == moodValue.green) green_increment = 0;
  if (blue == moodValue.blue) blue_increment = 0;
  red = red + red_increment;
  green = green + green_increment;
  blue = blue + blue_increment;
  rgbLed->setColorRGB(0, red, green, blue);
  delay(10);
}
```

Notice how we determine the increment. You may think an `if/else` clause would work where if the previous value was lower, we increment by +1; else, use -1. But that won't work if they are equal. Thus, we set the increment for each color to 0 first and then use separate `if` statements to set the increment for each color.

Notice also we set the color of the RGB LED with each pass through the loop. This is the fade effect where the color changes slowly.

Finally, notice what we do if the color is the same as the previous. In this case, we simply set the color on the RGB LED again since it is possible it is turned off.

Listing 14-6 shows the completed code for the class (documentation omitted for brevity). Take some time to read through the code to see how each function works. Take a bit longer in the `begin()` function to see how we use each of the hardware modules.

Listing 14-6. MoodLamp Code File

```
#include "MoodLamp.h"

void MoodLamp::begin() {
  // Setup light sensor
  pinMode(LIGHT_SENSOR, INPUT);

  // Setup the temperature sensor
  Wire.begin(); //Join I2C Bus
  // The TMP102 uses the default settings with the address 0x48
  if(!temperatureSensor.begin())
  {
    Serial.println("ERROR: Cannot connect to TMP102.");
    Serial.println("Is the board connected? Is the device ID
    correct?");
    while(1);
  }

  // Setup the LCD
  // Setup number of columns and rows:
  lcd.begin(16, 2);
  // Set background color?
  lcd.setRGB(127, 127, 127);

  // Setup RGB LED
  rgbLed = new ChainableLED(7, 8, NUM_LEDS);

  // Set the led default color
  moods = new Moods(); // Set default mood
  prevMood = {0, 0, 0, ""};
  clearMood();

  delay(1000);
}
```

```
bool MoodLamp::detectMood() {
  int lightValue = analogRead(LIGHT_SENSOR);
  if (lightValue <= LIGHT_TRIGGER_LEVEL) {
    delay(15);
    int temperature = readTemperature();
    // Do not proceed if temperature is too low
    if (temperature < MINIMUM_TEMPERATURE) {
      Serial.println("ERROR: temperature too low.");
      return false;
    }
    Serial.print("Temperature read: ");
    Serial.println(temperature);
    MoodValue moodValue = moods->getMood(temperature);
    Serial.print("Mood = ");
    Serial.println(moodValue.mood);
    setMood(moodValue);
    prevMood = moodValue;
    return true;
  }
  return false;
}

void MoodLamp::reset() {
  lcd.clear();
  lcd.setCursor(0, 0);
  lcd.print("Mood Detector");
  lcd.setCursor(0, 1); // column 1, row 2
  lcd.print("Cover sensors");
}

void MoodLamp::clearMood() {
  reset();
```

```
  // Return to default colors
  MoodValue moodValue = moods->getDefaultMood();
  setMood(moodValue, false);
  prevMood = moodValue;
}

void MoodLamp::off()  {
  lcd.clear();
  rgbLed->setColorRGB(0, 0, 0, 0);
  setMood({0, 0, 0, ""}, false);
}

int MoodLamp::readTemperature() {
  int temperature;

  // Turn sensor on to start temperature measurement.
  temperatureSensor.wakeup();
  delay(100);
  // read temperature data
  temperature = (int)temperatureSensor.readTempF();
  // Place sensor in sleep mode to save power.
  temperatureSensor.sleep();

  return temperature;
}

void MoodLamp::setMood(MoodValue moodValue, boolean show_
message) {
  int red;
  int green;
  int blue;
  int red_increment = 0;
  int green_increment = 0;
  int blue_increment = 0;f
```

```
// Display mood on LCD
if (show_message) {
  lcd.clear();
  lcd.setCursor(0, 0);
  lcd.print("You are feeling");
  lcd.setCursor(0, 1);
  lcd.print(moodValue.mood);
}

// Prepare starting position and increment values for R,G,B
red = prevMood.red;
green = prevMood.green;
blue = prevMood.blue;
if (red < moodValue.red) red_increment = 1;
if (red > moodValue.red) red_increment = -1;
if (green < moodValue.green) green_increment = 1;
if (green > moodValue.green) green_increment = -1;
if (blue < moodValue.blue) blue_increment = 1;
if (blue > moodValue.blue) blue_increment = -1;

// If the colors are the same, just reset the display
if ((red == moodValue.red) && (green == moodValue.green)
   && (blue == moodValue.blue)) {
  rgbLed->setColorRGB(0, red, green, blue);
  delay(10);
}
// Implement a smooth transition to the next values
while ((red != moodValue.red) || (green != moodValue.green)
|| (blue != moodValue.blue)) {
  // Check for incement stops
  if (red == moodValue.red) red_increment = 0;
  if (green == moodValue.green) green_increment = 0;
```

```
        if (blue == moodValue.blue) blue_increment = 0;
        red = red + red_increment;
        green = green + green_increment;
        blue = blue + blue_increment;
        rgbLed->setColorRGB(0, red, green, blue);
        delay(10);
    }
}
```

As you can see, the code is long, but except for a few places not difficult. This example shows you how you can encapsulate more than one module in a single class. In fact, what we are modeling here is the mood lamp itself, not the individual components. Thus, we hide all of the complex operations with the hardware from the user. Neat!

Compile the Sketch

The last step is to compile the sketch before uploading it to your board. If you encounter any errors, be sure to fix them and recompile to ensure the sketch compiles without errors or serious warnings.

Once everything compiles, we're ready to start testing. But first, let's look at the code for the Raspberry Pi. You can skip to the "Sketch on the Arduino" section if you're curious to see how the project works. While the code will execute the same on both platforms, the values differ due to the differences in how the sensors are read (the range of values differs).

Raspberry Pi

This section presents a walk-through of the Python code you will write to create a mood lamp and moods class and use them in the main script. But first, there are a couple of libraries we must install on our Raspberry Pi.

Note If you have not installed the GrovePi libraries, please see Chapter 12 for complete details and install them before you begin. If you encounter problems, see the "GrovePi/GrovePi+ Troubleshooting Tips" section in Chapter 12.

Install a Software Library

Aside from the GrovePi libraries, we need only one more software library. Specifically, we need a library for the TMP102. SparkFun does not have a Python library on their own, but there is a third-party library we can use. Unfortunately, we must download it and use it manually because it does not have an installer.

You can download it from `https://github.com/n8many/TMP102py` using the code download button or clone it directly to your project folder with the following command:

```
$ git clone https://github.com/n8many/TMP102py.git
```

What that means is you should download and unzip the file from GitHub into a folder that contains the Python files for this project. For example, if you create a folder named `mood_detector`, the TMP102 folder should reside in `mood_detector/TMP102py`.

Once you have that library downloaded (or cloned) and in your project folder, we're ready to write the code.

Write the Code

The code for the Python version of this project is a bit different than the Arduino code. We will still create the same classes (`MoodLamp`, `Moods`), but the libraries for the LCD and RGB LED are different and not so easy to use.

The reason is the Python code for those modules is written using code modules that contain no classes – only functions. Furthermore, the initialization code is done for us via importing the library making it much harder to set up.

While the code works and we can make it do all the things the Arduino version does, the code for these modules is not written to current Python standards and to a Python eye will seem primitive and a bit hinky.[2] Sadly, unless you use the newer Grove modules, you may find more situations where you will have to use functions instead of classes. That's too bad because there are a lot of Grove modules available!

Once again, we will not dive into every line of code, but we will see some of the more complex code and those areas that differ significantly from the Arduino version. You can read through the code and learn more about how it works at your leisure.

Let's start with writing the classes starting with the `Moods` class.

Moods Class

Open the Thonny Python IDE under the *Main* ➤ *Programming* submenu. Create a new file `moods.py`. We will use the same functions as the Arduino version. The only difference is the `MoodValue` structure in the Arduino version is replaced with a tuple. Recall a Python tuple is immutable (you cannot change its internal values) and you access the parts of the tuple with an index in the same way you do an array.

While that may seem like a big change, if you recall from reading the Arduino code, we don't change the values returned from this class. We simply use them. To make things easier, we define constants to represent the parts of the tuple as follows. This makes it easy to use the tuple. For example, `myMood[GREEN]` references the green value for the RGB values.

---

[2] A highly technical term that means "this fish smells bad."

```
# Index for mood value tuple
RED = 0
GREEN = 1
BLUE = 2
MOOD = 3
```

Other than that, the rest of the code is similar to the Arduino version just rewritten in Python.

Listing 14-7 shows the complete code for the class with documentation removed for brevity. Take a few moments to read through the code so that you understand all of the parts of the code. As you will see, it is not nearly as complicated as the Arduino class, thanks to the helpful class library and utility class from Adafruit.

Listing 14-7. Moods Class (Python)

```
Mood definitions by temperature
MOOD_TROUBLED = 72
MOOD_ALERT = 74
MOOD_CALM = 78
MOOD_HAPPY = 82
MOOD_ROMANTIC = 86
MOOD_NERVOUS = 90
MOOD_STRESSED = 94
# Index for mood value tuple
RED = 0
GREEN = 1
BLUE = 2
MOOD = 3

class Moods:
    """Moods Class"""

    default_mood = (102, 178, 255, "Resting")
```

```
    def get_default_mood(self):
        """Get Default Mood"""
        return self.default_mood

    def get_mood(self, temperature):
        """Get Mood"""
        if temperature == 0:
            return (0, 0, 0, "")
        if temperature < MOOD_TROUBLED:
            return (self.default_mood[RED], self.default_
            mood[GREEN],
                    self.default_mood[BLUE], "")
        if MOOD_TROUBLED == temperature <= MOOD_ALERT:
            return (255, 128, 0, "Troubled")
        if MOOD_ALERT >= temperature < MOOD_CALM:
            return (153, 0, 153, "Alert")
        if MOOD_CALM >= temperature < MOOD_HAPPY:
            return (0, 102, 0, "Calm")
        if  MOOD_HAPPY >= temperature < MOOD_ROMANTIC:
            return (255, 255, 0, "Happy")
        if MOOD_ROMANTIC >= temperature < MOOD_NERVOUS:
            return (204, 0, 0, "Romantic")
        if MOOD_NERVOUS >= temperature <= MOOD_STRESSED:
            return (255, 0, 255, "Nervous")
        return (255, 255, 255, "Stressed!")
```

Now, let's look at the MoodLamp class file.

MoodLamp Class

Open the Thonny Python IDE under the *Main* ➤ *Programming* submenu. Create a new file moodlamp.py. We will use the same functions as we did in the Arduino version renamed to a more Python-friendly naming scheme.

What differ are the functions for interacting with the RGB LED, TMP102, and LCD. For these, we must import the grovepi module and several functions as follows:

```
from grovepi import (pinMode, analogRead, chainableRgbLed_init,
                     storeColor, chainableRgbLed_pattern)
```

Here we see we will use (in addition to the pinMode and analogRead for the light sensor) the chainableRgbLed_init() function to initialize the RGB LED and the strangely named chainableRgbLed_pattern() function to set the color for the RGB LED.

The TMP102 sensor uses a different Python library, and we import it as follows:

```
from TMP102py.tmp102 import TMP102
```

Fortunately, the function names used are very similar to the Arduino version of the SparkFun library. For example, to read the temperature, we use readTemperature() - the same name as the Arduino version.

The LCD on the other hand uses a different library provided in the GrovePi libraries, but it is written using different functions. Rather than use the existing GrovePi library, we can create our own class that encapsulates the library and make the functions work in a similar manner as we saw in the Arduino version.

Specifically, we would expect to see clear(), print(), set_cursor(), and set_rgb() functions, but the GrovePi library doesn't use them. Listing 14-8 shows a new class named GroveLcdRgb that imports the GrovePi library named grove_rgb_lcd wrapping those functions into the ones we expect.

Listing 14-8. GroveLcdRgb Class (Python)

```
from grove_rgb_lcd import *

class GroveLcdRgb:
    def clear(self):
        textCommand(0x01) # clear display
        setRGB(0, 0, 0)
        setText("")

    def print(self, message):
        for symbol in message:
            bus.write_byte_data(DISPLAY_TEXT_ADDR, 0x40,
            ord(symbol))

    def set_cursor(self, col, row):
        if row == 0:
            dest_col = col | 0x80
        else:
            dest_col = col | 0xc0
        bus.write_byte_data(DISPLAY_TEXT_ADDR, 0x80, dest_col)

    def set_rgb(self, red, green, blue):
        setRGB(red, green, blue)
```

We can save this file as grove_lcd_rgb.py so as not to confuse it with the GrovePi library. When we want to use it in our `MoodLamp` class, we import it as follows. The functions this class provides are also (and thankfully) similar to the library we used in the Arduino version:

Note The import statement in this class is considered bad form. You should not use global imports. Rather, a better form would be `from grove_rgb_lcd import (textCommand, setRGB, setText, bus, DISPLAY_TEXT_ADDR)`. Try it yourself and see that it works.

```
from grove_lcd_rgb import GroveLcdRgb
```

Finally, we import the class and index constants from the Moods class as follows:

```
from moods import Moods, RED, GREEN, BLUE, MOOD
```

Those are the major differences in the Python code. Everything else is similar to the Arduino version. Listing 14-9 shows the complete code for the class with documentation removed for brevity. Take a few moments to read through the code so that you understand all of the parts of the code. As you will see, it is not nearly as complicated as the Arduino class, thanks to the helpful class library and utility class from Adafruit.

Listing 14-9. MoodLamp Class (Python)

```
import time
from grovepi import (pinMode, analogRead, chainableRgbLed_init,
                     storeColor, chainableRgbLed_pattern)
from TMP102py.tmp102 import TMP102
from grove_lcd_rgb import GroveLcdRgb
from moods import Moods, RED, GREEN, BLUE, MOOD

# Default value read from light sensor to initiate a mood
  detection.
LIGHT_TRIGGER_LEVEL = 300
LIGHT_SENSOR = 0
RGB_LED = 7
NUM_LEDS = 1
MINIMUM_TEMPERATURE = 72 # Degrees Fahrenheit

class MoodLamp:
    """Mood Lamp Class"""

    moods = Moods()
```

```
    lcd = GroveLcdRgb()                  # Grove LCD
    temperature_sensor = TMP102()        # Qwiic TMP102
    prev_mood = ()

    def begin(self):
        """Begin"""
        # Setup light sensor
        pinMode(LIGHT_SENSOR, "INPUT")

        # Setup the temperature sensor
        self.temperature_sensor.setUnits('F')

        # Set background color
        self.lcd.clear()
        self.lcd.set_rgb(127, 127, 127)

        # Setup the RGB LED
        chainableRgbLed_init(RGB_LED, NUM_LEDS)

        # Set the led to default color starting from off
        self.prev_mood = (0, 0, 0, "")
        self.clear_mood()

        time.sleep(1)

    def detect_mood(self):
        """Detect Mood"""
        if analogRead(LIGHT_SENSOR) <= LIGHT_TRIGGER_LEVEL:
            time.sleep(0.015)
            temperature = self.read_temperature()
            # Do not proceed if temperature is too low
            if temperature < MINIMUM_TEMPERATURE:
                print("ERROR: temperature too low.")
                return False
```

```
            print("Temperature read: {}".format(temperature))
            mood_value = self.moods.get_mood(temperature)
            print("Mood = {}".format(mood_value[MOOD]))
            self.set_mood(mood_value)
            self.prev_mood = mood_value
            return True
        return False

    def reset(self):
        self.lcd.clear()
        self.lcd.set_rgb(127, 127, 127)
        self.lcd.set_cursor(0, 0)
        self.lcd.print("Mood Detector")
        self.lcd.set_cursor(0, 1) # column 1, row 2
        self.lcd.print("Cover sensors")

    def clear_mood(self):
        """Clear Mood"""
        self.reset()
        # Return to default colors
        mood_value = self.moods.get_default_mood()
        self.set_mood(mood_value, False)
        self.prev_mood = mood_value

    def off(self):
        self.lcd.clear()
        self.lcd.set_rgb(0, 0, 0)
        self.set_mood((0, 0, 0, ""), False)

    def read_temperature(self):
        """Read Temperature"""
        temperature = 0
```

```
        # Turn sensor on to start temperature measurement.
        self.temperature_sensor.wakeup()
        time.sleep(0.500)
        # read temperature data
        temperature = int(self.temperature_sensor.
        readTemperature())
        # Place sensor in sleep mode to save power.
        self.temperature_sensor.sleep()
        return temperature

    def set_mood(self, mood_value, show_message=True):
        """Set Mood"""
        red = 0
        green = 0
        blue = 0
        red_increment = 0
        green_increment = 0
        blue_increment = 0

        # Display mood on LCD
        if show_message:
            self.lcd.clear()
            self.lcd.set_rgb(127, 127, 127)
            self.lcd.set_cursor(0, 0)
            self.lcd.print("You are feeling")
            self.lcd.set_cursor(0, 1)
            self.lcd.print(mood_value[MOOD])

        # Prepare starting position and increment values for
          R,G,B
        red = self.prev_mood[RED]
        green = self.prev_mood[GREEN]
        blue = self.prev_mood[BLUE]
```

```
if red < mood_value[RED]:
    red_increment = 1
if red > mood_value[RED]:
    red_increment = -1
if green < mood_value[GREEN]:
    green_increment = 1
if green > mood_value[GREEN]:
    green_increment = -1
if blue < mood_value[BLUE]:
    blue_increment = 1
if blue > mood_value[BLUE]:
    blue_increment = -1

# If the colors are the same, just reset the display
if (red == mood_value[RED]) and (green == mood_
value[GREEN])
    and (blue == mood_value[BLUE]):
    storeColor(red, green, blue)
    chainableRgbLed_pattern(RGB_LED, 0, 0)
    time.sleep(0.010)
    return

# Implement a smooth transition to the next values
while ((red != mood_value[RED]) or (green != mood_
value[GREEN]) or
       (blue != mood_value[BLUE])):
    # Check for incement stops
    if red == mood_value[RED]:
        red_increment = 0
    if green == mood_value[GREEN]:
        green_increment = 0
    if blue == mood_value[BLUE]:
        blue_increment = 0
```

```
            red = red + red_increment
            green = green + green_increment
            blue = blue + blue_increment
            storeColor(red, green, blue)
            chainableRgbLed_pattern(RGB_LED, 0, 0)
            time.sleep(0.001)
```

Now we can write our main script.

Main Script (Python)

Open the Thonny Python IDE under the *Main* ➤ *Programming* submenu. Create a new file `mood_detector.py`. There is nothing new in this code as it follows the same flow as the Arduino example. Specifically, we need only one instance of our new mood lamp class, and we call the `detect_mood()` to detect a new mood and display it on the RGB LED and LCD and then `clear_mood()` to return the RGB LED to the default color.

Once again, making a class to handle all of the hardware makes the main script far less complicated. Listing 14-10 shows the complete code for the main script for this project. You can read through it to see how all of the code works.

Listing 14-10. Main Script (Python)

```
# Import libraries
import sys
import time

from mood_lamp import MoodLamp

# Constants
MOOD_TIMEOUT = 10

# Global variables
mood_lamp = MoodLamp()
```

```
#
# main()
#
# Main script to respond to the user when she covers the
# temperature and light sensors with her fingers.
#
def main():
    """Main"""
    print("Welcome to the mood detector!")
    mood_lamp.begin()
    print("Mood detector is ready.")
    print("Place fingers over the light and temperature
    sensor.")
    print("Reading takes about 15 seconds.")
    while True:
        if mood_lamp.detect_mood():
            time.sleep(MOOD_TIMEOUT)
            mood_lamp.clear_mood()
        time.sleep(2)

if __name__ == '__main__':
    try:
        main()
    except (KeyboardInterrupt, SystemExit) as err:
        mood_lamp.off()
        print("\nbye!\n")
sys.exit(0)
```

OK, that's it! We've written the code. Unlike the Arduino, we do not need to compile the Python code. So we're now ready to execute the project!

Execute the Project

Now that we've spent many pages exploring the Grove modules and writing the code to interact with them, it is time to test the project by executing (running) it.

When the project runs (executes), you will see some diagnostic messages written to the serial monitor (Arduino) or the terminal (Raspberry Pi). You will also see a welcome message appear on the LCD. When you place a finger on each of the light and temperature sensors, you will see the mood detection results, and the RGB LED will change colors and stay that color for a period of time. Figure 14-11 shows an example of the default screen on the LCD. Notice the instructions are rather terse as there isn't much room to print characters.

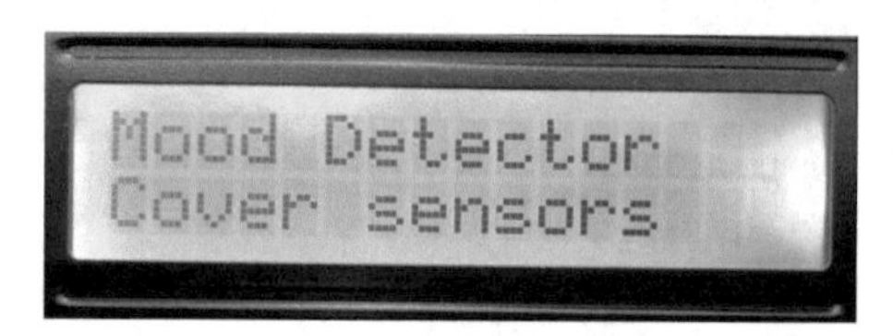

***Figure 14-11.** Executing the mood detector project*

Figure 14-12 shows an example of what you should see on the LCD when a mood detection/presentation is in progress.

***Figure 14-12.** Detecting a mood with the mood detector project*

Executing the code depends on which platform you're using. Let's look at the Arduino first.

Sketch on the Arduino

Executing the sketch on the Arduino requires connecting our board to our PC and then uploading the sketch to the Arduino. Recall the sketch will run so long as the USB cable is connected to our PC (and the Arduino).

Execute the Sketch

To execute the sketch, be sure your Arduino is connected and you've selected the correct board under the *Tools* ➤ *Board* menu. You also need to ensure you have the correct port selected under the *Tools* ➤ *Port* menu.

Once those items are set, you can click the *Upload* button or choose *Sketch* ➤ *Upload* from the menu. The Arduino IDE will compile the sketch and then upload it to your Arduino. Once you see the `Done uploading...` message, you can open the serial monitor. You should see the output begin momentarily that is the same as that on the LCD. Go ahead, and try it out! You should see values similar to the following:

```
Welcome to the mood detector!
Mood detector is ready.
Place fingers over the light and temperature sensor.
Reading takes about 15 seconds.
Temperature read: 81
Mood = Calm
Temperature read: 87
Mood = Happy
...
```

If something isn't working, check your connections or refer to Chapter 13 for troubleshooting tips.

Python Code on the Raspberry Pi

Executing the sketch on the Raspberry Pi requires running the Python code in a terminal after connecting your modules and powering on the Raspberry Pi. Recall the code will run until you stop it with *CTRL+C* on the keyboard.

Execute the Python Code

To run the Python code on the Raspberry Pi, you can issue the command `python3 ./mood_detector.py` from the same folder where the file was saved as shown in the following. You should get results similar to the following:

```
$ python3 ./mood_detector.py
Welcome to the mood detector!
Mood detector is ready.
Place fingers over the light and temperature sensor.
Reading takes about 15 seconds.
Temperature read: 83
Mood = Happy
Temperature read: 86
Mood = Romantic
...
```

If everything worked as executed, congratulations! You've just built your second Grove project. If something isn't working, check your connections or refer to Chapter 13 for troubleshooting tips.

Going Further

While we didn't discuss them in this chapter, there are some ideas where you could make this project into an IoT project. Here are just a few suggestions you can try once we have learned how to take our projects to the cloud. Put your skills to work!

- *Mood portal*: You can display the values of the last moods detected on a web page to allow you to see a progression of your mood from anywhere in the world. If you also add the date and time, you can see how your typical day or week goes.
- *Alternative hardware*: Implement the hardware used in the Python version in Arduino.
- *Sensor power management*: Add a relay to turn the power on and off to the light and temperature sensors to extend their life.
- *Bigger, better*: One of the ways you can enhance this project is to find a small white paper bag or opaque glass cylinder or dome to place over the RGB LED. Not only will this help diffuse the light, but it will also make it seem to glow. You can also add more RGB LEDs to make a larger, brighter lamp.

Summary

In this chapter, we got more hands-on experience making projects with Grove analog and digital modules as well as several I2C devices and even a Qwiic module thrown in for good measure. We used these modules to create a mood lamp that detects your mood when you place fingers on the light and temperature sensors.

Along the way, we learned more about how to work with Grove modules including how to write our own classes for managing multiple modules and sensors. We also saw how to use alternative software libraries in our Python project. Finally, we saw some potential to make this project better as well as some ideas for how to adapt the project for practical uses.

In the next chapter, we will see another project that demonstrates how to use more Grove modules to create a weather application to measure air temperature, humidity, and air quality – just the thing for the pollen season.

CHAPTER 15

Monitoring Your Environment

One of the most common examples of IoT projects is a weather project. Given the current health crisis, let's look a little closer at our indoor environment. There are several products you can buy to monitor indoor air quality, and for those with severe allergies and similar health conditions (some can be life threatening), an indoor air monitor may be a requirement to treat their condition.

In this chapter, we will see how to create a simple indoor environment monitor that detects air quality (the presence of harmful gases), dust concentration, barometric pressure, and temperature displaying the data on a small OLED. We'll see more analog and digital modules as well as the use of multiple I2C Grove modules.

Like the last chapter, we will see how to implement this project on the Arduino and Raspberry Pi but with an interesting challenge for the Python version. Let's get started.

Project Overview

The project for this chapter is designed to demonstrate how to use analog, digital, and multiple I2C devices on the same Grove host adapter to build an indoor environment monitor. It uses several sensors to sample the air for gases and dust as well as sampling the temperature and barometric pressure.

C. Bell, *Beginning IoT Projects*, https://doi.org/10.1007/978-1-4842-7234-3_15

Caution The project for this chapter should not be used for treating life-threatening health disorders. It is meant to be a demonstration of what is possible and should not be relied upon for critical health choices.

We will use a simple loop to sample the sensors every minute. For most uses, that is actually too frequent as indoor air quality may not change quickly. If you choose to install this project for long-term use, you may want to experiment with longer sampling times especially if you plan to log the data.

Note The Arduino version of this project is limited to Arduino AVR boards such as the Uno because one of the software libraries uses AVR-specific code. The code for this project may not work for other Arduino boards.

What Will We Learn?

By implementing this project, we will get more practice in using analog and digital Grove modules and how to connect Grove modules to the various protocol connectors on the host adapter. We will also see how to use multiple I2C sensors in the same project. Rather than a typical educational project, this project is fun to use as well as suited for learning all of the nuances of building Grove projects that use a mixture of analog, digital, and I2C modules.

The programming tasks will reveal how to read values from the sensors using a variety of methods and libraries to display the data on a small OLED.

Hardware Required

The hardware needed for this project is listed in Table 15-1. URLs for each component are included for ease of ordering including duplicate entries for alternative vendors. We will use the Grove OLED 0.96 and Buzzer along with Grove I2C High Accuracy Temperature, Temperature and Barometer, Air Quality, and Dust sensors. While this project doesn't include any Qwiic components, three of these sensors use I2C.

Table 15-1. *Hardware Needed for the Environment Monitor Project*

Component	URL	Qty	Cost
Grove OLED 0.96 v1.3	seeedstudio.com/Grove-OLED-Display-0-96.html	1	$16.40
Grove Buzzer	seeedstudio.com/Grove-Buzzer.html	1	$2.10
Grove I2C High Accuracy Temperature Sensor (MCP9808)	seeedstudio.com/Grove-I2C-High-Accuracy-Temperature-Sensor-MCP9808.html	1	$5.20
Grove Temperature and Barometer Sensor (BMP280)	seeedstudio.com/Grove-Barometer-Sensor-BMP280.html	1	$9.80
Grove Air Quality Sensor	www.seeedstudio.com/Grove-Air-Quality-Sensor-v1-3-Arduino-Compatible.html	1	$10.90
Grove Dust Sensor	www.seeedstudio.com/Grove-Dust-Sensor-PPD42NS.html	1	$12.70

(*continued*)

Table 15-1. (*continued*)

Component	URL	Qty	Cost
Grove cables(any length can be used)	Included with each preceding module	6	
Arduino MKR 1010 WiFi	www.sparkfun.com/products/15251	1	$35.95
Raspberry Pi 3B or later	www.sparkfun.com/categories/233 www.adafruit.com/category/176	1	$35.00+
Grove Base Shield V2.0 for Arduino	www.seeedstudio.com/Base-Shield-V2.html	1	$4.45
GrovePi+	www.sparkfun.com/products/15945	1	$5.95

Note At print, the Grove Dust Sensor is on back order from Seeed Studio due to high demand during the pandemic. You may be able to find it from another vendor or an online auction site.

About the Hardware

Let's discuss these components briefly. We will discover how to work with the hardware in more detail later in the chapter. If you implement the Python version, be sure to note the changes in the libraries used.

Grove OLED 0.96

Since we have more data than can fit on two short lines, we must change our display of choice to use a small OLED module. The Grove OLED 0.96 is a monochrome 128×64 dot matrix display with high brightness and contrast ratio and low power consumption. You can address all of the pixels (dots) on the screen too. Note that there are several versions of this module. We will be using the version that uses the SSD1308 chip. If you use a different version, you may need to use a different software library. Figure 15-1 shows the Grove Light Sensor.

Figure 15-1. *Grove Light Sensor (courtesy of seeedstudio.com)*

Grove Buzzer

A new option for this project is the use of sound so that we can play a tone as a warning "beep."[1] We will keep it simple and use the Grove Buzzer module, which is a simple Piezo buzzer that is normally used to make beep sounds by turning it on and off with a digital connection. Figure 15-2 shows the Grove Buzzer module.

[1] As you will hear, it is reminiscent of a typical, annoying smoke alarm low battery signal.

Figure 15-2. *Grove Buzzer (courtesy of seeedstudio.com)*

Grove I2C High Accuracy Temperature Sensor (MCP9808)

The Grove I2C High Accuracy Temperature Sensor (or simply MCP9808) is a high-accuracy digital module based on the MCP9808 microchip. It features high accuracy measuring temperatures ranging from –40 to 125 degrees Celsius. While there are other temperature sensors available for use, this module is not only reliable and accurate, but it also uses I2C for easy integration into our environment monitor. Figure 15-3 shows the Grove I2C High Accuracy Temperature Sensor (MCP9808).

Figure 15-3. *Grove I2C High Accuracy Temperature Sensor (courtesy of seeedstudio.com)*

If you recall from our Qwiic modules, most permit you to alter the I2C address and other features using jumpers. This module is similar, and you can change the I2C address by soldering the jumpers on the back of the module. Figure 15-4 shows what the jumpers look like. Notice the labels for each.

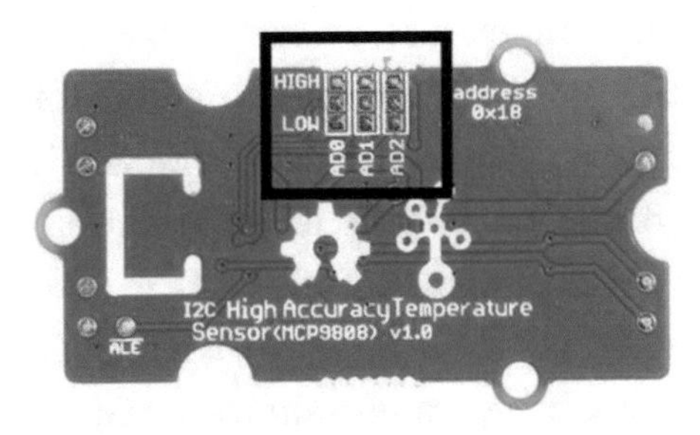

Figure 15-4. *Grove I2C jumpers – temperature sensor (courtesy of seeedstudio.com)*

You can change the I2C address by soldering across the jumpers as shown in Table 15-2.

Table 15-2. *I2C Address Map for the Grove I2C High Accuracy Temperature Sensor*

A0	A1	A2	Address
0	0	0	0x18
0	0	1	0x19
0	1	0	0x1A
0	1	1	0x1B
1	0	0	0x1C
1	0	1	0x1D
1	1	0	0x1E
1	1	1	0x1F

You may need to change the address if you add another I2C module with the same address or if you want to use multiple Grove I2C High Accuracy Temperature Sensor modules.

Grove Temperature and Barometer Sensor (BMP280)

Since we are capturing temperature, we may also want to measure the barometric pressure. The Grove Temperature and Barometer Sensor (or simply BMP280) is an excellent choice for that data. While it can also measure temperature and can be used to determine altitude, we will use it solely for the barometric pressure. If you'd like to see how to do that, visit `https://www.seeedstudio.com/Grove-Barometer-Sensor-BMP280.html` for more information. Figure 15-5 shows the Grove Temperature and Barometer Sensor.

Figure 15-5. *Grove Temperature and Barometer Sensor (courtesy of seeedstudio.com)*

Like the High Accuracy Temperature Sensor, you can also change the I2C address for this module using the jumpers on the back as shown in Figure 15-6.

Figure 15-6. *Grove I2C jumpers – barometric pressure sensor (courtesy of seeedstudio.com)*

Here, our choices are a bit narrower. We can use the jumpers to change the address from 0x76 (default) to 0x77.

Grove Air Quality Sensor

The Grove Air Quality Sensor is an analog sensor designed for indoor air quality testing and measures certain gases including carbon monoxide, alcohol, acetone, thinner, formaldehyde, and similar slightly toxic gases. While it does not differentiate among the gases, it provides a general value that you can use to determine thresholds for "safe" air quality. In fact, we will write the code to determine ranges for good, fair, and poor air quality. Figure 15-7 shows the Grove Air Quality Sensor.

Figure 15-7. *Grove Air Quality Sensor (courtesy of seeedstudio.com)*

Grove Dust Sensor

We will also be measuring the dust or particles in the air. The Grove Dust Sensor is a digital module and an excellent choice because it provides a percentage of particles found in the air. We can therefore write our code to test for a threshold of particulates in the air to determine dusty or even smoky conditions. Figure 15-8 shows the Grove Dust Sensor.

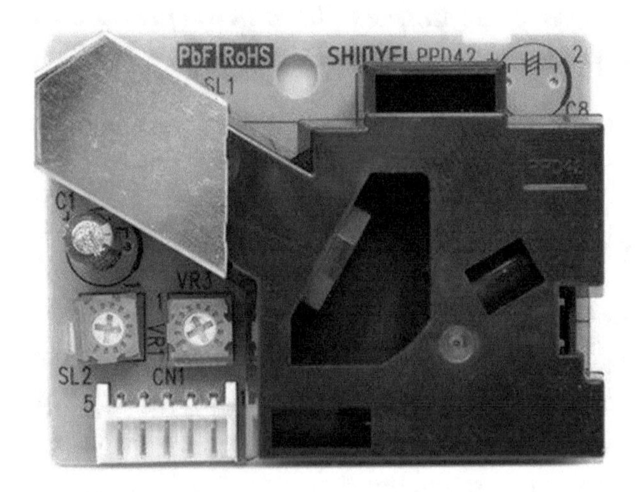

Figure 15-8. *Grove Dust Sensor (courtesy of seeedstudio.com)*

Tip See Chapter 12 for more details on the Grove host adapters.

Connect the Grove Modules

Recall from Chapter 12 we can use a single Grove cable to connect each Grove module separately to a specific connector on the host adapter on our host board. The host adapter for the Raspberry Pi has a different layout but has the same connectors we will use. Table 15-3 includes the details of each connection on the host adapter to help you make the right connections. Simply use the table to connect a Grove cable from the module to the Grove connector on the host adapter as marked in the table.

Table 15-3. *Grove Connections*

Module	Protocol	Grove Connector on the Host Adapter
OLED 0.96	I2C	I2C1
Buzzer	Digital	D6
High Accuracy Temperature	I2C	I2C2
Barometer	I2C	I2C3
Air Quality	Analog	A0
Dust	Digital	D7

Figure 15-9 shows an example of how you should connect your modules for this project. Notice the figure shows the Arduino host adapter, but the connections are labeled the same for the GrovePi on the Raspberry Pi.

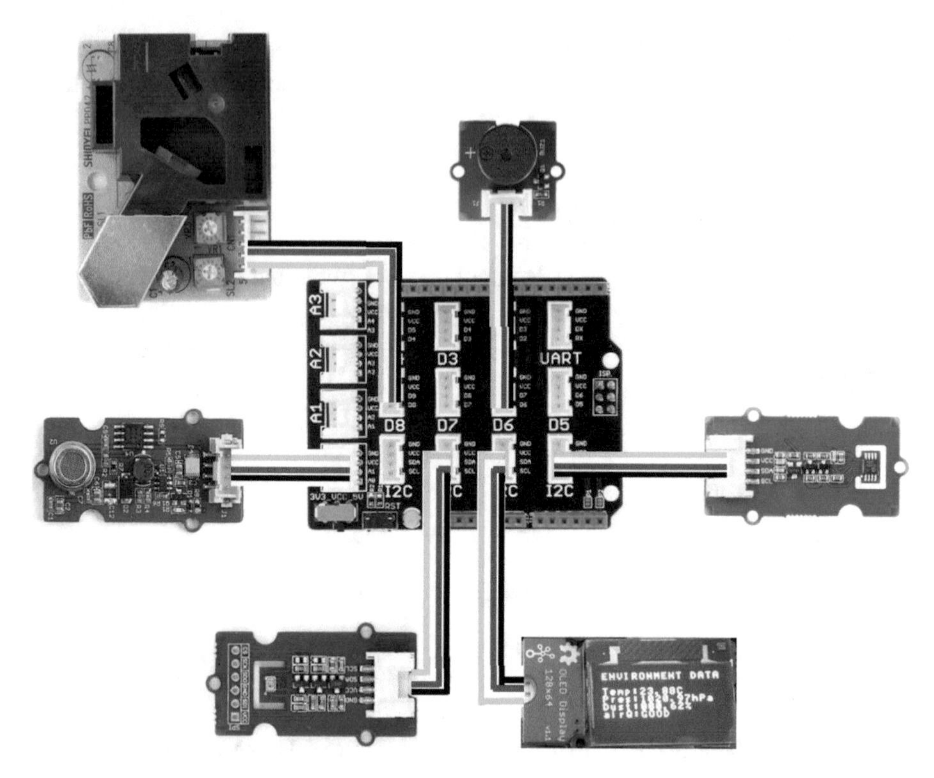

Figure 15-9. *Environment monitor project Grove connections*

Using an Enclosure

Since we have so many components and a bunch of cables connecting them all together, using the project can take a little bit of space. With all of those modules dangling by their cables tethered only to the Grove host adapter, you run the risk of accidentally unplugging a module, or, worse, the electronics on the module may come into contact with conductive material. You can mitigate this somewhat by using double-sided tape to tape them to your desk, but a better solution is to create a mounting plate. We could create a full enclosure, but as you will see, leaving the modules exposed gives the project a genuine cool factor.

If you have your own or access to a 3D printer, you can print a mounting plate. The source code for this chapter includes the 3D printing files you need to create a simple enclosure to mount the modules arranged in a manner that enables experimentation. Figure 15-10 shows the mounting plate.

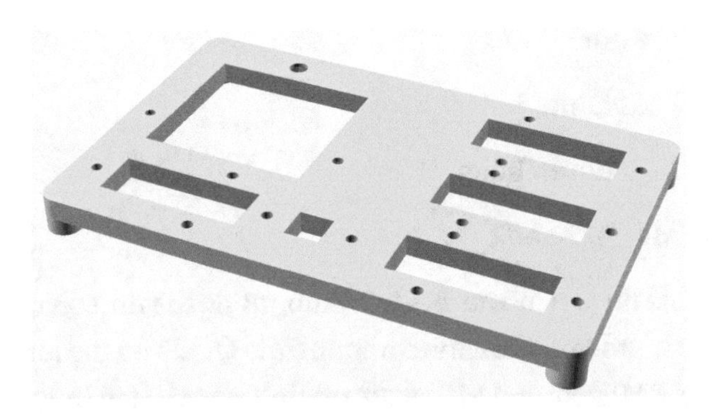

Figure 15-10. *3D mounting plate design for the environment monitor project*

If you're thinking this resembles a simple plank of wood (which would work equally as well), there are feet on the bottom of the plate and places for the nuts on the bottom as well. In fact, you will need to print this plate upside down.

There is also a set of spacers you will need to print as shown in Figure 15-11.

Figure 15-11. *3D spacer design for the environment monitor project*

Notice from left to right there are 11 short M2 spacers for the MCP9808, BMP280, air quality, and buzzer modules. There are three long M2 spacers for the OLED module. Finally, there is one M4 spacer for the dust sensor.

To mount the modules, you will need the following hardware:

- (14) M2 nuts
- (1) M4 nut
- (11) M2×8mm bolts
- (3) M2×19mm bolts
- (1) M4×5mm bolt

To assemble the enclosure, begin by mounting the dust sensor on the upper left, the buzzer on the center bottom, the OLED on the lower left, and the air quality, BMP280, and MCP9808 modules on the left (any order is fine). Figure 15-12 shows the completed project with the cables routed to the top.

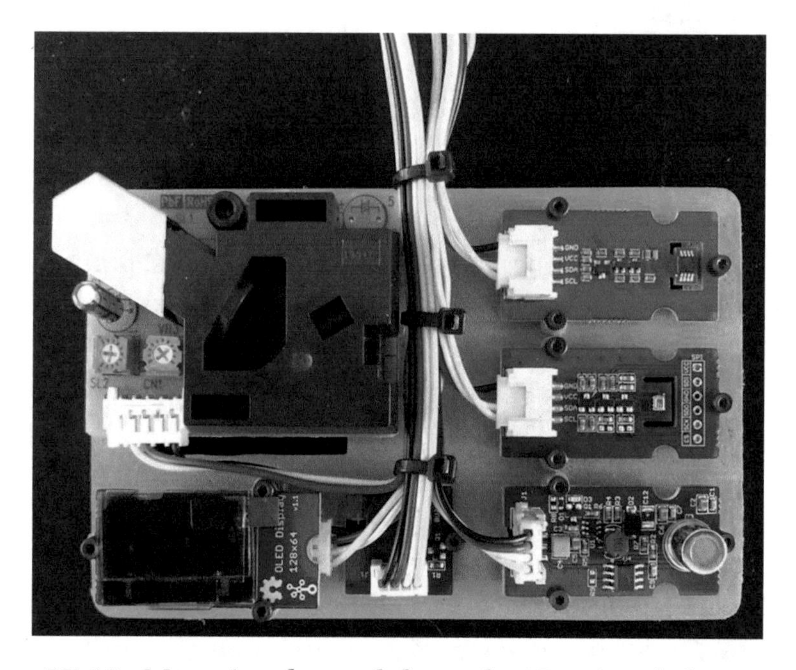

Figure 15-12. *Mounting the modules to the 3D printed plate*

Note The OLED will display text rotated 180 degrees from the Arduino version. If you prefer the orientation used in the Arduino version, you may want to open the 3D printer file and rotate the OLED mount accordingly. Or simply turn the unit around on your desk!

Before you celebrate by plugging all of your modules into your host adapter, take a few moments to carefully label each of the cables using a piece of masking or painter's tape. Write the connector label on the tape as shown in Figure 15-13. You don't have to worry about the I2C connections because they can be plugged into any of the I2C connectors.

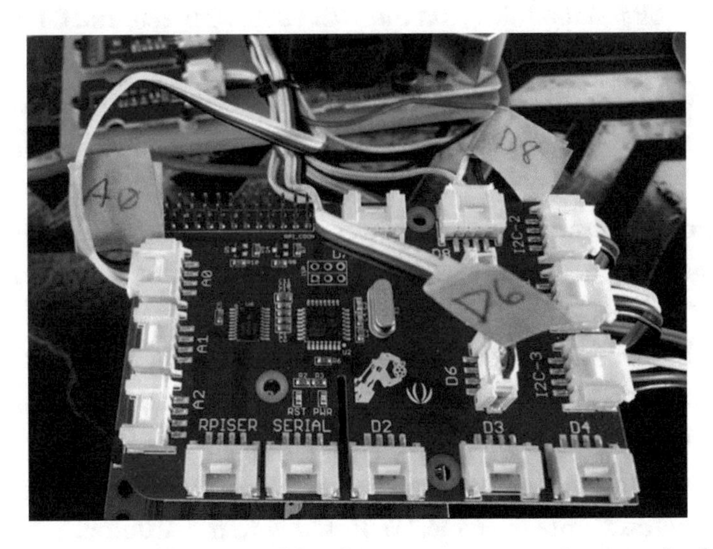

Figure 15-13. *Label your cables for easier connections to the host adapter*

If you have experience creating 3D models for printing, feel free to experiment with creating your own enclosure - perhaps one that also includes a battery and a small form factor host board. If you decide to build a complete enclosure, make sure to place holes or a grid opening over the sensors for airflow. The dust and air quality sensors are the modules that need openings most.

Now that we know more about the hardware for this chapter, let's write the code!

Write the Code

The code for this project involves following the usual pattern. For this project, that means using analog and digital modules as well as multiple I2C devices. The air quality sensor is an analog sensor, the buzzer and dust sensors are digital modules, and the MCP9808, BMP280, and OLED are I2C devices.

As you will see, the code isn't overly complicated for the Arduino version, but we will have some more work to do for the Python version. This is because some of the Grove modules used are not directly supported by the GrovePi library and one requires doing things a little unorthodox. Never fear, though. The code we will see is fully functional albeit not "standard" Python.

Like the previous projects, we will use a class to wrap our functionality. In this case, we will put reading of all of the sensors in the new class and control the OLED and buzzer from the main sketch.

Let's walk through how to prepare our computers to use the components and write the code. We'll start with the Arduino.

Arduino

This section presents a walk-through of the sketch and class you will write to read values from the sensors and display the values on the OLED. But first, there are a couple of libraries we must install on our PCs.

Install Software Libraries

We will need to install the Arduino libraries for the OLED, MCP9808, BMP280, and air quality modules separately. As you will see, some of these libraries are not from Seeed Studio directly, but they will work with our project.

Open the Library Manager from the Arduino IDE menu (*Sketch* ➤ *Include Library* ➤ *Library Manager...*). Then search for `U8G2` and install the latest version of the U8G2 library as shown in Figure 15-14. This library works with a host of different graphics chips including the one for our Grove 0.96 OLED.

Figure 15-14. *Installing the U8G2 OLED library (Arduino IDE)*

Similarly, we need to install the library for the Grove High Accuracy Temperature Sensor (MCP9808). The library for this module is not part of the Arduino Library Manager. Rather, we must download the library from GitHub (`https://github.com/Seeed-Studio/Grove_Temperature_sensor_MCP9808`) and install it manually. This is an excellent opportunity to learn a different method to install a library. Rather than download it, unzip, and then copy the module to the Arduino `libraries` folder, we will use the Arduino IDE to install it from the `.zip` file. First, download the `.zip` file from GitHub by clicking the *Code* button and then *Download ZIP* as shown in Figure 15-15.

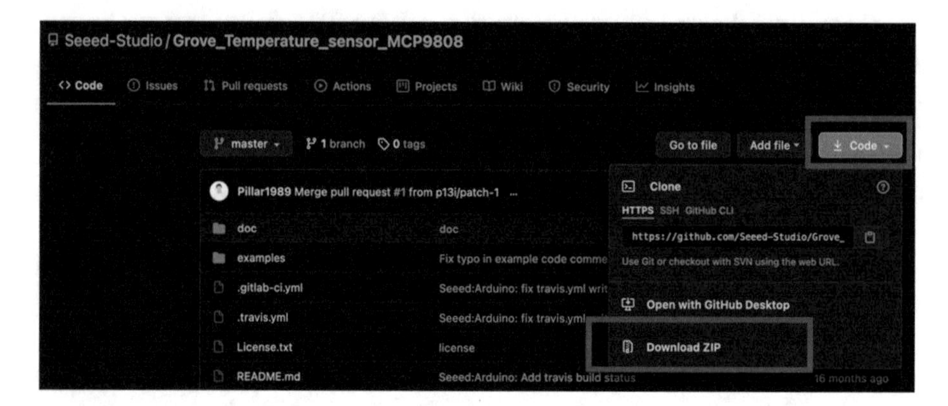

***Figure 15-15.** Downloading the Grove MCP9808 library (GitHub)*

Once the file has downloaded, go back to your Arduino IDE and click Sketch ➤ *Add .ZIP Library...* as shown in Figure 15-16. This will install the library from the zipped file. Cool, eh?

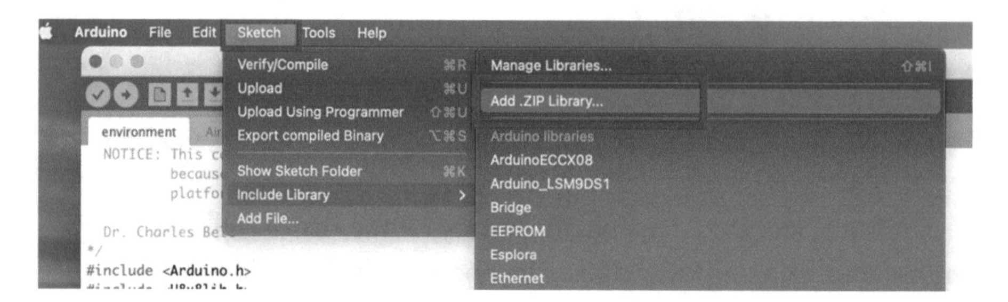

***Figure 15-16.** Installing a library from a .zip file*

We also need to install the library for the BMP280 sensor. Open the Library Manager and search for `Seeed BMP280` and then install the latest version as shown in Figure 15-17.

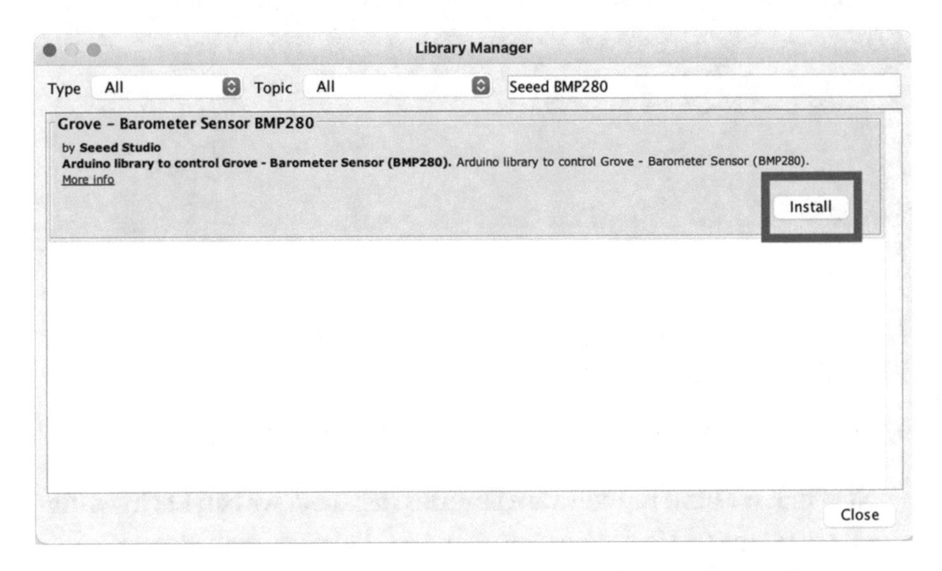

Figure 15-17. *Installing the BMP280 library (Arduino IDE)*

Finally, we need to install the library for the air quality sensor. Open the Library Manager and search for `AirQuality` and then install the latest version as shown in Figure 15-18.

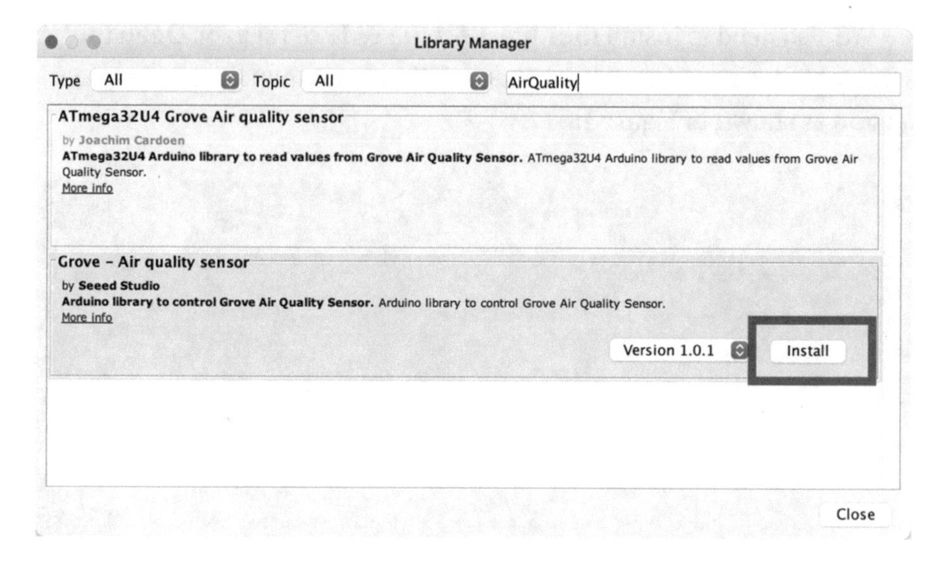

Figure 15-18. *Installing the air quality library (Arduino IDE)*

Now that we have the software libraries installed, we can begin writing our sketch. Since this is not our first Arduino sketch, we will discuss the code at a high level and skip the line-by-line details focusing on the mechanics of how the code works. You can study the code at your leisure to ensure you understand the sketch in more detail.

Write the Sketch

Recall we are going to use a class to encapsulate the sensors. We will write the class header first and then the main sketch and finally complete the code for the class. This is typically how programmers develop code with class modules (but not always). By creating the header first, we can understand how to use the class making writing the main sketch easier. The sensor class will be named `AirMonitor`.

Recall there is no way (currently) to create and add new files to a sketch (but you can add existing files by clicking *Sketch* ➤ *Add File...*). So we will once again create the main sketch and add the code header and source files manually.

Open a new sketch and name it environment.ino or whatever you'd like to use. Save the file and then close the project in the Arduino IDE.

To create the class files, navigate with your File Explorer (Finder) to the folder where you stored your main sketch (environment.ino). Then, use your File Explorer or a text file editor to create two new files named AirMonitor.h and AirMonitor.cpp. Or you can use a terminal to navigate to the folder and issue these commands to create the empty files:

```
environment % touch AirMonitor.h
environment % touch AirMonitor.cpp
```

Now, let's see the code for each file starting with the header file.

Class Header File

Click the tab named AirMonitor.h to open the blank file. Here, we will add the header or blueprint for the class. The module provides a mechanism to retrieve values from the sensors.

Aside from the sensor setup, which we will accomplish in the constructor, we will adopt a common set of functions that most libraries that manage multiple sensors use. We will use a function to read the data from all of the sensors and separate get functions to retrieve the values from each sensor (or data type).

For example, we first call readData() and then getTemperature(), getPressure(), getDust(), and getAirQuality(). An interesting side effect is if you don't call the read function, subsequent calls to the get functions return the same values. Thus, the read plus get functions may help in situations where some processing must be done after the read but before the get calls. This allows you to know precisely when the data was sampled, but delay consuming the data. Another possible solution is to provide a single function that reads the data from all sensors and returns the data in a structure. But we will keep it simple and use the read and get functions.

The `readData()` function returns a Boolean where true means the read was successful. The get functions all return a float for the data item except the `getAirQuality()`, which returns an `enum`. Recall the Grove Air Quality Sensor gives us a general value to examine rather than a specific value. According to the Seeed Studio documentation, we can set thresholds for three values of poor, fair, and good air quality. We also need to include two error conditions: one where the air quality detected is above the maximum threshold and another for a read error, which results in a value that is below the minimum threshold. The following shows the `enum` values we will use:

```
enum air_quality { ERROR_POOR, POOR, FAIR, GOOD, NO_READ };
```

Aside from those functions, we will create some private variables for storing the data. However, for the dust sensor, we need to do something a bit different. Recall the dust sensor has a read time of 30 seconds. So, while we can call the `readData()` function whenever we want, we must keep track of when the last successful read of the dust sensor has occurred. For example, if we call `readData()` every 10 seconds, only the first and third calls will refresh the dust sensor because it can only be read every 30 seconds. Thus, the dust value may be stale if the `readData()` function is called more frequently than every 30 seconds. This is a good example of how sensor refresh or minimal read times can affect your sampling rate.

CHOOSING A SAMPLE RATE

One of the things that you must consider when writing IoT solutions is how often you need to read data called the sample rate (or sampling rate). There are several factors you must consider, all of which should help you determine how often you should read data.

First, you must consider how often you can get data from the sensors. Some sensors may require as much as several minutes to refresh values. Most of those either let you read stale data (the last value read) or emit an error if you read the data too frequently.

Aside from the sensors, you also need to consider how often the data changes or how often you need to check/retrieve the data. The application will play a big factor in determining an optimum rate. For example, if you are sampling a sensor for data that doesn't change often, there is no point in reading it more frequently.

Another factor to consider concerns storing the data. If you are planning to store the data, reading the data every second could generate more data than your storage mechanism can handle.

Finally, the criticality of the data may also be a factor. More specifically, if the data is used to make critical decisions for industrial, mechanical, or health decisions, the sample rate may need to be high (fast). For example, it would be far too late to detect oncoming vehicles every 30 seconds.

When choosing a sample rate, you must consider all of these elements: refresh rates of your sensors, how often the data will change, how much data you want to store, and the criticality of the data.

Since the header code is not difficult to read, let's look at the completed code for the header file. Listing 15-1 shows the file.

Listing 15-1. AirMonitor Header File

```
#include <Arduino.h>
#include "Seeed_BMP280.h"
#include "Seeed_MCP9808.h"

// Constants
#define DUST_PIN 8
#define DUST_SAMPLE_RATE 30000

class AirMonitor {
public:
  enum air_quality { ERROR_POOR, POOR, FAIR, GOOD, NO_READ };
  AirMonitor();
```

```
  bool readData();
  float getTemperature() { return temperature; }
  float getPressure() { return pressure; }
  float getDust() { return dustConcentration; }
  air_quality getAirQuality() { return airQuality; }

private:
  // Environment variables
  float temperature {0.0};
  float pressure {0.0};
  air_quality airQuality {GOOD};
  float dustConcentration {0.0};
  bool initOk {true};

  // Variables for dust sensor
  unsigned long starttime;

  BMP280 *bmp280;
  MCP9808 *tempSensor;
};
```

OK, let's return to the main sketch to see how we can use this class.

Main Sketch

Now click the `environment.ino` tab to return to the main sketch. Since we are placing all of the sensor work in the `AirMonitor` class, all we need to do here is instantiate a new class instance (stored in a variable named `airQuality`), set up the serial class, initialize the OLED and buzzer, and then print the greeting. All of these are done in the `setup()` function.

The `loop()` function simply calls the `readData()` method, and if it returns true, we get the data and display it on the OLED. The only extra work we need to do is determine what the air monitor is returning and print the correct value and examine the data to ensure it is below established levels (to determine air quality). If the air quality is low, we display a message and play an alarm sequence on the buzzer.

We will also use a helper function to sound a tone on the buzzer. This isn't absolutely necessary, but it does help reduce the amount of code, and this concept will help you understand the way we use the buzzer in the next chapter.

While the main sketch has more code than some of the other example projects, it is still fairly simple to write. The added code are the calls to display data on the OLED. Listing 15-2 shows the code for the main sketch. Take a few moments and read through the code. Notice the sampling rate defined in the SAMPLING_RATE constant. This is set to 60 seconds, but could be set to a lower rate by increasing the value. Since this project is an air quality monitor, a sample rate of once every 5–10 minutes may be fine, but 60 seconds is an acceptable rate for an experimental project.

Listing 15-2. Main Sketch

```
#include <Arduino.h>
#include <U8x8lib.h>
#include <Wire.h>
#include "AirMonitor.h"

// Constants
#define SAMPLING_RATE 60000 // 60 seconds
#define BUZZER_PIN 6
#define WARNING_BEEPS 5

// Constants for environmental quality
#define MAX_TEMP 30.0
#define MAX_DUST 40.0

// Global variables
U8X8_SSD1306_128X64_NONAME_HW_I2C *oled;
AirMonitor *airQuality;
```

```
void beep(int duration=150) {
  digitalWrite(BUZZER_PIN, HIGH);
  delay(duration);
  digitalWrite(BUZZER_PIN, LOW);
}

void setup() {
  // Setup buzzer
  pinMode(BUZZER_PIN, OUTPUT);
  Serial.begin(115200);
  while (!Serial);
  // Setup OLED
  oled = new U8X8_SSD1306_128X64_NONAME_HW_I2C(U8X8_PIN_NONE);
  oled->begin();
  oled->setFont(u8x8_font_chroma48medium8_r);
  Serial.println("Welcome to the Environment Monitor!");
  Serial.print("Starting....");
  oled->drawString(0, 1, "Environment");
  oled->drawString(0, 2, "Monitor");
  oled->drawString(0, 4, "Starting...");
  airQuality = new AirMonitor();
  delay(3000);
  Serial.println("done.");
  oled->drawString(11, 4, "done.");
  beep();
  delay(3000);
  oled->clear();
}

void loop(void) {
  if (airQuality->readData()) {
    // Retrieve the data
```

```
float tempC = airQuality->getTemperature();
float pressure = airQuality->getPressure();
float dust = airQuality->getDust();
AirMonitor::air_quality air = airQuality->getAirQuality();

oled->drawString(0, 0, "ENVIRONMENT DATA");
oled->drawString(0, 3, "Temp: ");
oled->drawString(5, 3, String(tempC, 2).c_str());
oled->drawString(11, 3, "C    ");
oled->drawString(0, 4, "Pres: ");
oled->drawString(5, 4, String(pressure, 1).c_str());
oled->drawString(14, 4, "Pa");
oled->drawString(0, 5, "Dust: ");
if (dust == 0.0) {
  oled->drawString(5, 5, "--         ");
} else {
  oled->drawString(5, 5, String(dust, 2).c_str());
  oled->drawString(10, 5, "%     ");
}
oled->drawString(0, 6, "AirQ: ");
switch (air) {
  case AirMonitor::air_quality::ERROR_POOR:
  case AirMonitor::air_quality::POOR:
    oled->drawString(5, 6, "POOR");
    break;
  case AirMonitor::air_quality::FAIR:
    oled->drawString(5, 6, "FAIR");
    break;
  case AirMonitor::air_quality::GOOD:
    oled->drawString(5, 6, "GOOD");
    break;
```

```
      default:
        oled->drawString(5, 6, "--         ");
    }

    // Check for environmental quality
    if ((dust > MAX_DUST) or (tempC > MAX_TEMP) or
        (air == AirMonitor::air_quality::POOR) or
        (air == AirMonitor::air_quality::ERROR_POOR)) {
      for (int x = 0; x < WARNING_BEEPS; x++) {
        oled->drawString(3, 7, "ENV NOT OK");
        beep(250);
        delay(250);
        oled->drawString(3, 7, "           ");
        delay(250);
      }
    }
  } else {
    oled->clear();
    oled->drawString(0, 2, "ERROR! CANNOT");
    oled->drawString(0, 3, "READ DATA");
  }
  delay(SAMPLING_RATE);
}
```

Now we can write the final portion of our project – the code for the class.

Class Code File

Click the tab named `AirMonitor.cpp` to open the blank file. Here, we will add the code for the class. There are only two public functions to write: the constructor and `readData()`. Note that we defined the get functions in the class header file.

The code for both of the public functions is not difficult, and we leave the explanation of that code as an exercise. Likewise, using the software libraries to read the temperature and barometric pressure is not complicated.

The code for the dust sensor is a bit more complex because we cannot read the data faster than every 30 seconds. As you will see, the code to do this simply records a start time and a duration calculation. If 30 seconds or more has expired since the last read, we can read the sensor. Even so, the code isn't overly complicated.

However, the code for the air quality sensor requires some explanation. The air quality sensor software library uses an interrupt mechanism that is only available for the AVR-based Arduino boards. This is because it is a part of the hardware library itself and thus is hardware specific.

Fortunately, we do not need to learn about interrupts and AVR coding because Seeed Studio has provided the function we need. In short, this function is fired (called) at a specific time based on an internal hardware timer. When called, the function is designed to wait a certain period of time (approximately 2 seconds) before reading the value. Thus, this exposes a refresh cycle of this sensor of every 2 seconds. That isn't a problem for this project, but a project where you are using the sensor for industrial or health applications may need to consider this refresh rate. The following shows the interrupt function:

```
// Interrupt timer for air quality sensor taken from
// https://wiki.seeedstudio.com/Grove-Air_Quality_Sensor_v1.3/
ISR(TIMER2_OVF_vect)
{
  // Set 2 seconds as a detected duty
  if(airqualitysensor.counter == 122)
  {
    airqualitysensor.last_vol=airqualitysensor.first_vol;
    airqualitysensor.first_vol=analogRead(A0);
```

```
    airqualitysensor.counter=0;
    airqualitysensor.timer_index=1;
    PORTB=PORTB^0x20;
  }
  else
  {
    airqualitysensor.counter++;
  }
}
```

Aside from the interrupt function we need to add, we also have to treat the air quality sensor differently. Specifically, we cannot create a private variable in the class like we did for the other sensors. Rather, we must declare it as a global variable. This, along with the interrupt, makes the code to communicate with the air quality sensor a bit odd when you read the code.

Listing 15-3 shows the completed code for the class (documentation omitted for brevity). Take some time to read through the code to see the features we've discussed.

Listing 15-3. AirMonitor Code File

```
#include "AirMonitor.h"
#include "AirQuality.h"

// Global variables
AirQuality airqualitysensor;
int current_quality =-1;

AirMonitor::AirMonitor() {
  // Setup BMP280 sensor
  bmp280 = new BMP280();
```

```
  if(!bmp280->init()){
    Serial.println("ERROR: Cannot read BMP280!");
    initOk = false;
  }
  // Setup MCP9808 sensor
  tempSensor = new MCP9808();
  if(tempSensor->init()) {
    Serial.println("ERROR: Cannot read MCP9808!");
    initOk = false;
  }
  // Setup air quality sensor
  airqualitysensor.init(A0);
}

bool AirMonitor::readData() {
  // Variables for dust sensor
  unsigned long duration;
  unsigned long sampleTime = DUST_SAMPLE_RATE;
  unsigned long lowPulse = 0;
  float ratio = 0;
  float particleConcentration {0.0};

  // Check to see if initialization is Ok.
  // Don't read data if initialization fails.
  if (!initOk) return false;

  // Read barometer
  Serial.print("Pressure: ");
  Serial.print(pressure = bmp280->getPressure());
  Serial.println("Pa");
```

```
  // Read temperature
  tempSensor->get_temp(&temperature);
  Serial.print("Temperature: ");
  Serial.println(temperature);

  duration = pulseIn(DUST_PIN, LOW);
  lowPulse = lowPulse + duration;
  if ((millis()-starttime) > sampleTime) {
    ratio = lowPulse/(sampleTime*10.0);
    particleConcentration = 1.1 * pow(ratio, 3) - 3.8 *
    pow(ratio, 2) +
      520 * ratio + 0.62; // using spec sheet curve
    Serial.print(lowPulse);
    Serial.print(",");
    Serial.print(ratio);
    Serial.print(",");
    Serial.println(particleConcentration);
    // Guard against spurious values
    if (particleConcentration < 100.0) {
      dustConcentration = particleConcentration;
    }
    lowPulse = 0;
    starttime = millis();
  } else {
    particleConcentration = 0.0;
  }
  Serial.print("Dust %: ");
  Serial.println(particleConcentration);

  // Read air quality sensor
  current_quality=airqualitysensor.slope();
  if (current_quality >= 0) {
```

```
      if (current_quality == 0) {
        airQuality = ERROR_POOR;
        Serial.println("High pollution! ERROR");
      } else if (current_quality == 1) {
        airQuality = POOR;
        Serial.println("High pollution!");
      } else if (current_quality == 2) {
        airQuality = FAIR;
        Serial.println("Low pollution!");
      } else if (current_quality >= 3) {
        airQuality = GOOD;
        Serial.println("Fresh air");
      }
  } else {
    airQuality = NO_READ;
  }
  return true;
}

// Interrupt timer for air quality sensor taken from
// https://wiki.seeedstudio.com/Grove-Air_Quality_Sensor_v1.3/
ISR(TIMER2_OVF_vect)
{
  // Set 2 seconds as a detected duty
  if(airqualitysensor.counter == 122)
  {
    airqualitysensor.last_vol=airqualitysensor.first_vol;
    airqualitysensor.first_vol=analogRead(A0);
    airqualitysensor.counter=0;
    airqualitysensor.timer_index=1;
    PORTB=PORTB^0x20;
  }
```

```
  else
  {
    airqualitysensor.counter++;
  }
}
```

As you can see, the code is long, but except for a few places not difficult to read or learn how it works.

Compile the Sketch

The last step is to compile the sketch before uploading it to your board. If you encounter any errors, be sure to fix them and recompile to ensure the sketch compiles without errors or serious warnings.

Remember you can only compile this code with AVR-based Arduino boards. These include the Uno, Leonardo, Mega, etc. You can see a complete list by clicking the *Tools* ➤ *Board* ➤ *Arduino AVR Boards* menu item. If your chosen board is not in the list, you may not be able to compile the code for that board.

Once everything compiles, we're ready to start testing. But first, let's look at the code for the Raspberry Pi. You can skip to the "Sketch on the Arduino" section if you're curious to see how the project works. While the code will execute the same on both platforms, the values differ due to the differences in how the sensors are read (the range of values differs).

Raspberry Pi

This section presents a walk-through of the Python code you will write to create an AirMonitor class and use it in the main script. On a higher level, the Python code follows the same design as the Arduino code, but the libraries differ greatly. Let's begin with installing the libraries we will need on our Raspberry Pi.

Note If you have not installed the GrovePi libraries, please see Chapter 13 for complete details and install them before you begin. If you encounter problems, see the "GrovePi/GrovePi+ Troubleshooting Tips" section in Chapter 13.

Install Software Libraries

Aside from the GrovePi libraries, we need two more software libraries. Specifically, we need a library for the BMP280 and MCP9808. There are no such libraries from Seeed Studio for these sensors, but fortunately Pimoroni (`shop.pimoroni.com`) has a nice BMP280 library, and Adafruit (`www.adafruit.com`) has a nice MCP9808 library we can use. You can install both with the following commands:

```
$ pip3 install bmp280
$ pip3 install adafruit-circuitpython-mcp9808
```

Once you have those libraries installed, we're ready to write the code.

Write the Code

The code for the Python version of this project is a bit different than the Arduino code. We will still create the same class (`AirMonitor`), but the code to read the temperature, barometer, and air quality sensors differs.

Once again, we will not dive into every line of code, but we will see some of the more complex code and those areas that differ significantly from the Arduino version. You can read through the code and learn more about how it works at your leisure.

Let's start with writing the `AirMonitor` class.

AirMonitor Class

Open the Thonny Python IDE under the *Main* ➤ *Programming* submenu. Create a new file AirMonitor.py. We will use a similar `read_data()` function as the Arduino version, but we will return the data using a single get function. In addition, we will be using different libraries to read from the sensors. Let's look at the ways the code differs from the Arduino version.

Firstly, we do not have an `enum` in Python, but we do have an `enum` class that we can use to create our own `enum` for the air quality sensor values. The following shows the new class. To use the class values, we use a dotted notation like `AirQualityEnum.POOR`, but otherwise it represents the analogous values in the Arduino version:

```
class AirQualityEnum(enum.Enum):
    """Air Quality Enum"""
    POOR = 0
    FAIR = 1
    GOOD = 2
    ERROR = 3
```

Next, we use a dictionary in the class to store the sensor data and return it with the `get_data()` function. The following shows the new dictionary. Once returned to the main sketch, we simply use the key to retrieve the sensor data. For example, `data["temperature"]` fetches the temperature value:

```
data = {
    "temperature": 0.0,
    "pressure": 0.0,
    "dust_concentration": 0.0,
    "air_quality": AirQualityEnum.GOOD,
}
```

With regard to the sensors, we simply substitute a different software library for the temperature and barometric pressure sensors. For the air quality sensor, we use a library that is part of the GrovePi suite, which is much easier to use than the Arduino version's AVR interrupt.

Finally, the dust collector sensor is read via another GrovePi library, but we will use a private (in Python, that's a function that starts with an underscore) function to limit reads of the dust sensor data to no sooner than 30 seconds. The following shows the code for the new `_read_dust()` function. Notice how the code returns an unknown value if the sensor is read for the first time (e.g., at startup) and how it returns the old value until the timeout (30 seconds) has expired:

```
def _read_dust(self):
    """_read_dust()"""
    # If this is the first reading...
    if not self.start_time:
        self.start_time = time.time()
        # return 0.0 or "unknown"
        return 0.0
    if (time.time() - self.start_time) < DUST_SAMPLE_RATE:
        # return last value stored
        return self.data["dust_concentration"]
    # Threshold reached, reset timer
    self.start_time = time.time()
    # Sefault update period is 30000 ms
    grovepi.dust_sensor_en(pin=DUST_PIN)
    new_val = grovepi.dust_sensor_read(pin=DUST_PIN)
    print("> LPO time = {:3d} | LPO% = {:5.2f} | "
          "pcs/0.01cf = {:6.1f}".format(*new_val))
    grovepi.dust_sensor_dis(pin=DUST_PIN)
    if new_val[2] > 100.00:
        return 0.0
    return new_val[2]
```

Listing 15-4 shows the complete code for the class with documentation removed for brevity. Take a few moments to read through the code so that you understand all of the parts of the code. As you will see, it is not nearly as complicated as the Arduino class, thanks to the helpful libraries from Pimoroni, Adafruit, and GrovePi.

Listing 15-4. AirMonitor Class (Python)

```
import enum
import time
import board
import smbus
import adafruit_mcp9808
from bmp280 import BMP280
import grovepi

# Constants
DUST_PIN = 8
DUST_SAMPLE_RATE = 30 # 30 seconds
AIR_SENSOR_PIN = 0

class AirQualityEnum(enum.Enum):
    """Air Quality Enum"""
    POOR = 0
    FAIR = 1
    GOOD = 2
    ERROR = 3

class AirMonitor:
    data = {
        "temperature": 0.0,
        "pressure": 0.0,
        "dust_concentration": 0.0,
        "air_quality": AirQualityEnum.GOOD,
    }
```

```
    mcp9808 = None
    bmp280 = None
    start_time = None

    def __init__(self):
        # Setup MCP9808 sensor
        i2c = board.I2C()  # uses board.SCL and board.SDA
        self.mcp9808 = adafruit_mcp9808.MCP9808(i2c)

        # Setup the BMP280
        bus = smbus.SMBus(1)
        self.bmp280 = BMP280(i2c_dev=bus, i2c_addr=0x77)

        # Setup air quality
        grovepi.pinMode(AIR_SENSOR_PIN, "INPUT")

    def read_data(self):
        """read_data"""
        print("\n>> Reading Data <<")
        # Read temperature
        try:
            print("> Reading temperature = ", end="")
            self.data["temperature"] = self.mcp9808.temperature
            print(self.data["temperature"])
        except Exception as err:
            print("ERROR: Cannot read temperature: {}".
            format(err))
            return False

        # Read pressure
        try:
            print("> Reading pressure = ", end="")
            self.data["pressure"] = self.bmp280.get_pressure()
            print(self.data["pressure"])
```

```
        except Exception as err:
            print("ERROR: Cannot read pressure: {}".format(err))
            return False

        # Read dust
        try:
            print("> Reading dust concentration")
            self.data["dust_concentration"] = self._read_dust()
            print("> Dust concentration =
            {}".format(self.data["dust_concentration"]))
        except Exception as err:
            print("ERROR: Cannot read dust concentration: {}".
            format(err))
            return False

        # Read air quality
        try:
            print("> Reading air quality = ", end="")
            sensor_value = grovepi.analogRead(AIR_SENSOR_PIN)
            if sensor_value > 700:
                self.data["AirQualityEnum"] = AirQualityEnum.
                POOR
            elif sensor_value > 300:
                self.data["AirQualityEnum"] = AirQualityEnum.
                FAIR
            else:
                self.data["AirQualityEnum"] = AirQualityEnum.
                GOOD
            print(self.data["AirQualityEnum"])
        except IOError as err:
            print("ERROR: cannot read air quality: {0}".
            format(err))
```

```
            self.data["AirQualityEnum"] = AirQualityEnum.ERROR
            return False

        return True

    def get_data(self):
        """get_data"""
        return self.data

    def _read_dust(self):
        """_read_dust()"""
        # If this is the first reading...
        if not self.start_time:
            self.start_time = time.time()
            # return 0.0 or "unknown"
            return 0.0
        if (time.time() - self.start_time) < DUST_SAMPLE_RATE:
            # return last value stored
            return self.data["dust_concentration"]
        # Threshold reached, reset timer
        self.start_time = time.time()
        # Sefault update period is 30000 ms
        grovepi.dust_sensor_en(pin=DUST_PIN)
        new_val = grovepi.dust_sensor_read(pin=DUST_PIN)
        print("> LPO time = {:3d} | LPO% = {:5.2f} | "
              "pcs/0.01cf = {:6.1f}".format(*new_val))
        grovepi.dust_sensor_dis(pin=DUST_PIN)
        if new_val[2] > 100.00:
            return 0.0
        return new_val[2]
```

Now we can write our main script.

Main Script (Python)

Open the Thonny Python IDE under the *Main* ➤ *Programming* submenu. Create a new file `environment.py`. There is nothing new in this code as it follows the same flow as the Arduino example. Specifically, we will control the OLED and buzzer, and the flow of the code is a similar loop that reads and displays the data.

However, we will be using a different software library for the OLED. We will use another of the GrovePi libraries, which has different functions than the Arduino version. For example, rather than write a string to the OLED memory (for display) at a given row and column (X and Y), the GrovePi uses two functions: one to position the cursor (start of string) and another to write the string. To make the code read more like the Arduino version, we will use a helper function as follows:

```
def oled_write(column, row, message):
    """oled_write"""
    oled.setTextXY(column, row)
    oled.putString(message)
```

And since the setup for the OLED GrovePi library differs, we will use another helper function to isolate those lines of code as follows:

```
def setup_oled():
    """setup_oled"""
    oled.init()                 # initialize SEEED OLED display
    oled.clearDisplay()         # clear the screen
    oled.setNormalDisplay()     # set display to normal mode
    oled.setPageMode()          # set addressing mode to Page Mode
    oled_write(0, 1, "Environment")
    oled_write(0, 2, "Monitor")
    oled_write(0, 4, "Starting...")
```

The beep() helper function remains from the Arduino version. Listing 15-5 shows the complete code for the main script for this project. You can read through it to see how all of the code works.

Listing 15-5. Main Script (Python)

```
import sys
import time

from grovepi import pinMode, digitalWrite
import grove_128_64_oled as oled
from air_monitor import AirMonitor, AirQualityEnum

# Constants
SAMPLING_RATE = 5 # 60 seconds
BUZZER_PIN = 6
WARNING_BEEPS = 5
HIGH = 1
LOW = 0

# Constants for environmental quality
MAX_TEMP = 30.0
MAX_DUST = 40.0

def beep(duration=0.150):
    """beep"""
    digitalWrite(BUZZER_PIN, HIGH)
    time.sleep(duration)
    digitalWrite(BUZZER_PIN, LOW)

def oled_write(column, row, message):
    """oled_write"""
    oled.setTextXY(column, row)
    oled.putString(message)
```

```
def setup_oled():
    """setup_oled"""
    oled.init()                 # initialize SEEED OLED display
    oled.clearDisplay()         # clear the screen
    oled.setNormalDisplay()     # set display to normal mode
    oled.setPageMode()          # set addressing mode to Page Mode
    oled_write(0, 1, "Environment")
    oled_write(0, 2, "Monitor")
    oled_write(0, 4, "Starting...")

def main():
    """Main"""
    print("Welcome to the Environment Monitor!")
    # Setup the buzzer
    pinMode(BUZZER_PIN, "OUTPUT")
    # Setup the OLED
    setup_oled()
    # Start the AirMonitor
    air_quality = AirMonitor()
    time.sleep(3)
    oled_write(11, 4, "done")
    beep()
    oled.clearDisplay()

    while True:
        if air_quality.read_data():
            # Retrieve the data
            env_data = air_quality.get_data()

            oled_write(0, 0, "ENVIRONMENT DATA")
            oled_write(0, 2, "Temp: ")
            oled_write(5, 2, "{:3.2f}C".format(env_
            data["temperature"]))
```

```
oled_write(0, 3, "Pres: ")
oled_write(5, 3, "{:05.2f}hPa".format(env_
data["pressure"]))
oled_write(0, 4, "Dust: ")
if env_data["dust_concentration"] == 0.0:
    oled_write(5, 4, "--          ")
else:
    oled_write(5, 4,
        "{:06.2f}%".format(env_data["dust_
        concentration"]))
oled_write(0, 5, "airQ: ")
if env_data["air_quality"] in
    {AirQualityEnum.ERROR, AirQualityEnum.POOR}:
    oled_write(5, 5, "POOR")
elif env_data["air_quality"] == AirQualityEnum.FAIR:
    oled_write(5, 5, "FAIR")
elif env_data["air_quality"] == AirQualityEnum.GOOD:
    oled_write(5, 5, "GOOD")
else:
    oled_write(5, 5, "--        ")

# Check for environmental quality
if ((env_data["dust_concentration"] > MAX_DUST) or
        (env_data["temperature"] > MAX_TEMP) or
        (env_data["air_quality"] == AirQualityEnum.
        POOR) or
        (env_data["air_quality"] == AirQualityEnum.
        ERROR)):
    #pylint: disable=unused-variable
    for i in range(0, WARNING_BEEPS):
        oled_write(3, 7, "ENV NOT OK")
        beep(0.250)
```

```
                    time.sleep(0.250)
                    oled_write(3, 7, "          ")
                    time.sleep(0.250)

        else:
            oled.clearDisplay()
            oled_write(0, 2, "ERROR! CANNOT")
            oled_write(0, 3, "READ DATA")

        time.sleep(SAMPLING_RATE)

if __name__ == '__main__':
    try:
        main()
    except (KeyboardInterrupt, SystemExit) as err:
        print("\nbye!\n")
sys.exit(0)
```

OK, that's it! We've written the code, and we're now ready to execute the project!

Execute the Project

Now that we've spent many pages exploring the Grove modules and writing the code to interact with them, it is time to test the project by executing (running) it.

When the project runs (executes), you will see some diagnostic messages written to the serial monitor (Arduino) or the terminal (Raspberry Pi). You will also see a welcome message appear on the OLED. The code will start displaying data, but you won't see any values for the dust sensor until 30 seconds has passed. Figure 15-19 shows an example of the project running (Python version). Recall the Arduino version has the OLED display rotated 180 degrees.

***Figure 15-19.** Executing the environment monitor project*

Executing the code shows similar output in the serial monitor (Arduino) and terminal (Python). The differences are mainly in how the data is shown. In the Arduino version, some of the software libraries contain print statements that add to the output, and in the Python version, we print the dictionary with the data. Let's look at the Arduino first.

Sketch on the Arduino

Executing the sketch on the Arduino requires connecting our board to our PC and then uploading the sketch to the Arduino. Recall the sketch will run so long as the USB cable is connected to our PC (and the Arduino).

Execute the Sketch

To execute the sketch, be sure your Arduino is connected and you've selected the correct board under the *Tools* ➤ *Board* menu. You also need to ensure you have the correct port selected under the *Tools* ➤ *Port* menu.

Once those items are set, you can click the *Upload* button or choose *Sketch* ➤ *Upload* from the menu. The Arduino IDE will compile the sketch and then upload it to your Arduino. Once you see the `Done uploading...` message, you can open the serial monitor. You should see the output begin momentarily that is the same as that on the OLED. Go ahead, and try it out! You should see values similar to the following:

```
Welcome to the Environment Monitor!
Starting....sys_starting...
The init voltage is ...
94
Sensor ready.
Test begin...
done.
Pressure: 101931.00Pa
Temperature: 21.12
0,0.00,0.62
Dust %: 0.62
sensor_value:83      Air fresh
Fresh air
Pressure: 101934.00Pa
Temperature: 21.12
Dust %: 0.00
sensor_value:77      Air fresh
...
```

If something isn't working, check your connections or refer to Chapter 13 for troubleshooting tips.

Python Code on the Raspberry Pi

Executing the sketch on the Raspberry Pi requires running the Python code in a terminal after connecting your modules and powering on the Raspberry Pi. Recall the code will run until you stop it with *CTRL+C* on the keyboard.

Execute the Python Code

To run the Python code on the Raspberry Pi, you can issue the command `python3 ./environment.py` from the same folder where the file was saved as shown in the following. You should get results similar to the following:

```
$ python3 ./environment.py
Welcome to the Environment Monitor!

>> Reading Data <<
> Reading temperature = 23.75
> Reading pressure = 1021.2385746409756
> Reading dust concentration
> Dust concentration = 0.0
> Reading air quality = AIR_QUALITY.GOOD

>> Reading Data <<
> Reading temperature = 23.75
> Reading pressure = 1021.2504208636452
> Reading dust concentration
> Dust concentration = 0.0
> Reading air quality = AIR_QUALITY.GOOD
...
```

If everything worked as executed, congratulations! You've just built your third Grove project. If something isn't working, check your connections or refer to Chapter 13 for troubleshooting tips.

Going Further

While we didn't discuss them in this chapter, there are some ideas where you could make this project into an IoT project. Here are just a few suggestions you can try once we have learned how to take our projects to the cloud. Put your skills to work!

- *Environment portal*: You can display the values of the last sensor(s) read on a web page to allow you to see the condition of your environment from anywhere in the world. If you also add the date and time, you can see how your environment changes over time.
- *Additional sensors*: Implement additional sensors to read more data such as specific gases such as CO2 and O2 and a light sensor to detect day and night cycles. You could also include a vibration sensor if you live in areas prone to seismic tremors. Interestingly, you can use vibration sensors to detect when someone walks into the room.
- *Sampling rate*: Adjust the sampling rate to match your environmental needs. For example, if you live in a very clean apartment or house with good climate control, your sampling rate may be lower than if you live in a dusty area prone to temperature changes such as an RV or typical rustic cabin.

Summary

In this chapter, we got more hands-on experience making projects with Grove analog and digital modules as well as multiple I2C devices. We used these modules to create an environment monitor that displays the temperature, barometric pressure, air quality, and dust (particle) concentration in the air – in other words, an indoor air monitoring solution.

Along the way, we learned more about how to work with Grove modules including how to write our own class for managing multiple sensors. We also saw how to use alternative software libraries in both the Arduino and Python versions of our project. Finally, we saw some potential to make this project better as well as some ideas for how to adapt the project for practical uses.

In the next chapter, we will see another project that demonstrates how to use more Grove modules to create a classic electronic game called Simon Says. It's time to have some fun!

CHAPTER 16

Simon Says

If you like early, vintage electronic games, you most likely have played a game named Simon.[1] It is a round tabletop game that has four large colored buttons on top. One or more players can play with the objective to repeat a sequence from memory. The game presents the player with a sequence of colored lights in a random pattern. The player's goal is to press the buttons for each color in the sequence before time runs out. If the player repeats the sequence correctly, the game adds another light to the sequence. The game starts with a single light, so early levels are pretty easy, but as the sequence gets longer, it becomes harder to play. Throw in several players, and you've got a cool, Internet-free game party!

In this chapter, we will see how to create a version of the Simon game using Grove modules including analog, digital, and I2C protocols. We will also see how to incorporate a set of Qwiic modules to make things more interesting.

Note Unlike previous projects, this project requires some soldering to assemble the Qwiic components. Fortunately, the soldering isn't difficult, so it would be a good time to learn, or you can find a friend to assemble the components for you.

---

[1] `https://en.wikipedia.org/wiki/Simon_(game)`

C. Bell, *Beginning IoT Projects*, https://doi.org/10.1007/978-1-4842-7234-3_16

Like the last chapter, we will see how to implement this project on the Arduino and Raspberry Pi but with an interesting twist. Let's get started.

Project Overview

The project for this chapter is designed to demonstrate how to use analog, digital, and I2C devices on the same Grove host adapter to build a Simon game. It works very much like the original game but with an LCD for displaying messages. We will use a Grove buzzer for sound and a set of Qwiic LED buttons.

While this seems like a simple project build, the number of modules in use and integrating all of the code for those modules make this project the most ambitious in the book. If you haven't read through and worked on the other projects, you may want to work on the earlier chapters first and save this one until you've mastered a few of the others.

For those with access to a 3D printer, we will also see a simple mounting plate you can print to install the modules to protect the modules and make them easier to use in playing the game.

What Will We Learn?

By implementing this project, we will get more practice in using analog and digital Grove modules and how to connect Grove modules to the various protocol connectors on the host adapter. We will also integrate multiple I2C modules including integrating both Grove and Qwiic I2C components.

The programming tasks will reveal how to read button press events, turn LEDs on and off, play sound (tones) on a small speaker, and display messages to the players on an LCD. As you will see, the sound component adds an interesting challenge for both the Arduino and Python versions.

Let's see what hardware we will need.

Hardware Required

The hardware needed for this project is listed in Table 16-1. URLs for each component are included for ease of ordering including duplicate entries for alternative vendors. We will use the Grove Dual Button, Grove Buzzer, Grove LCD RGB Backlight, Grove Qwiic Hub, and a set of Qwiic LED buttons.

***Table 16-1.** Hardware Needed for the Simon Project*

Component	URL	Qty	Cost
Grove Dual Button	www.seeedstudio.com/Grove-Dual-Button-p-4529.html	1	$2.20
Grove Buzzer	www.seeedstudio.com/Grove-Buzzer.html	1	$1.90
Grove LCD RGB Backlight	www.seeedstudio.com/Grove-LCD-RGB-Backlight.html	1	$11.90
Grove Qwiic Hub	www.seeedstudio.com/Grove-Qwiic-Hub-p-4531.html	1	$1.90
SparkFun Qwiic LED Button Breakout	www.sparkfun.com/products/15931	4	$3.10
LED Tactile Button - White	www.sparkfun.com/products/10439	1	$2.10
LED Tactile Button - Green	www.sparkfun.com/products/10440	1	$2.10
LED Tactile Button - Red	www.sparkfun.com/products/10442	1	$2.10

(*continued*)

Table 16-1. (*continued*)

Component	URL	Qty	Cost
LED Tactile Button - Blue	www.sparkfun.com/products/10443	1	$2.10
Grove cables (any length can be used but longer may be best)	Included with each preceding Grove module	3	
Grove female breakout (Python version only)	www.seeedstudio.com/Grove-4-pin-Female-Jumper-to-Grove-4-pin-Conversion-Cable-5-PCs-per-PAck.html	1	$3.90
Qwiic cable	www.sparkfun.com/products/14426	4	$0.95
Arduino MKR 1010 WiFi	www.sparkfun.com/products/15251	1	$35.95
Raspberry Pi 3B or later	www.sparkfun.com/categories/233 www.adafruit.com/category/176	1	$35.00+
Grove Base Shield V2.0 for Arduino	www.seeedstudio.com/Base-Shield-V2.html	1	$4.45
GrovePi+	www.sparkfun.com/products/15945	1	$5.95

About the Hardware

Let's discuss these components briefly. We will discover how to work with the hardware in more detail later in the chapter. If you implement the Python version, be sure to note the changes in the libraries used as well as different connections used for the buzzer module. We saw the LCD RGB Backlight and Grove Qwiic Hub modules in Chapter 14, so we won't cover those here. See Chapter 14 for explanations of those components.

Grove Dual Button

Thus far in the book, we've only worked with one type of button. In this chapter, we will work with two types. The first is the Grove Dual Button. This is a digital module that has two momentary buttons. While there are two buttons on the module, we need only a single Grove cable to connect to the host adapter.

This is because digital modules use only three wires: ground, 5V, and one for signal. Since we have four cables available, we can use the extra wire for the second button.

The button module comes with a set of colored button caps that you can use to help color-code your button choices, which is a nice option.

Figure 16-1 shows the Grove Dual Button.

Figure 16-1. *Grove Dual Button (courtesy of seeedstudio.com)*

Grove Buzzer

We will be adding a sound feature for this project so that we can play tones (notes) just like the old electronic games. We will keep it simple and use the same Grove Buzzer module from Chapter 15.

However, there is one issue with using this module with Python on the Raspberry Pi. The GrovePi/GrovePi+ does not permit you to communicate directly with the module over a digital protocol. This is because all communication to the Grove modules is processed through a chip on the GrovePi/GrovePi+, which does not support this option.

Fortunately, we can use an analog protocol for the Python version and use pulse wave modulation (PWM) to control it.[2] PWM simply uses a loop to modulate how much power is sent to the component. PWM is often used to dim LEDs like we did in Chapter 14, but with a buzzer it permits you to change the tone of the sound produced.

We can use a Grove female breakout cable to connect to the Raspberry Pi GPIO directly using the GPIO pass-through pins on top of the GrovePi. We will see how to make these connections in the next section.

Qwiic LED Buttons

The Simon game uses four large colored buttons. Underneath the case, this is simply four momentary buttons and four (or eight) small lights (newer versions use white LEDs) that shine through the opaque button cover.

Since we don't have anything like that in the Grove, Qwiic, or STEMMA QT component systems, we will use the best option available - the Qwiic LED Button Breakout. Unfortunately, as of this writing, SparkFun doesn't sell assembled LED modules in four colors. So we will have to purchase four breakout boards (boards without the LED button soldered) and four different LED buttons (white, red, green, and blue).

[2] https://en.wikipedia.org/wiki/Pulse-width_modulation

Once again, this will require some soldering, but the soldering isn't difficult. If you don't know how to solder, you can use this opportunity to learn or find a friend and have them teach you how to solder.

Assembly requires placing the LED button onto the breakout board and soldering six connections: four for the button and two for the LED that is built into the button. We will see connections that need to be made and learn how to orient the LED button correctly, but first let's see the components. Figure 16-2 shows the Qwiic LED Button Breakout.

Figure 16-2. *Qwiic LED Button Breakout (courtesy of sparkfun.com)*

The LED buttons are discrete components that have an LED built into the top of the button. So there are two components in a single package. You can turn the LED on or off without pressing the button or, through code, turn on the button when it is pressed, which is what we will do in this project. Figure 16-3 shows the four LED buttons we will use for this project.

Figure 16-3. *Qwiic LED buttons (courtesy of sparkfun.com)*

If you look closely at the breakout board and buttons, you will see six holes (and six pins) where two of the holes/pins are located on the side of the button. If you look closer, you will see these "ears" are marked with a + sign on one and either a - or nothing on the other. These are the pins for the LED. The other four are for the button. While placement of the button doesn't matter for the connections, the pins for the LED must be oriented correctly.

That is, the + pin must be inserted into the + hole on the breakout board for the LED. Figure 16-4 shows where the pins are located on the breakout board. Once you identify the positive pin on the LED button, you can then solder all six pins. Remember to double- and triple-check your connections so you don't have to unsolder and resolder the button.

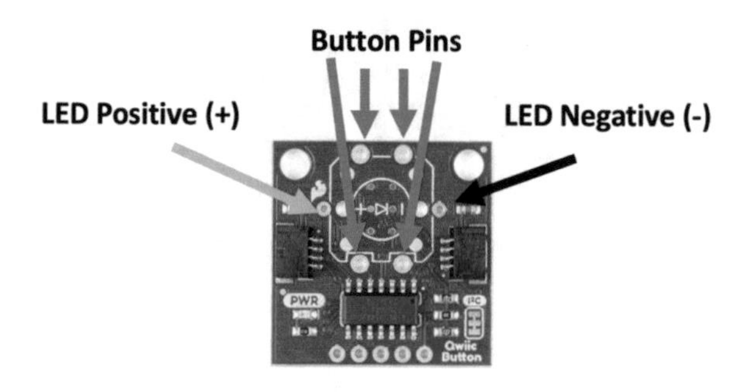

Figure 16-4. *Qwiic LED and button pins*

Caution Some of the LED buttons may come with a mark made by a marker. Don't assume this is the positive leg of the LED.[3] Always double-check by examining the button ears with a magnifier.

---

[3] Yep. That particular snake bit me, and I had to unassemble and resolder! Unsoldering is much harder than soldering. Do yourself a big favor and check it twice before you solder!

There is one other thing you must do to use these modules in our project. Since we will be using six I2C modules (LCD RGB Backlight, four buttons, and the Grove Qwiic Hub), we need to disable some of the I2C resistors on at least one module. This may not be necessary for the Arduino version, but is required for the Python version. This is because there are too many resistors on the I2C bus and that can result in erratic or dropped (failed) I2C communication.

To disable the resistors on the assembled Qwiic LED button modules, break the connections on the front of the board labeled I2C as shown in Figure 16-5. You can use a sharp knife to carefully break the two small connections.

Figure 16-5. *I2C resistors on the Qwiic LED Button Breakout (courtesy of sparkfun.com)*

We also need to change the I2C addresses of at least three of the buttons. Why? Because they are all set to use address 0x6F. Fortunately, like most Qwiic modules, we can change the I2C address by soldering across the pads on the back as shown in Figure 16-6.

Figure 16-6. *I2C address jumpers (pads) on the Qwiic LED Button Breakout (courtesy of sparkfun.com)*

Since we have four I2C jumpers, we can address up to 16 buttons. We simply apply solder across the pads as shown in Table 16-2 where C means the pad is soldered (closed) and O means the pad is not soldered (open).

Table 16-2. *Qwiic LED Button I2C Addresses*

Binary	Hex	A0	A1	A2	A3
1111	0x60	S	S	S	S
1110	0x61	0	S	S	S
1101	0x62	S	0	S	S
1100	0x63	0	0	S	S
1011	0x64	S	S	0	S
1010	0x65	0	S	0	S
1001	0x66	S	0	0	S
1000	0x67	0	0	0	S

(*continued*)

Table 16-2. (*continued*)

Binary	Hex	A0	A1	A2	A3
0111	0x68	S	S	S	0
0110	0x69	0	S	S	0
0101	0x6A	S	0	S	0
0100	0x6B	0	0	S	0
0011	0x6C	S	S	0	0
0010	0x6D	0	S	0	0
0001	0x6E	S	0	0	0
0000	0x6F	0	0	0	0

For this project, we're going to keep it simple and leave one button with all jumpers open (0x6G), one with a jumper on A0 (0x6E), one with a jumper on A1 (0x6D), and the last with a jumper on A2 (0x6B). If you'd rather use other addresses, you can. Just make sure to change the code accordingly.

Once you have assembled four Qwiic LED buttons and set the I2C addresses, we can assemble our project.

Connect the Grove Modules

Recall from Chapter 12 we can use a single Grove cable to connect each Grove module separately to a specific connector on the host adapter on our host board. The host adapter for the Raspberry Pi has a different layout but has the same connectors we will use. Table 16-3 includes the details of each connection on the host adapter to help you make the right connections. Simply use the table to connect a Grove cable from the module to the Grove connector on the host adapter as marked in the table.

Table 16-3. Grove Connections (Arduino)

Module	Protocol	Grove Connector on the Host Adapter
Buzzer	Digital	D6
Dual Button	Digital	D7
LCD RGB Backlight	I2C	I2C1
Qwiic Hub + Qwiic LED buttons	I2C	I2C2 (and a Qwiic daisy chain)

As mentioned, the connections for the Python version of the project are the same except for the buzzer module. Recall we will connect this module using a female Grove breakout cable. Table 16-4 shows the connections we need to make. The pins shown are the physical pin numbers, and the Raspberry Pi GPIO values are shown in parentheses.

Table 16-4. Grove Buzzer Connections (Python)

Wire Color	Connection	GPIO Pin#
Black	Ground	4 (GND)
Red	5V	6 (5V)
White	No connection	N/A
Orange/brown	Signal	16 (GPIO23)

Let's look at a closeup of these connections. Figure 16-7 shows a closeup of the connections for the buzzer on the Raspberry Pi. Notice the white cable is not used, so you can simply secure it so that it doesn't come into contact with anything.

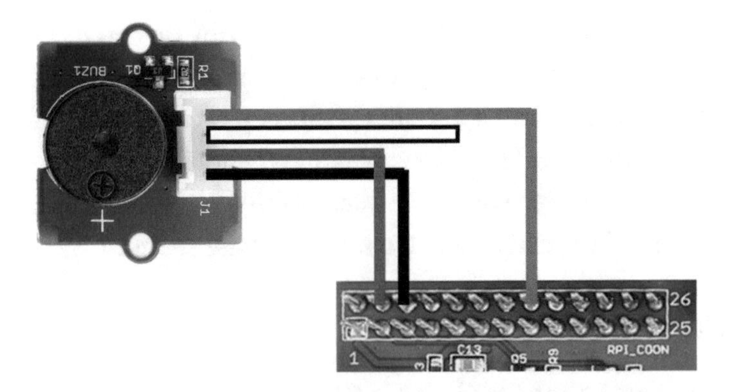

***Figure 16-7.** Closeup of buzzer connections for the Raspberry Pi*

Now, let's look at all of the connections we need to make for each of the project versions. As you will see, there are significant differences for the buzzer module.

Connections for Arduino

Figure 16-8 shows an example of how you should connect your modules for this project for the Arduino version.

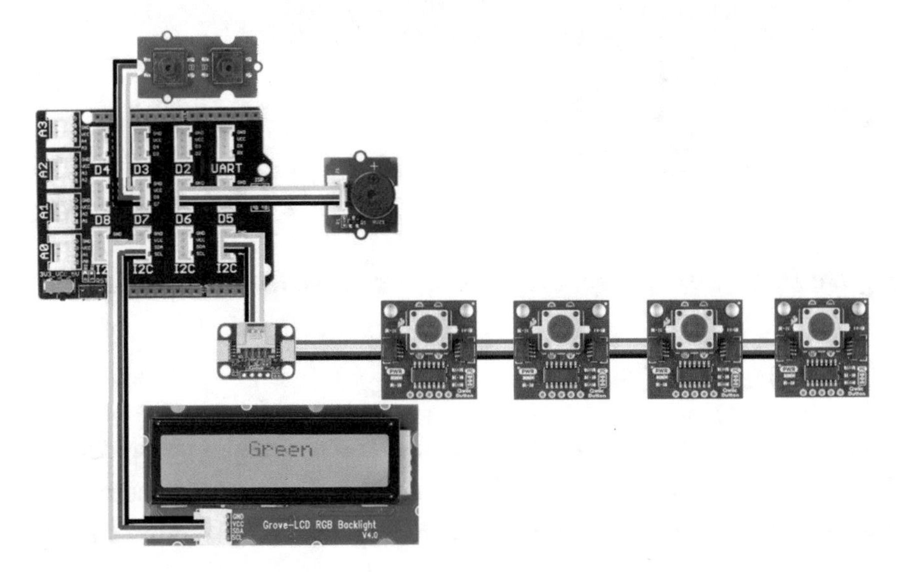

***Figure 16-8.** Simon Says project Grove connections (Arduino only)*

Wow, that's a lot of cables! Now you can see why this project is the most complex of the projects in this book. Once you make all of these connections, you may find it a challenge to keep everything together as they tend to slide around and the cables aren't very flexible. You may find it necessary to use a piece of wood to screw the modules to in order to play the game. We will see a 3D printed option in a later section.

Connections for Raspberry Pi

The connections for the Raspberry Pi (Python) version differ in two ways. First, the buzzer is connected to an analog pin to the GPIO, and it uses the GrovePi/GrovePi+ host adapter that has the Grove connections in a different location. Recall these are the same connections as the Arduino host adapter but put together on a map look quite different. Figure 16-9 shows the connections for the Python version.

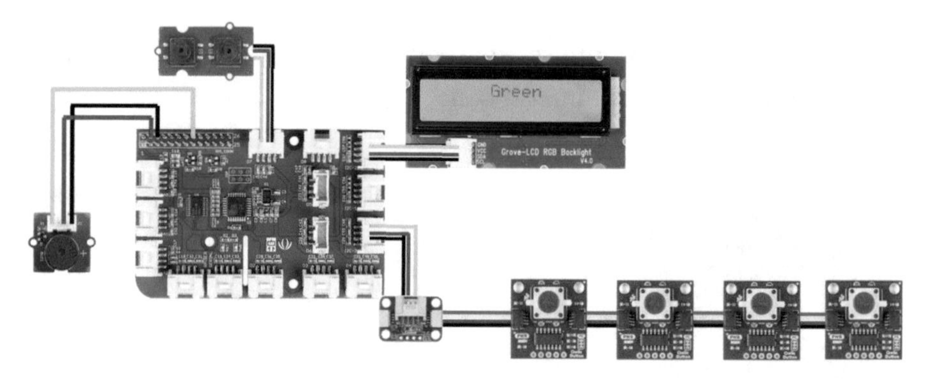

Figure 16-9. *Simon Says project Grove connections (Python only)*

Using an Enclosure

Since we have so many components and a bunch of cables connecting them all together, using them to play a game like Simon can become a lesson in patience (either that or you can use some double-sided tape to tape them to your desk). We can fix this with a mounting plate like we did in Chapter 15. We could create a full enclosure, but as you will see leaving the modules exposed gives the project a genuine cool factor.

If you have your own or access to a 3D printer, you can print a mounting plate. The source code for this chapter includes the 3D printing files you need to create a simple enclosure to mount the modules arranged in a manner that enables game play. Figure 16-10 shows the mounting plate.

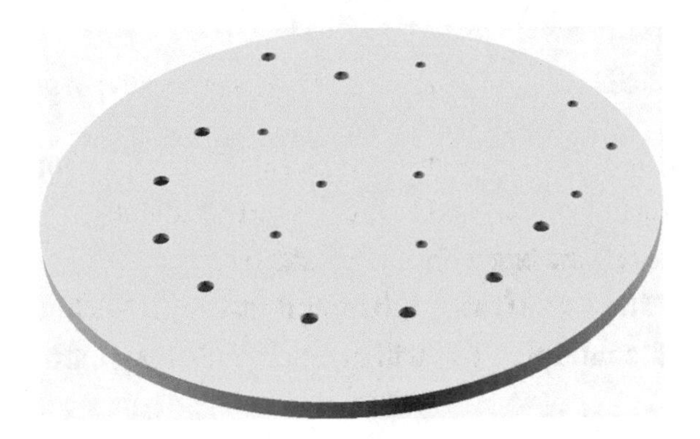

Figure 16-10. *3D mounting plate design for the Simon Says project*

While this looks like nothing more than a coaster, there are feet on the bottom of the plate and places for M2 and M3 nuts. In fact, you will need to print this plate upside down.

There is also a set of spacers you will need to print as shown in Figure 16-11.

Figure 16-11. *3D spacer design for the Simon Says project*

Notice from left to right there are eight short M3 spacers for the LED buttons, three medium-length M2 spacers for the dual button module, two short M2 spacers for the buzzer module, and four long M2 spacers for the LCD RGB Backlight. The Grove Qwiic Hub mounts to the plate without spacers.

To mount the modules, you will need the following hardware:

- (9) M2 nuts
- (10) M3 nuts
- (8) M3×7mm bolts
- (2) M3×5mm bolts
- (2) M2×10mm bolts
- (3) M2×12mm bolts
- (4) M2×19mm bolts

To assemble the enclosure, begin by mounting all modules except the LCD RGB Backlight as shown in Figure 16-12. You should also make all of the cable connections as well since we will route all wiring under the LCD

RGB Backlight. Notice we used different-length Qwiic cables to keep the wiring away from the buttons.

Figure 16-12. *Mounting the modules to the 3D printed plate (part 1)*

Next, mount the LCD RGB Backlight routing all wiring under the LCD as shown in Figure 16-13.

Tip You may want to use the longest Grove cables you can find to ensure your plate can be placed far enough away from the host board to allow players to gather around it to play.

Figure 16-13. *Mounting the modules to the 3D printed plate (part 2)*

Finally, attach the Qwiic cables to your host board. Figure 16-14 shows a completed example of the mounting plate.

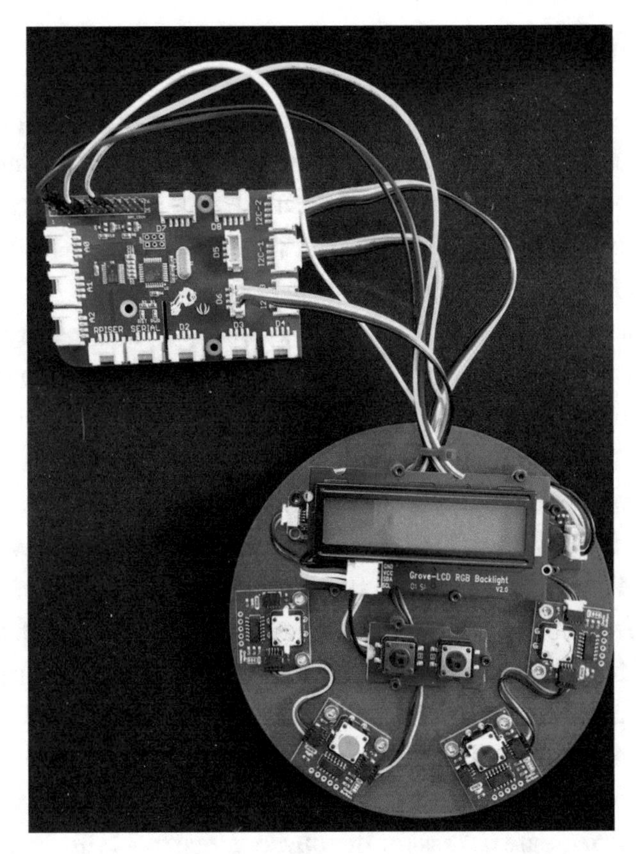

***Figure 16-14.** Mounting the modules to the 3D printed plate (part 3)*

If you have experience creating 3D models for printing, feel free to experiment with creating your own enclosure - perhaps one that also includes a battery and a small form factor host board.

Now that we know more about the hardware for this chapter, let's write the code!

Write the Code

The code for this project is once again written using the usual pattern. For this project, that means using analog and digital modules as well as a host of I2C devices. The dual button and buzzer are digital (the buzzer is an analog module for the Python version). The LCD, Grove Qwiic Hub, and four Qwiic LED buttons are all I2C devices.

As you will see, the code isn't overly complicated for the Arduino version, but we will have some more work to do for the Python version. This is because the buzzer module is not directly supported by the GrovePi library. Never fear, though. The code we will see is fully functional in the Python version by switching the buzzer from a digital to an analog connection, which shows how versatile components that can work with PWM can be controlled via an analog connection.

Like the previous projects, we will use classes to wrap our functionality. Like we did in Chapter 14, we'll focus on making the Simon game its own class and general control of the game system in the main code. That is, we will use the dual button and LCD in the main code and everything else in the Simon code module (class).

We will also create a class for the buzzer since this is likely to be a core component for future projects. That's one of the greatest benefits of object-oriented design - we can reuse the code in other projects without modification. Plus, it makes it easier if you need to substitute radically different code like we need to do for the Python version.

Let's walk through how to prepare our computers to use the components and write the code. We'll start with the Arduino.

Arduino

This section presents a walk-through of the sketch and classes you will write to read values from the buttons, control the LEDs on the Qwiic LED buttons, and play sounds on the buzzer. We will also display short messages on the LCD module. But first, there are a couple of libraries we must install on our PCs.

Install Software Libraries

We will need to install the Arduino libraries for the LCD and Qwiic LED modules. We do not need any libraries for the digital modules as they are supported directly by the Arduino platform. And we installed the code library we need for the LCD in Chapter 14, so if you haven't done that already, you will want to see how to do that in Chapter 14. That leaves the Qwiic LED button library.

To install the Qwiic LED software library, open the Library Manager from the Arduino IDE menu (*Sketch ➤ Include Library ➤ Library Manager...*). Then search for `qwiic button` and install the latest version of the Qwiic LED library as shown in Figure 16-15.

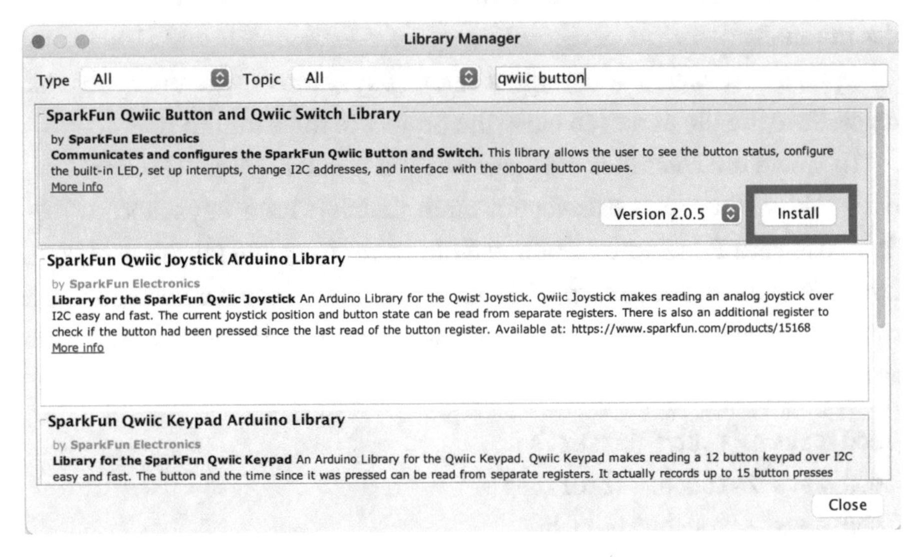

***Figure 16-15.** Installing the Qwiic button Library (Arduino IDE)*

Now that we have the software libraries installed, we can begin writing our sketch. Since this is not our first Arduino sketch, we will discuss the code at a high level and skip the line-by-line details focusing on the mechanics of how the code works. You can study the code at your leisure to ensure you understand the sketch in more detail.

Write the Sketch

Recall we are going to use a class to encapsulate the Simon game and another for the buzzer. Thus, we will create two classes: one to define the Simon game that we can control from the main code and another for the buzzer that we will control from the Simon game class.

We will write the class headers first and then the main sketch and finally complete the code for the classes. The Simon game class will be named `Simon`, and the buzzer class will be named `Buzzer`. Yes, not very imaginative, but sometimes simplicity is best.

Recall there is no way (currently) to create and add new files to a sketch (but you can add existing files by clicking *Sketch* ➤ *Add File...*). So we will once again create the main sketch and add the code header and source files manually.

Open a new sketch and name it `simon_says.ino` or whatever you'd like to use. Save the file and then close the project in the Arduino IDE.

To create the class files, navigate with your File Explorer (Finder) to the folder where you stored your main sketch (`simon_says.ino`). Then, use your File Explorer or a text file editor to create four new files named `Buzzer.h`, `Buzzer.cpp`, `Simon.h`, and `Simon.cpp`. Or you can use a terminal to navigate to the folder and issue these commands to create the empty files:

```
simon_says % touch Buzzer.h
simon_says % touch Buzzer.cpp
simon_says % touch Simon.h
simon_says % touch Simon.cpp
```

Now, let's see the code for each file starting with the header files.

Class Header File: Buzzer

Click the tab named `Buzzer.h` to open the blank file. Here, we will add the header or blueprint for the class. The module provides a mechanism to play tones. Specifically, we will create functions for each of the sounds that the Simon game uses. In this case, we need the following tones (or tone sequences):

- *playThemeSong()*: An introductory song played when the game starts.
- *playReadySetGo()*: A tone to indicate the player can begin entering the sequence of buttons.
- *playSuccess()*: A tone to indicate the sequence entered matches the challenge sequence.
- *playFailure()*: A set of tones to indicate the sequence is not correct and the player's turn ends.
- *playColor()*: Play a unique tone for each of the four LED buttons.

Aside from those functions, we will also create a constructor so we can set up the class and two private functions: one for playing a set of tones (song) and another to get the frequency for the tone.

We will also use a scale of notes and their frequencies stored in private variables. In this way, we can record the notes in variables for each of the preceding tones/sounds and then use the frequency function to retrieve the frequency of the note. The frequency defines how long the buzzer will sound. By varying the frequency, we can get different notes.

Tip For more information about how to use a buzzer (Piezo speaker) to play notes, see `www.arduino.cc/en/Tutorial/BuiltInExamples/toneMelody` or `https://learn.digilentinc.com/Documents/392` for more details on how PWM can help.

We can also define how long to hold (play) each tone, which will help us determine a cadence or primitive rhythm. We will call these "beats" where each beat is a quarter note (so we'll be using 4/4 time). Thus, a 1 is one quarter, 2 is half, etc. We also use a tempo to determine the speed, which we will set globally, but you could easily modify the code to allow different tones to be played at different speeds. This way, we can make the song faster or slower depending on our aesthetic requirements.

Let's look at the completed code for the header file. Listing 16-1 shows the file.

Listing 16-1. Buzzer Header File

```
#include <Arduino.h>

#define DEFAULT_TEMPO 113
#define NOTES_IN_SCALE 8
#define BUZZER_PIN 6

typedef struct {
  // The speed to play the notes
  int tempo = DEFAULT_TEMPO;
  // Number of notes in song
  int numNotes;
  // Notes is an array of text characters corresponding to
     the notes
  // in your song. A space represents a rest (no tone)
```

```
  String *notes;
  // Beats is an array of values for each note and rest.
  // A "1" represents a quarter-note, 2 a half-note, etc.
  int *beats;
} SongStruct;

class Buzzer {
public:
  Buzzer();
  void playThemeSong() {playSong(&theme); }
  void playSuccess() { playSong(&success); }
  void playFailure() { playSong(&failure); }
  void playColor(int color) { playSong(&colors[color]); }
  void playReadySetGo() { playSong(&go); }

private:
  SongStruct failure;
  SongStruct success;
  SongStruct theme;
  SongStruct go;
  SongStruct colors[4];

  // The following arrays hold the note characters and their
  // corresponding frequencies. The last "C" note is uppercase
  // to separate it from the first lowercase "c". If you want to
  // add more notes, you'll need to use unique characters.
  char noteNames[NOTES_IN_SCALE] = { 'c', 'd', 'e', 'f', 'g',
  'a', 'b', 'C' };
  int frequencies[NOTES_IN_SCALE] = {262, 294, 330, 349, 392,
  440, 494, 523};

  void playSong(SongStruct *song);
  int frequency(char note);
};
```

Notice the public functions are simply calls to the private `play()` function with the song specified. The songs (sequences of tones) are stored as structures that contain variables and an array to contain the tones (song). These are generated dynamically in the constructor. Notice also the notes scale and frequency arrays. This is where we define the notes we will play. You can change these as you see fit to make the class usable for playing more complex songs.

Now, let's look at the header file for the Simon game.

Class Header File: Simon

Click the tab named `Simon.h` to open the blank file. Recall this module controls the four Qwiic LED buttons and the buzzer.

The public functions are not difficult. We will use functions for the setup routine where we can change the number of players (`setupMode()`), start the game (`startGame()`), show the number of players (`showPlayers()`), and play the game (`play()`).

Aside from that, we will also need a number of private functions that are a bit more complicated. We need functions to control the LCD, play a challenge sequence, read a sequence of buttons from the player, generate the challenge sequence using the `randint()` function to generate a random integer from 0 to 3 to correspond to the button array index, and even determine a winner for the multiplayer mode. The following lists the private functions and their uses:

- `numAlive()`: Determine the number of players still active (alive).
- `resetScreen()`: Reset the LCD and display a new message.
- `showWinner()`: Show the winner on the LCD.
- `readButton()`: Determine if a button is pressed.

- `generateSequence()`: Generate a challenge sequence.
- `playSequence()`: Play a challenge sequence by turning on the corresponding LED and playing the tone for the button.
- `readSequence()`: Read a sequence from the player.

Finally, we will need a number of variables to store information including an instance of the LCD class and the Qwiic buttons (stored as an array). The constructor will also need to be added to set up the hardware.

Since the header code is not difficult to read, let's look at the completed code for the header file. Listing 16-2 shows the file.

Listing 16-2. Simon Header File

```
#include <Arduino.h>
#include <SparkFun_Qwiic_Button.h>
#include <rgb_lcd.h>
#include "Buzzer.h"

#define MIN_BEATS 2       // Starting number of beats
#define MAX_PLAYERS 4     // Max number of players
#define MAX_TIMEOUT 5000 // Seconds to wait to abort read
#define KEY_INTERVAL 500 // Interval between button playback

typedef struct {
  int highScore {0};
  bool isAlive {true};
} PlayerStruct;

class Simon {
public:
  Simon();
  void startGame(int players);
```

```
  void play();
  void setupMode();
  void showPlayers(int players);

private:
  int numPlayers = 1;
  PlayerStruct **playerScores {NULL};

  Buzzer *buzzer = new Buzzer();     // Buzzer
  rgb_lcd lcd;                       // Grove LCD
  QwiicButton buttons[4];

  int numAlive();
  void resetScreen(const char *message);
  void showWinner();
  bool readButton(int button);
  int *generateSequence(int numNotes);
  void playSequence(int *challengeSequence, int numNotes);
  bool readSequence(int *challengeSequence, int numNotes);
};
```

OK, let's return to the main sketch to see how we can use the classes.

Main Sketch

Now click the `simon_says.ino` tab to return to the main sketch. Since we are placing most of the hardware work in the Simon game class, all we need to do here is set up the Grove Dual Button and write code to interact with the Simon game class.

We set up the dual button in the `setup()` function, which requires initializing each button separately. This is because, while we're using only one Grove connector, the module uses the extra wire for the second button. Specifically, when we connect the module to the D7 connector, which is wired to the digital 7 pin, we can access the second button via the digital 8 pin. The only other thing we do in this function is create an instance of the Simon game class.

The `loop()` function is responsible for controlling the *mode* button to set the number of players and the *start* button to start the game. We will make the code allow use of the mode button so long as a game is not in process.

Note The *mode* button is closest to the Grove connector. If your orient the module with the Grove connector on the left, the *mode* button is the left button, and *start* is the right button.

Recording the number of players is done using a variable where we allow up to four players. So pressing the *mode* button continually will cycle through the options (e.g., 1, 2, 3, 4, 1, 2, 3, 4...). We will use this value when the player presses the *start* button.

When the *start* button is pressed, we use the `Simon` class to start a new game with the `startGame()` method passing in the number of players selected. Then we call the `play()` function turning control over to the Simon class. Once the game ends, we place the Simon instance back to the setup mode with the `setupMode()` function. A few short delays are added to make the game flow better.[4]

By placing all of the game control in its own class, we've simplified the main code. Listing 16-3 shows the code for the main sketch. Take a few moments and read through the code.

---

[4] Purists may say the use of delays or sleep is a poor replacement for excellent code, but they are handy for controlling flow and execution speed. Plus, there are (good) side effect benefits in some languages such as Python related to threading.

Listing 16-3. Main Sketch

```
#include "Simon.h"

// Constants
#define START_BUTTON 7
#define MODE_BUTTON 8

// Global variables
Simon *simon;
bool gameStarted = false;
int numPlayers = 1;

void setup()
{
  pinMode(START_BUTTON, INPUT);
  pinMode(MODE_BUTTON, INPUT);

  Serial.begin(115200);
  while (!Serial);

  // Initialize the game
  Serial.println("Welcome to the Simon Says game!");
  simon = new Simon();
}

void loop()
{
  bool startButton = false;
  bool modeButton = false;

  if (!gameStarted) {
    startButton = digitalRead(START_BUTTON) == LOW;
    modeButton = digitalRead(MODE_BUTTON) == LOW;
```

```
    if (startButton) {
      Serial.println("Start button pressed.");
      simon->startGame(numPlayers);
      delay(1000);
      simon->play();
      delay(1000);
      simon->setupMode();
    } else if (modeButton) {
      numPlayers = numPlayers + 1;
      if (numPlayers > MAX_PLAYERS) numPlayers = 1;
      Serial.print("Mode button pressed - ");
      Serial.print(numPlayers);
      Serial.println(" players.");
      simon->showPlayers(numPlayers);
      delay(2000);
      simon->setupMode();
      delay(50);
    }
  }
}
```

Now we can write the final portion of our project – the code for the classes.

Class Code File: Buzzer

Click the tab named `Buzzer.cpp` to open the blank file. Here, we will add the code for the class. Aside from the constructor, there are two functions we need to write: the `frequency()` and `playSong()` functions. Let's talk about each of these.

As mentioned previously, the constructor is where we set up all of the sequences of tones (songs) for each of the sounds we will play for the Simon game. We use a statically allocated structure for each sound to store the number of notes, the notes (as a string), and the beats as an integer array as shown in the following. We do this for each of the four tones:

```
// Success tones
success.numNotes = 3;
success.notes = new String("CCC");
success.beats = new int[success.numNotes] {1,1,1};
```

For the button tones, we use an array, which makes things a bit easier since we can loop over the array and set the properties since all will be the same number of tones and beats. However, we will need to set up each of the notes separately as shown in the following:

```
// Tones for the colors
for (int i = 0; i < 4; i++) {
  colors[i].numNotes = 1;
  colors[i].beats = new int[1] {1};
}
colors[0].notes = new String("a");
colors[1].notes = new String("g");
colors[2].notes = new String("C");
colors[3].notes = new String("f");
```

The `frequency()` function takes as a parameter the note we want to play and then uses the constants defined in the header file to locate the frequency for that note. We use a loop here in the Arduino version, but we will see a slightly more efficient way to access the frequency in the Python version. The reason this is inefficient is because it has to look at each note until it finds a match. Yes, there are only eight notes and processing time is insignificant, but for the sake of better coding, we should always strive to find more efficient methods.

The playSong() function takes as a parameter a pointer named song to the SongStruct we used to save the song (number of notes, notes, and beats). We use a loop to loop through each of the notes using the number of notes to limit the for loop. For each note, we calculate the duration by multiplying the number of beats by the tempo. If there is a space in the sequence, we use that as a quarter note rest (we don't play anything). Otherwise, we get the frequency for the note from the frequency() function and pass that and the duration to the Arduino tone() function to play the note. Cool! Listing 16-4 shows the code for the playSong() function.

Listing 16-4. The playSong() Function (Buzzer.cpp)

```
void Buzzer::playSong(SongStruct *song) {
  int i, duration;

  for (i = 0; i < song->numNotes; i++) {
    duration = song->beats[i] * song->tempo;  // length of note/
                                                 rest in ms
    if (song->notes->charAt(i) == ' ') {     // is this a rest?
      delay(duration);                        // then pause for
                                                 a moment
    } else {                                  // otherwise,
    play the note
      tone(BUZZER_PIN, frequency(song->notes->charAt(i)),
      duration);
      delay(duration);                        // wait for tone
                                                 to finish
    }
    delay(song->tempo/10);                    // brief pause
                                                 between notes
  }
}
```

Recall Python doesn't have a tone() function, so we will need to substitute a different mechanism for playing notes for the Python version of the project.

Listing 16-5 shows the complete Buzzer code for completeness.

Listing 16-5. Buzzer Code File

```
#include "Buzzer.h"

Buzzer::Buzzer() {
  // Failure tones
  failure.numNotes = 5;
  failure.notes = new String("g g c");
  failure.beats = new int[failure.numNotes] {4,1,4,1,12};
  // Success tones
  success.numNotes = 3;
  success.notes = new String("CCC");
  success.beats = new int[success.numNotes] {1,1,1};
  // Theme song
  theme.numNotes = 18;
  theme.notes = new String("cdfda ag cdfdg gf ");
  theme.beats =
    new int[theme.numNotes] {1,1,1,1,1,1,4,4,2,1,1,1,1,1,1,
    4,4,2};
  // Start signal
  go.numNotes = 3;
  go.notes = new String("aaa");
  go.beats = new int[failure.numNotes] {1,1,1};
  // Tones for the colors
  for (int i = 0; i < 4; i++) {
    colors[i].numNotes = 1;
    colors[i].beats = new int[1] {1};
  }
```

```
  colors[0].notes = new String("a");
  colors[1].notes = new String("g");
  colors[2].notes = new String("C");
  colors[3].notes = new String("f");
}

int Buzzer::frequency(char note) {
  int i;
  // Search through the letters in the array, and return the
  frequency for that note.
  for (i = 0; i < NOTES_IN_SCALE; i++) {
    if (noteNames[i] == note) {
      return(frequencies[i]);
    }
  }
  return(0);
}

void Buzzer::playSong(SongStruct *song) {
  int i, duration;

  for (i = 0; i < song->numNotes; i++) {
    duration = song->beats[i] * song->tempo;  // length of
                                                 note/rest in ms
    if (song->notes->charAt(i) == ' ') {      // is this a
                                                 rest?
      delay(duration);                        // then pause for
                                                 a moment
    } else {                                  // otherwise,
                                                 play the note
      tone(BUZZER_PIN, frequency(song->notes->charAt(i)),
      duration);
```

```
      delay(duration);              // wait for tone to finish
    }
    delay(song->tempo/10);          // brief pause between notes
  }
}
```

Now let's look at the code file for the Simon class. As you will see, this class contains a lot of code, but most is rather easy to read and understand.

Class Code File: Simon

Click the tab named `Simon.cpp` to open the blank file. Here, we will add the code for the class. Since there are a lot of functions in the class, we will first discuss each function in overview and then highlight some of the more complex ones in more detail, but none are overly difficult. You can discover how the other functions work as an exercise. We will begin with the public functions.

The constructor, `Simon()`, is where we set up the hardware for the class, which includes the Qwiic LED buttons, LCD, and the Buzzer class. The only tricky part is how to deal with the button addresses. We stored the Qwiic LED buttons in an array so we can instantiate a separate instance of the Qwiic LED button class for each passing in the correct address and then easily loop through them to turn off the LEDs as shown in the following:

```
// Setup the buttons
buttons[0].begin(0x6f);
buttons[1].begin(0x6E);
buttons[2].begin(0x6D);
buttons[3].begin(0x6B);
```

```
// Turn off all buttons
for (int button = 0; button < 4; button++) {
  buttons[button].LEDoff();
}
```

We also initialize the random number generator using a read from an analog pin as the seed. This will simulate using a different seed each time because reading an uninitialized pin will generate an unpredictable value. We can place the game in setup mode with the `setupMode()` function, which simply resets the LCD to indicate we are in the setup mode.

The `startGame()` function takes an integer for the number of players and simply zeros out the player scores, plays the theme song, and sets the LCD for start of game play.

The `play()` function is a bit more complicated. Here is where the game play is coded. At the highest level, the function loops generating a challenge sequence, playing it to the user, and then reading the player's response. If the challenge is met, the loop continues with an extra button added and a new random sequence generated.[5]

When there are more than one player, the loop cycles through each player in turn. If a player misses the sequence, that player is removed from the cycle (considered no longer playing or "alive"). Play continues until there are no more players alive, and a winner is determined, and the game ends. When the game ends, the code pauses and then resets the game class for the next game.

Next, let's look at the private functions. Recall private functions are used internally to the class and not visible to the caller.

The `numAlive()` function loops through the player scores to determine how many players are still playing. It is used in the `play()` function to determine when the game ends.

[5] I've seen many examples of the Simon game for Arduino and other platforms that use the same sequence adding a new button each time. To me, that's nowhere near as challenging as having a new sequence each turn.

The resetScreen() takes as a parameter a message to be displayed on the LCD. The function clears the display and then adds the message. It is used to control the LCD during game play.

The showWinner() function loops through the player scores to determine which player has the highest score. Since the play() function is designed to keep going until all players have failed to complete a sequence, it is possible for two or more players to have the same score. This is an intentional omission that you are encouraged to solve as an exercise. Hint: You can simply declare a tie.

The readButton() function takes a parameter for the button number (index) to determine if the button was pressed. If it was pressed, the code also turns on the LED so the player can get instant feedback. It simply returns true if the button was pressed or false if not.

The code for this function is shown in the following and demonstrates how to deal with latency or sampling challenges (also called debouncing) to ensure you don't get multiple press events over a short time. That is, it is possible the contacts inside the button make momentary contact several times before becoming a steady contact. We can use code to help with that situation. Study the following code to see how it was solved (there are many such methods):

```
bool Simon::readButton(int button) {
  if (buttons[button].isPressed() == true) {
    buttons[button].LEDon();
    // Wait for user to stop pressing
    while (buttons[button].isPressed() == true)
      delay(10);
    buzzer->playColor(button);
    buttons[button].LEDoff();
    Serial.print("Button ");
    Serial.print(button);
```

```
    Serial.println(" pressed.");
    return true;
  }
  return false;
}
```

The generateSequence() function takes as a parameter the number of buttons and returns an integer array allocated from memory that includes a set of random integers in range 0–3 to represent the buttons in the sequence. To create a random integer in that range, we call randint(4), which returns the correct range.

The playSequence() function uses two parameters: one for the button (challenge) sequence and another for the number of notes. It simply loops through the array turning on the LED and playing the tone for each button using a delay between each. This is used by the play() function to present the challenge sequence to the player.

The readSequence() function also uses two parameters: one for the button (challenge) sequence and another for the number of notes. It simply loops through the array reading the button presses from the player. If the correct button is pressed, the next button is read and so on. If all buttons were pressed in the correct order (the sequence pressed equals the challenge sequence), the function returns true, or false is returned on the first incorrect button press in the sequence. This is used by the play() function to read the player's response.

OK, that's a lot of functions! Let's now look at the complete code for the class. Take a few moments to read through it (there's a lot of code) to ensure you understand how it all works. Listing 16-6 shows the completed code for the class (documentation omitted for brevity).

Listing 16-6. Simon Code File

```
#include "Simon.h"

Simon::Simon() {
  Wire.begin();

  // Setup the LCD
  lcd.begin(16, 2);
  // Set background color?
  lcd.setRGB(127, 127, 127);

  randomSeed(analogRead(0));

  // Setup the buttons
  buttons[0].begin(0x6f);
  buttons[1].begin(0x6E);
  buttons[2].begin(0x6D);
  buttons[3].begin(0x6B);
  // Turn off all buttons
  for (int button = 0; button < 4; button++) {
    buttons[button].LEDoff();
  }

  // Put game in setup mode
  setupMode();
}

void Simon::startGame(int players) {
  numPlayers = players;
  playerScores = new PlayerStruct*[numPlayers];
  for (int player = 0; player < numPlayers; player++) {
    playerScores[player] = new PlayerStruct;
    playerScores[player]->isAlive = true;
    playerScores[player]->highScore = 0;
  }
```

```
  Serial.print("Playing theme...");
  buzzer->playThemeSong();
  Serial.println("done.");
  resetScreen("Press START");
}

void Simon::play() {
  bool gameOver = false;
  int numNotes = 1;
  String message("Player ");

  // Main game loop
  while (!gameOver) {
    // For each player, generate a new sequence and test skills
    for (int player = 0; player < numPlayers; player++) {
      if (playerScores[player]->isAlive) {
        numNotes = playerScores[player]->highScore + 1;
        Serial.print("Challenge sequence size ");
        Serial.println(numNotes);
        resetScreen((String(message) += (player + 1)).c_str());
        int *challengeSequence = generateSequence(numNotes);
        playSequence(challengeSequence, numNotes);
        delay(250);
        buzzer->playReadySetGo();
        Serial.println("Go!");
        delay(250);
        if (readSequence(challengeSequence, numNotes)) {
          buzzer->playSuccess();
          Serial.println("Success!");
          delay(500);
          playerScores[player]->highScore = numNotes;
        } else {
```

```
          Serial.println("Fail");
          playerScores[player]->isAlive = false;
        }
        delete challengeSequence;
      }
    }
    // Check to see if any players remain alive
    // and show winner if multiple players
    int playersRemaining = numAlive();
    if (playersRemaining == 0) {
      resetScreen("GAME OVER");
      if (numPlayers > 1) showWinner();
      gameOver = true;
      Serial.println("Game over...");
      delay(2000);
    }
  }
  // Delete the player memory
  for (int player = 0; player < numPlayers; player++) {
    delete playerScores[player];
  }
  delete playerScores;
}

void Simon::setupMode() {
  lcd.clear();
  lcd.setCursor(0, 0);
  lcd.print("Simon Says!");
  lcd.setCursor(0, 1); // column 1, row 2
  lcd.print("Setup Mode");
}
```

```
void Simon::showPlayers(int players) {
  lcd.clear();
  lcd.setCursor(0, 0);
  lcd.print("Simon Says!");
  lcd.setCursor(0, 1); // column 1, row 2
  if (players == 1) {
    lcd.print("single player");
  } else  {
    lcd.print(players);
    lcd.print(" players");
  }
}

void Simon::showWinner() {
  int winner = -1;
  int score = 0;
  for (int player = 0; player < numPlayers; player++) {
    if (playerScores[player]->highScore > score) {
      winner = player;
      score = playerScores[player]->highScore;
    }
  }
  lcd.setCursor(0, 0);
  lcd.print("Player ");
  lcd.print(winner + 1);
  lcd.print("WON!");
  lcd.setCursor(0, 1); // column 1, row 2
  lcd.print("Score = ");
  lcd.print(score);
}
```

```
int Simon::numAlive() {
  int count = 0;
  if (playerScores != NULL) {
    for (int player = 0; player < numPlayers; player++) {
      if (playerScores[player]->isAlive) count++;
    }
  }
  return count;
}

void Simon::resetScreen(const char *message) {
  lcd.clear();
  lcd.setCursor(0, 0);
  lcd.print("Simon Says! (");
  lcd.print(numPlayers);
  lcd.print(")");
  lcd.setCursor(0, 1); // column 1, row 2
  lcd.print(message);
}

bool Simon::readButton(int button) {
  if (buttons[button].isPressed() == true) {
    buttons[button].LEDon();
    // Wait for user to stop pressing
    while (buttons[button].isPressed() == true)
      delay(10);
    buzzer->playColor(button);
    buttons[button].LEDoff();
    Serial.print("Button ");
    Serial.print(button);
    Serial.println(" pressed.");
    return true;
```

```
  }
  return false;
}

bool Simon::readSequence(int *challengeSequence, int numNotes) {
  int buttonRead = -1;
  int index = 0;
  unsigned long startTime = millis();

  // Loop reading buttons and compare to stored sequence
  while (index < numNotes) {
    if (readButton(0)) {
      buttonRead = 0;
    } else if (readButton(1)) {
      buttonRead = 1;
    } else if (readButton(2)) {
      buttonRead = 2;
    } else if (readButton(3)) {
      buttonRead = 3;
    }
    // if a button is pressed, check the sequence
    if (buttonRead >= 0) {
      if (challengeSequence[index] != buttonRead) {
        buzzer->playFailure();
        resetScreen("FAIL SEQUENCE");
        delay(5000);
        return false;
      } else {
        Serial.println("MATCH!");
        startTime = millis();
```

```
        index++;
        buttonRead = -1;
      }
    }
    if (millis() > (startTime + MAX_TIMEOUT)) {
      buzzer->playFailure();
      resetScreen("FAIL TIMEOUT");
      delay(5000);
      return false;
    }
    delay(50);
  }
  return true;
}

void Simon::playSequence(int *challengeSequence, int numNotes) {
  for (int note = 0; note < numNotes; note++) {
    int buttonIndex = challengeSequence[note];
    buttons[buttonIndex].LEDon();
    buzzer->playColor(buttonIndex);
    delay(KEY_INTERVAL);
    buttons[buttonIndex].LEDoff();
    delay(KEY_INTERVAL);
  }
}

int *Simon::generateSequence(int numNotes) {
  if (numNotes == 0) return NULL;
  // Create a new sequence adding a new note
  int *challengeSequence = new int[numNotes];
```

```
  for (int note = 0; note < numNotes; note++) {
    challengeSequence[note] = (int)random(4);
  }
  return challengeSequence;
}
```

As you can see, the code is long, but except for a few places not difficult. It is a lot of code, but demonstrates how to work with multiple hardware components including mixing Grove and Qwiic in the same project.

Compile the Sketch

The last step is to compile the sketch before uploading it to your board. If you encounter any errors, be sure to fix them and recompile to ensure the sketch compiles without errors or serious warnings.

Once everything compiles, we're ready to start testing. But first, let's look at the code for the Raspberry Pi. You can skip to the "Sketch on the Arduino" section if you're curious to see how the project works. While the code will execute the same on both platforms, the values differ due to the differences in how the sensors are read (the range of values differs).

Raspberry Pi

This section presents a walk-through of the Python code you will write to create the Simon game. But first, there are a couple of libraries we must install on our Raspberry Pi.

Note If you have not installed the GrovePi libraries, please see Chapter 13 for complete details and install them before you begin. If you encounter problems, see the "GrovePi/GrovePi+ Troubleshooting Tips" section in Chapter 13.

Install a Software Library

Aside from the GrovePi libraries, we need only one more software library. Specifically, we need a library for the Qwiic LED Button Breakout. Fortunately, SparkFun has provided an excellent library we can use, and you can install it as shown in the following:

```
$ pip3 install sparkfun-qwiic-button
```

We also need the custom library we created in Chapter 14 to work with the LCD RGB Backlight. Simply copy that library named `grove_lcd_rgb.py` to your Simon game folder.

Once you have that library installed and the grove LCD library from Chapter 14 copied, we're ready to write the code.

Write the Code

The code for the Python version of this project is a bit different than the Arduino code. We will still create the same classes (`Simon`, `Buzzer`), but the code for controlling the buzzer is very different.

The reason is there is no Python code for the Grove buzzer and there is no tone() function in the Python platform or even the GrovePi libraries. We are on our own to create a substitute. The solution works, but it is very different, so we will take a detailed walk-through for that part of the code.

Other than that, the other differences are minor and most due to differences in the language. We will see each of the nuances to ensure you understand how the Python code works in the same manner as the Arduino code.

While we will not dive into every line of code, we will see some of the more complex code and those areas discussed that differ significantly from the Arduino version. You can read through the code and learn more about how it works at your leisure.

Let's start with writing the classes starting with the `Buzzer` class.

Buzzer Class

Open the Thonny Python IDE under the *Main* ➤ *Programming* submenu. Create a new file `buzzer.py`. We will use the same functions as the Arduino version except for how the `play_song()` function works. However, before we look at that, let's discuss how we are going to substitute a Python data mechanism for the C++ structure.

Rather than use a tuple, we will use one of the more powerful Python data storage called a dictionary. A dictionary allows us to create a structure where we can store one or more key/value pairs where we can store all of the parts of the song: the number of notes, notes, and beats. We will also see how to store the tempo for each song.[6] Interestingly, the tempo for the Python version needed to be adjusted to make the same sounds. This is largely due to how we implement the PWM for the notes.

The following shows the layout of the dictionary we will use for each song. Here, we use the keys `tempo`, `num_notes`, `notes`, and `beats`, which will be used in the code to reference the value for each:

```
# Success tones dictionary
self.success = {
    'tempo': DEFAULT_TEMPO,
    'num_notes': 3,
    'notes': "CCC",
    'beats': [1, 1, 1]
}
```

While that may seem like a minor change, it is quite different in how the data is accessed in the rest of the code.

[6] See the challenge in the "Arduino" section regarding storing the tempo with each song.

Now, let's look at the `play_song()` function. This is where things get interesting. Recall we must use an analog connection here since we are going to use a PWM to produce a sound on the buzzer. Thus, we must initialize the buzzer pin using an analog connection. However, the GrovePi does not permit direct connections to analog pins. Rather, we must choose a different analog pin and use the GPIO functions directly. The following shows the code differences for initializing the GPIO pin:

```
import RPi.GPIO as GPIO
...
BUZZER_PIN = 23
...
class Buzzer:
...
    def __init__(self):
        ...
        # Setup the buzzer
        GPIO.setmode(GPIO.BCM)
        GPIO.setwarnings(False)
        GPIO.setup(BUZZER_PIN, GPIO.OUT)
```

Here we see that we import a different library: the base Raspberry Pi GPIO library, which should already be on your Raspberry Pi. We then define the button as pin 23 (the drawings in the "Connect the Grove Modules" section show where this pin is physically located), then use the GPIO library to set up the GPIO using the Broadcom (BCM) numbering scheme,[7] and set up the pin as an output pin.

Next, we will create our own `tone()` function so that the `play_song()` function remains largely intact semantically from the Arduino version. The following shows the new `tone()` function:

[7] See `https://pinout.xyz/#` for more details about pin layouts.

```
def tone(frequency, duration):
    half_wave = 1 / (frequency * 2)
    waves = int(duration * frequency)
    for i in range(waves):
        GPIO.output(BUZZER_PIN, True)
        time.sleep(half_wave)
        GPIO.output(BUZZER_PIN, False)
        time.sleep(half_wave)
```

Notice here we do some math first where we get the half wave of the frequency. We are getting one half of the sine wave so that we can turn the buzzer for half the wave and off for half the wave, which is the frequency times the duration and, hence, a pulse. Again, there are other ways to generate a PWM, but this works well for the buzzer.

Other than that significant change, the rest of the code is similar to the Arduino version just rewritten in Python.

Listing 16-7 shows the complete code for the class with documentation removed for brevity. Take a few moments to read through the code so that you understand all of the parts of the code. As you will see, it is not nearly as complicated as the Arduino class, thanks to the helpful class library and utility class from Adafruit.

Listing 16-7. Buzzer Class (Python)

```
import time
import RPi.GPIO as GPIO

# CONSTANTS
DEFAULT_TEMPO = 0.095
BUZZER_PIN = 23
NOTES_IN_SCALE = 8
HIGH = 1
LOW = 0
```

```
def tone(frequency, duration):
    half_wave = 1 / (frequency * 2)
    waves = int(duration * frequency)
    for i in range(waves):
        GPIO.output(BUZZER_PIN, True)
        time.sleep(half_wave)
        GPIO.output(BUZZER_PIN, False)
        time.sleep(half_wave)

class Buzzer:
    note_names = ['c', 'd', 'e', 'f', 'g', 'a', 'b', 'C']
    frequencies = [262, 294, 330, 349, 392, 440, 494, 523]
    failure = {}
    success = {}
    theme_song = {}
    ready_set_go = {}
    colors = [{}, {}, {}, {}]

    def __init__(self):
        """Constructor"""
        # Failure tones dictionary
        self.failure = {
            'tempo': DEFAULT_TEMPO,
            'num_notes': 5,
            'notes': "g g c",
            'beats': [4, 1, 4, 1, 10]
        }
        # Success tones dictionary
        self.success = {
            'tempo': DEFAULT_TEMPO,
            'num_notes': 3,
            'notes': "CCC",
```

```
        'beats': [1, 1, 1]
    }
    # Theme song dictionary
    self.theme_song = {
        'tempo': DEFAULT_TEMPO,
        'num_notes': 18,
        'notes': "cdfda ag cdfdg gf ",
        'beats': [1, 1, 1, 1, 1, 1, 4, 4, 2,
                  1, 1, 1, 1, 1, 1, 4, 4, 2]
    }
    # Start signal dictionary
    self.ready_set_go = {
        'tempo': DEFAULT_TEMPO,
        'num_notes': 3,
        'notes': "aaa",
        'beats': [1, 1, 1]
    }
    # Tones for the colors
    for i in range(0, 4):
        self.colors[i]['tempo'] = DEFAULT_TEMPO
        self.colors[i]['num_notes'] = 1
        self.colors[i]['beats'] = [1]
    self.colors[0]['notes'] = "a"
    self.colors[1]['notes'] = "g"
    self.colors[2]['notes'] = "C"
    self.colors[3]['notes'] = "f"

    # Setup the buzzer
    GPIO.setmode(GPIO.BCM)
    GPIO.setwarnings(False)
    GPIO.setup(BUZZER_PIN, GPIO.OUT)
```

```
    def play_theme_song(self):
        self.play_song(self.theme_song)

    def play_start(self):
        self.play_song(self.ready_set_go)

    def play_success(self):
        self.play_song(self.success)

    def play_failure(self):
        self.play_song(self.failure)

    def play_color(self, color):
        self.play_song(self.colors[color])

    def play_ready_set_go(self):
        self.play_song(self.ready_set_go)

    def frequency(self, note):
        # Search through the letters in the array, and
        # return the frequency for that note.
        for i in range(0, NOTES_IN_SCALE):
            if self.note_names[i] == note:
                return self.frequencies[i]
        return 0

    def play_song(self, song):
        for i in range(0, song['num_notes']):
            duration = song['beats'][i] * song['tempo']
            if song['notes'][i] == ' ':
                time.sleep(duration)
            else:
                freq = self.frequency(song['notes'][i])
                tone(freq, duration)
```

```
        time.sleep(duration)
        time.sleep(song['tempo']/10)
```

Now, let's look at the Simon class file.

Simon Class

Open the Thonny Python IDE under the *Main* ➤ *Programming* submenu. Create a new file Simon.py. We will use the same functions as we did in the Arduino version renamed to a more Python-friendly naming scheme.

What differs from the Arduino is the use of the LCD class from Chapter 14. Fortunately, the functions are similar to the library we used in the Arduino version. Recall we import the new class with the following code:

```
from grove_lcd_rgb import GroveLcdRgb
```

Another major difference is in the play() function. Here, the timing of the events is a bit different. That is, the sleeps in the code had to be adjusted so that game play followed the same flow from the Arduino version.

There are also necessitated delay changes in the read_button() function to control how fast the button presses are read. Recall we need to debounce the button press, and in this the Python version of the Qwiic LED button library differs.

On a very minor point, we moved the generate_sequence() function out of the class because it contains no references to the internal variables of functions of the class. This is a Python style thing. You can leave the function in the class, but if you run a tool like pylint to check Python style and best practices, it will complain.

Those are the major differences in the Python code. Everything else is similar to the Arduino version. Listing 16-8 shows the complete code for the class with documentation removed for brevity. Take a few moments to read through the code so that you understand all of the parts of the code. As you will see, it is not nearly as complicated as the Arduino class, thanks to the helpful class library and utility class from Adafruit.

Listing 16-8. Simon Class (Python)

```
import random
import time
import smbus
import qwiic_button
from grovepi import analogRead
from buzzer import Buzzer
from grove_lcd_rgb import GroveLcdRgb

# Constants
MIN_BEATS = 2          # Starting number of beats
MAX_PLAYERS = 4        # Max number of players
MAX_TIMEOUT = 5.0      # Seconds to wait to abort read
KEY_INTERVAL = 0.500 # Interval between button playback
LED_BRIGHTNESS = 200 # Button LED brightness

def generate_sequence(num_notes):
    if num_notes == 0:
        return []
    challenge_sequence = []
    i = 0
    while i < num_notes:
        challenge_sequence.append(random.randint(0, 3))
        i = i + 1
    return challenge_sequence

class Simon:
    bus = smbus.SMBus(1)
    num_players = 1
```

```
    player_scores = []
    buzzer = Buzzer()                    # Buzzer
    lcd = GroveLcdRgb()                  # Grove LCD
    buttons = [qwiic_button.QwiicButton(0x6f),
               qwiic_button.QwiicButton(0x6E),
               qwiic_button.QwiicButton(0x6D),
               qwiic_button.QwiicButton(0x6B)]

    def __init__(self):
        # Setup the LCD
        self.lcd.clear()
        # Set background color?
        self.lcd.set_rgb(127, 127, 127)

        random.seed(analogRead(0))

        # Setup the buttons
        # Turn off all buttons
        for button in range(0, 4):
            self.buttons[button].begin()
            self.buttons[button].LEDoff()
            self.buttons[button].setDebounceTime(500)
        # Put game in setup mode
        self.setup_mode()

    def start_game(self, players):
        self.num_players = players
        for player in range(0, players):
            player_score = {
                'number': player,
                'is_alive': True,
                'high_score': 0
            }
```

```
            self.player_scores.append(player_score)
        print("Playing theme...")
        self.buzzer.play_theme_song()
        print("done.")
        self.reset_screen("Press START")

    def play(self):
        game_over = False
        num_notes = 1

        while not game_over:
            for player in range(0, self.num_players):
                if self.player_scores[player]['is_alive']:
                    num_notes = self.player_scores[player]
                    ['high_score'] + 1
                    self.reset_screen("Player {0}".
                    format(player + 1))
                    challenge_sequence = generate_sequence(num_
                    notes)
                    self.play_sequence(challenge_sequence, num_
                    notes)
                    time.sleep(0.250)
                    self.buzzer.play_ready_set_go()
                    print("Go!")
                    time.sleep(0.250)
                    if self.read_sequence(challenge_sequence,
                    num_notes):
                        self.buzzer.play_success()
                        print("Success!")
                        time.sleep(0.500)
                        self.player_scores[player]['high_
                        score'] = num_notes
```

```
                else:
                    print("Fail")
                    self.player_scores[player]['is_alive']
                    = False
        players_remaining = self.num_alive()
        if players_remaining == 0:
            self.reset_screen("GAME OVER")
            if self.num_players > 1:
                self.show_winner()
            game_over = True
            print("Game over...")
            time.sleep(2)
    self.player_scores = []

def setup_mode(self):
    self.lcd.clear()
    self.lcd.set_cursor(0, 0)
    self.lcd.print("Simon Says!")
    self.lcd.set_cursor(0, 1) # column 1, row 2
    self.lcd.print("Setup Mode")

def show_players(self, num_players):
    self.lcd.clear()
    self.lcd.set_cursor(0, 0)
    self.lcd.print("Simon Says!")
    self.lcd.set_cursor(0, 1) # column 1, row 2
    if num_players == 1:
        self.lcd.print("single player")
    else:
        self.lcd.print(chr(num_players + 0x30))
        self.lcd.print(" players")
```

```
    def show_winner(self):
        winner = -1
        score = 0
        for player in range(0, self.num_players):
            if self.player_scores[player]['high_score'] >
            score:
                winner = player
                score = self.player_scores[player]['high_
                score']
        self.lcd.set_cursor(0, 0)
        self.lcd.print("Player ")
        self.lcd.print(winner + 1)
        self.lcd.print("WON!")
        self.lcd.set_cursor(0, 1) # column 1, row 2
        self.lcd.print("Score = ")
        self.lcd.print(score)

    def num_alive(self):
        count = 0
        for player in range(0, self.num_players):
            if self.player_scores[player]['is_alive']:
                count = count + 1
        return count

    def reset_screen(self, message):
        self.lcd.clear()
        self.lcd.set_cursor(0, 0)
        self.lcd.print("Simon Says! (")
        self.lcd.print("{0}".format(self.num_players))
        self.lcd.print(")")
        self.lcd.set_cursor(0, 1) # column 1, row 2
        self.lcd.print(message)
```

```
    def read_button(self, button):
        if self.buttons[button].isButtonPressed():
            self.buttons[button].LEDon(LED_BRIGHTNESS)
            time.sleep(0.10)
            self.buzzer.play_color(button)
            self.buttons[button].LEDoff()
            print("Button {0} pressed.".format(button))
            return True
        return False

    def read_sequence(self, challenge_sequence, num_notes):
        button_read = -1
        index = 0
        start_time = time.time()

        while index < num_notes:
            if self.read_button(0):
                button_read = 0
            elif self.read_button(1):
                button_read = 1
            elif self.read_button(2):
                button_read = 2
            elif self.read_button(3):
                button_read = 3
            # if a button is pressed, check the sequence
            if button_read >= 0:
                if challenge_sequence[index] != button_read:
                    self.buzzer.play_failure()
                    self.reset_screen("FAIL SEQUENCE")
                    time.sleep(5)
                    return False
                print("MATCH!")
```

```
                start_time = time.time()
                index = index + 1
                button_read = -1
            if (time.time() - start_time) > MAX_TIMEOUT:
                print("ERROR: Timeout!")
                self.buzzer.play_failure()
                self.reset_screen("FAIL TIMEOUT")
                time.sleep(5)
                return False
            time.sleep(0.050)
        return True

    def play_sequence(self, challenge_sequence, num_notes):
        for beat in range(0, num_notes):
            button_index = challenge_sequence[beat]
            self.buttons[button_index].LEDon(LED_BRIGHTNESS)
            self.buzzer.play_color(button_index)
            time.sleep(KEY_INTERVAL)
            self.buttons[button_index].LEDoff()
            time.sleep(KEY_INTERVAL)
```

Now we can write our main script.

Main Script (Python)

Open the Thonny Python IDE under the *Main* ➤ *Programming* submenu. Create a new file `simon_says.py`. There is nothing new in this code as it follows the same flow as the Arduino example. Listing 16-9 shows the complete code for the main script for this project. You can read through it to see how all of the code works.

Listing 16-9. Main Script (Python)

```
import sys
import time

from grovepi import pinMode, digitalRead
from simon import Simon, MAX_PLAYERS

# Constants
START_BUTTON = 7
MODE_BUTTON = 8
LOW = 0

def main():
    """Main"""
    print("Welcome to the Simon Says game!")
    simon = Simon()
    game_started = False
    start_button = False
    mode_button = False
    num_players = 1
    pinMode(START_BUTTON, "INPUT")
    pinMode(MODE_BUTTON, "INPUT")
    while True:
        if not game_started:
            # Show number of players
            start_button = digitalRead(START_BUTTON) == LOW
            mode_button = digitalRead(MODE_BUTTON) == LOW
            if start_button:
                print("Start button pressed.")
                simon.start_game(num_players)
                time.sleep(1)
                simon.play()
```

```
                time.sleep(1)
                simon.setup_mode()
            elif mode_button:
                num_players = num_players + 1
                if num_players > MAX_PLAYERS:
                    num_players = 1
                print("Mode button pressed - {0} players."
                      "".format(num_players))
                time.sleep(0.050)
                simon.show_players(num_players)
                time.sleep(2)
                simon.setup_mode()

if __name__ == '__main__':
    try:
        main()
    except (KeyboardInterrupt, SystemExit) as err:
        print("\nbye!\n")
sys.exit(0)
```

OK, that's it! We've written the code. Unlike the Arduino, we do not need to compile the Python code. So we're now ready to execute the project!

Execute the Project

Now that we've spent many pages exploring the Grove modules and writing the code to interact with them, it is time to test the project by executing (running) it.

When the project runs (executes), you will see some diagnostic messages written to the serial monitor (Arduino) or the terminal (Raspberry Pi). You will also see a welcome message appear on the LCD.

You can then press the *mode* button to set the number of players and, when you're ready, press the start button to *start* the game. Figure 16-16 shows examples of the LCD when in setup mode, X, and Y.

Figure 16-16. *Executing the Simon Says project*

Executing the code depends on which platform you're using, but the output in the serial monitor and the terminal is the same. Let's look at the Arduino first.

Sketch on the Arduino

Executing the sketch on the Arduino requires connecting our board to our PC and then uploading the sketch to the Arduino. Recall the sketch will run so long as the USB cable is connected to our PC (and the Arduino).

Execute the Sketch

To execute the sketch, be sure your Arduino is connected and you've selected the correct board under the *Tools* ➤ *Board* menu. You also need to ensure you have the correct port selected under the *Tools* ➤ *Port* menu.

Once those items are set, you can click the *Upload* button or choose *Sketch* ➤ *Upload* from the menu. The Arduino IDE will compile the sketch and then upload it to your Arduino. Once you see the `Done uploading...` message, you can open the serial monitor. You should see the output begin momentarily that is the same as that on the LCD. Go ahead, and try it out! You should see values similar to the following:

```
Welcome to the Simon Says game!
Mode button pressed - 2 players.
Mode button pressed - 3 players.
Mode button pressed - 4 players.
Mode button pressed - 1 players.
Start button pressed.
Playing theme...
done.
Go!
Button 0 pressed.
MATCH!
Success!
Go!
Button 1 pressed.
MATCH!
Button 2 pressed.
MATCH!
Success!
Go!
Button 3 pressed.
MATCH!
Button 0 pressed.
MATCH!
Button 3 pressed.
MATCH!
```

```
Success!
Go!
ERROR: Timeout!
Fail
Game over...
...
```

If something isn't working, check your connections or refer to Chapter 13 for troubleshooting tips.

Python Code on the Raspberry Pi

Executing the sketch on the Raspberry Pi requires running the Python code in a terminal after connecting your modules and powering on the Raspberry Pi. Recall the code will run until you stop it with *CTRL+C* on the keyboard.

Execute the Python Code

To run the Python code on the Raspberry Pi, you can issue the command `python3 ./simon_says.py` from the same folder where the file was saved as shown in the following. You should get results similar to the following:

```
$ python3 ./simon_says.py
Welcome to the Simon Says game!
Mode button pressed - 2 players.
Mode button pressed - 3 players.
Mode button pressed - 4 players.
Mode button pressed - 1 players.
Start button pressed.
Playing theme...
...
```

If everything worked as executed, congratulations! You've just built your fourth Grove project. If something isn't working, check your connections or refer to Chapter 13 for troubleshooting tips.

Going Further

While we didn't discuss them in this chapter, there are some ideas where you could make this project into an IoT project. Here are just a few suggestions you can try once we have learned how to take our projects to the cloud. Put your skills to work!

- *Simon Says portal*: You can display the high scores of your games to see how each player is fairing over time. Or, if you're playing it yourself, you can track your progress in becoming a Simon Says gaming guru.
- *Complete the enclosure*: Use the sample base plate and create a cover for the game.
- *Increase the difficulty*: One of the ways you can enhance game play is to make the timeout time for a player to enter a sequence shorter as game play continues. For example, for the first n sequences, use the default timeout; for the next n sequences, reduce the timeout by a portion; and so on until the timeout gets to a minimum timeout. If you do the same thing for the delay used in playing the challenge sequence, it will ensure the game will become much more difficult and possibly more fun to play.

Summary

In this chapter, we completed a set of example projects to explore building IoT projects with Grove modules. Along the way, we learned how to work with Grove modules including how to write our own classes for managing multiple modules and sensors (buttons are sensors after all).

Combined with the Qwiic projects, we took a long look at our component systems without the Internet component intentionally so that we can make the projects easier to implement so that we can learn how to select and connect the hardware as well as how to write the code to interact with it.

Along the way, we were presented with a list of optional enhancements, many of which encouraged you to expand your understanding of the hardware and software (code). We also saw some seeds to begin thinking about how to make the projects interact with the Internet. Not all of the projects were necessarily well suited for use over the Internet, but with some imagination, we can make it work.

Now that we've mastered how to build projects with Qwiic, STEMMA QT, and Grove modules, we can focus on the Internet portion. As you will see, it isn't as simple as adding a few methods - we will have to think about how to store and present the data.

In the next section, we will conclude our exploration of IoT projects by completing the missing piece - the Internet part! Yes, you've been waiting for that for 16 chapters, and it's finally here.

PART IV

Going Further: IoT and the Cloud

This part introduces an overview of cloud systems for the IoT. It features a tutorial on ThingSpeak that demonstrates how to store and share data from the example projects for the Arduino and Raspberry Pi.

CHAPTER 17

Introducing IoT for the Cloud

Now that you've seen a number of projects, ranging from very basic to advanced in difficulty, it is time to discuss how to make your IoT data viewable by others via the cloud. More specifically, you will get a small glimpse at what is possible with the more popular cloud computing services and solutions.

I say a glimpse because it is not possible to cover all cloud services and solutions available for IoT in a single chapter. Once again, this is a case where learning a little bit about something and seeing it in practice will help you get started. Like the other chapters where you've had a lightning tour, this chapter presents a few of the newer concepts and features of cloud solutions at a high level.

In this chapter, we will get an overview of what the cloud is and how it is used for IoT solutions. The chapter also presents a short overview of the popular cloud systems for IoT as well as a short example using one of the free options to give you a sense of what is possible and how some of our projects can be modified to use the Internet.

C. Bell, *Beginning IoT Projects*, https://doi.org/10.1007/978-1-4842-7234-3_17

THE CLOUD: ISN'T THAT JUST MARKETING HYPE?

Don't believe all the hype or sales talk about any product that includes "cloud" in its name. Cloud computing services and resources should be accessible via the Internet from anywhere, available to you via subscription (with fee or for free), and permit you to consume or produce and share the data involved. Also, consider the fact that you must have access to the cloud to get to your data. Thus, you have no alternative if the service is unreachable (or down).

Since the technologies presented are quite unique in implementation (but rather straightforward in concept), I keep the project hardware and programming to a minimal effort. In fact, you will reuse the weather project that you used in Chapter 8 and the secret knock example from Chapter 13. However, rather than saving the data in MySQL, you will send it to Microsoft Azure. You will also see how to present the data via a website connected to the data in Microsoft Azure.

Overview

Unless you live in a very isolated location, you have likely been bombarded with talk about the cloud and IoT. Perhaps you've seen advertisements in magazines and on television or read about it in other books or attended a seminar or conference. Unless you've spent time learning what cloud means, you are probably wondering what all the fuss is about.

What Is the Cloud?

Simply stated,[1] the cloud is a name tagged to services available via the Internet. These can be servers you can access (running as a virtual machine on a larger server), systems that provide access to a specific software or environment, or resources such as disks or IP addresses that you can attach to other resources. The technologies behind the cloud include grid computing (distributed processing), virtualization, and networking. The correct scientific term is cloud computing. Although a deep dive into cloud computing is beyond the scope of this book, it is enough to understand that you can use cloud computing services to store your sensor data.

What Is Cloud Computing Then?

The term *cloud computing* is sadly overused and has become a marketing term for some. True cloud computing solutions are services that are provided to subscribers (customers) via a combination of virtualization, grid computing (distributed processing and storage), and facilities to support virtualized hardware and software, such as IP addresses that are tied to the subscription rather than a physical device. Thus, you can use and discard resources on the fly to meet your needs.

These resources, services, and features are priced by usage patterns (called *subscription plans* or *tiers*), in which you can pay for as little or as much as you need. For example, if you need more processing power, you can move up to a subscription level that offers more CPU cores, more memory, and so forth. Thus, you only pay for what you need, which means that organizations can potentially save a great deal on infrastructure.

[1] Experienced cloud researchers will tell you there is a lot more to learn about the cloud.

A classic example of this benefit is a case where an organization experiences a brief and intense level of work that requires additional resources to keep their products and services viable. Using the cloud, organizations can temporarily increase their infrastructure capability and, once the peak has passed, scale things back to normal. Clearly, this is a lot better than having to rent or purchase a ton of hardware for that one event.

Sadly, there are some vendors that offer cloud solutions (typically worded as *cloud enabled* or simply *cloud*) that fall far short of being a complete solution. In most cases, they are nothing more than yesterday's Internet-based storage and visualization. Fortunately, Microsoft Azure is the real deal: a full cloud computing solution with an impressive array of features to support almost any cloud solution you can dream up.

Tip If you would like to know more about cloud computing and its many facets, see `https://en.wikipedia.org/wiki/Cloud_computing`.

How Does the Cloud Help IoT?

OK, so now that we know what cloud systems are, how do they help us with our IoT projects? There are a variety of ways, but most common are mechanisms for storing and presenting your data rather than storing it locally or even remotely on another system such as a dedicated database server. That is, you can send the data you collect from your sensors to the cloud for storage and even use additional cloud services to view the data using charts, graphs, or just plain text. The sky is the limit with respect to how you can present your data.

But storing data isn't the only feature you can leverage in the cloud. There are other services that you can use to link to yet other services to form a solution. For example, most paid IoT cloud systems provide features that can "talk" to each other allowing you to link them together to quickly build a solution. The features are often called components rather than services, but both terms apply.

For example, in Microsoft Azure, you can store your data with one of several components and then link it to others that allow you to modify the data via queries, to others to route the data to other places (even to another cloud service vendor), and ultimately to one of several components for displaying the data. Yes, it really is a set of building blocks like that.

Now that we've had a general overview of cloud systems, let's look at those that support IoT projects directly.

IoT Cloud Systems

There are a number of IoT cloud vendors that offer all manner of products, capacities, and features to match just about anything you can conjure for an IoT project. With so many vendors offering IoT solutions, it can be difficult to choose one. The following is a short list of the more popular IoT offerings from the top vendors in the cloud industry:

- *Oracle IoT*: `www.oracle.com/internet-of-things/`
- *Microsoft Azure IoT Hub*: `https://azure.microsoft.com/en-us/product-categories/iot/`
- *Google IoT Core*: `https://cloud.google.com/iot-core`
- *IBM IoT*: `www.ibm.com/internet-of-things`
- *Arduino IoT Cloud*: `www.arduino.cc/en/IoT/HomePage`
- *Adafruit IO*: `https://io.adafruit.com/`
- *If This Then That (IFTTT)*: `https://ifttt.com/`
- *MathWorks ThingSpeak*: `https://thingspeak.com/`

Most of the vendors offer commercial products, but a few like Google, Azure, Arduino, IFTTT, and ThingSpeak offer limited free accounts. A few are free like Adafruit IO and Arduino IoT Cloud, but may limit you to a particular platform or a smaller set of features. As you may surmise, some of the offerings are complex solutions with steep learning curve, but the IFTTT and ThingSpeak offerings are simple and easy to use. Since we want a solution that supports Arduino and Raspberry Pi (and other platforms), we will use IFTTT in this chapter as an example of what is possible when working with the cloud. We will then use ThingSpeak in the next chapter to round out our introduction to IoT cloud systems.

Tip If you want or need to use one of the other vendors, be sure to read all of the tutorials thoroughly before jumping into your code.

Let's look at some of the types of services available in cloud systems that support IoT projects.

IoT Cloud Services Available

IoT projects offer an amazing opportunity to expand our knowledge of the world around us and to observe events from all over the world no matter where we are located. To meet these needs, IoT cloud services are an array of services that you can leverage in your applications.

There are services for collecting data, managing your devices, performing analytics, and even application and processing extensions for you to exploit. For example, some vendors include complete user management where you can provide user accounts for people to log in and use your cloud solution and see your data.

The following lists a number of the types of services available. Some vendors may not offer all of the services, and a service common among the vendors may work very differently from one vendor to another.

However, this should give you an idea of what services are available and a general idea of the feature set:

- *Device management*: Allows you to set up, manage, and track devices in your IoT network.
- *Data storage*: Permits storage of your IoT data either on a temporary (typically free for a number of days) or permanent (paid) storage.
- *Data analytics*: Allows you to perform analysis on your data to find trends, outliers, or any form of analytical query.
- *Data query and filters*: You can perform queries or filter your IoT data after it has been sent to the cloud service for detailed presentations or transformations.
- *Big data*: Permits you to store vast amounts of data and perform operations on the data (think data warehousing).
- *Visualization tools*: Various dashboards and graphics you can use to help present your data in meaningful ways (spreadsheet, pie charts, etc.).
- *High availability*: Provides features that allow you to operate even if portions of your cloud servers (or the vendors) fail or go offline due to network issues.
- *Third-party integration*: Allows you to connect your IoT services to other IoT servers from other vendors, for example, connecting your Adafruit IO data to IFTTT for triggering an SMS message.
- *Security (data, user)*: Provides support for managing user accounts, security access, and more for your applications.

- *Encryption*: Allows you to encrypt your data either in the cloud or when transmitting the data from one service to another.
- *Deployment*: Similar to device management, but on a grander scale where you build IoT devices using common profiles, operating systems, configurations, etc.
- *Scalability*: The ability to scale from a small number of devices and services to many devices. This is often only available in the larger, paid vendor services.
- *APIs (Rest, programming)*: Allows you to write code to communicate directly with the services instead of issuing web requests. Often part of the larger, paid vendor services.

For our beginning IoT projects, we will be focusing on a subset of these services, which can be grouped into several categories. Let's look at a few of the most common services you may want to start using right away.

Data Storage

These services allow you to store your data in the cloud rather than on your local device. Some data such as alerts or notices do not need to be stored, and you should consider if you would need the data in the future or it will be project dependent. For example, if you want to create a weather alert project, you may not care what the temperature was a week or even a month ago. However, if you want to do some amateur weather forecasting, you will want to store data for some time (perhaps years). You may consider storing the data locally, which may be possible for some platforms such as the Raspberry Pi, but the Arduino and similar boards have very limited storage capabilities.

Thus, if you need to store your data for some period and storing it locally is not an option, you should consider this when selecting a cloud vendor. Look for how data will be stored, the mechanisms needed to send the data to the service, and how to get the data out of the service.

Data Transformation (Queries)

These services allow you to perform queries on the data as it flows to or through the cloud services. You may want to show only a subset of the data to your users, or you may want to filter the data so that data from certain devices, dates, etc. are shown for one of several views.

The case where you'd want to consider includes IoT projects that collect data from multiple sensors and multiple devices where the data is stored for a period of time. For example, if you have devices geographically distributed over a wide area, you may only want to see data from a subset of those devices. Similarly, if you have data from several time periods, hours, days, and weeks, you may only want to see data from a specific time.

Visualization Tools

These services along with routing and messaging are the most commonly used for beginning IoT projects. These are simply services that allow you to see your data on the Internet. It may be nothing more than a simple list of the data, or it may be an elaborate data dashboard complete with controls that users can use to manipulate the display. Fortunately, most cloud vendors provide a robust set of tools (some more than one) that you can use to present your data to yourself or your users.

Routing and Messaging

These services are the heart or the bones of the IoT cloud. They encompass the glue to bind different services together. More specifically, they provide mechanisms for you to connect your devices to services and those services to other services such as queries, filters, and visualization tools permitting you to build an IoT solution using several cloud services. We'll see an example of such a service in the next section.

Now that we've had an overview of the IoT cloud services and the most common services we will encounter, let's jump into a simple example using If This Then That (IFTTT).

Cloud Services Example: IFTTT

Getting Started

Now that you've seen a variety of project implementations and coding examples, you're ready to add that extra bit to connect them to the cloud.

As you will see, the code we will use to connect to IFTTT isn't difficult, but the concepts of using the service are quite different than what you may have seen in the past. This is the primary reason we wait until this chapter to connect our sample projects to the Internet.

But first, let's talk about basic networking capabilities.

Networking: Connecting Your Board to the Internet

The connecting of your board to the Internet is the missing piece in our pursuit of successful IoT projects. How you connect to the Internet depends on the platform. Connecting with an Arduino is very different than connecting with a Raspberry Pi. Even so, there are different ways you can connect each of these boards such as different WiFi options for Arduino, choosing Ethernet for either board, etc. We will concentrate on using WiFi connections.

Let's start with the Arduino.

Arduino

Depending on which Arduino board you are using, it may already support WiFi. For example, the Uno WiFi Rev2 and some of the MKR boards come with WiFi capabilities. Other boards can take advantage of a WiFi or Ethernet shield. However, be advised that vendors of Arduino products are moving away from these options as the newest boards have onboard WiFi support. Fortunately, the older shields will still work. All we need to know is how to get started.

All of the WiFi shields and boards that support WiFi have examples you can use to learn how to use the WiFi capabilities. As you will see, none are difficult to set up, but may appear strange at first. For example, you can find how to connect your Arduino Uno WiFi using the *File* ➤ *Examples* ➤ *WiFiNINA* ➤ *Tools* ➤ *ConnectWithWPA* sketch. You will find a lot of different examples for that board and others like the WiFi 101 shield. If you have an older WiFi shield, you may need to look under *File* ➤ *Examples* ➤ *RETIRED* ➤ *WiFi* to find them. As you will see, they are very similar in how you set up. The only difference is which WiFi library to use.

For those Arduino boards that support the Uno shield pinout, you have several options for a WiFi shield. You could find an older (retired) Arduino WiFi or WiFi 101 shield like the one shown in Figure 17-1 or look into the options from Adafruit and SparkFun.

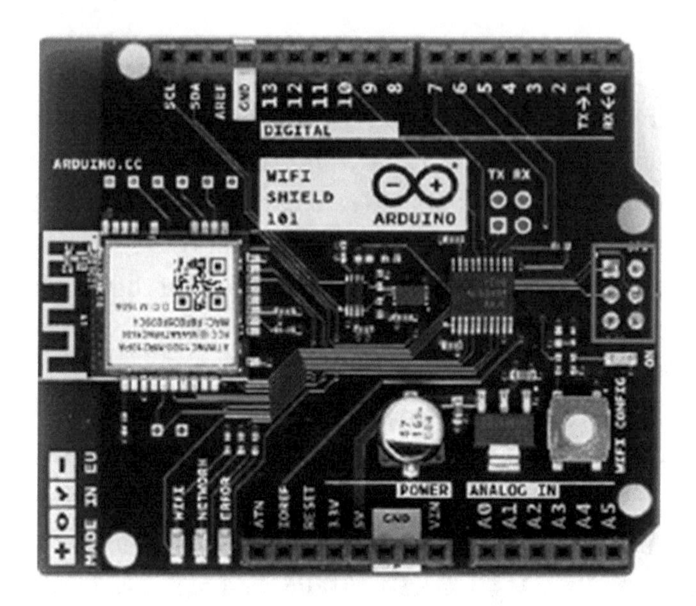

Figure 17-1. *Arduino WiFi 101 shield (courtesy of Arduino.cc)*

Let's open an example sketch to see how to get started. We'll use our Arduino Uno WiFi Rev 2. Open the Arduino IDE and open a new project and name it `wifi_example.ino`. Or, if you prefer, you can open the `ConnectWithWPA` example sketch and rename it.

The first thing we will do is set up our include files. For most WiFi uses, we will need the SPI and WiFi libraries (not all WiFi libraries require the SPI library, but there is no penalty if you include it).

The WiFi library you will use depends on the WiFi shield or WiFi capability of your Arduino board. For older WiFi shields, you include the `WiFi.h` library; for older WiFi 101 shields and boards, you include `WiFi101.h`; and for newer Arduino Uno WiFi Rev2 and similar boards, you include `WiFiNINA.h`. The following shows the complete list of includes we will need for this example. Notice we include all of the libraries, but comment out the ones we don't need. Recall we will be using an Arduino Uno WiFi Rev2:

```
#include <SPI.h>
#include <WiFiNINA.h>      // Arduino Uno, MKR-boards, etc.
//#include <WiFi101.h>     // Arduino WiFi 101 shield
//#include <WiFi.h>        // (retired) Arduino WiFi shields
```

Next, we need to declare variables for our SSID and password (assuming you're using WPA or similar authentication). Newer Arduino examples use a separate file to store the SSID and password. The file is typically named arduino_secrets.h. You can add the file by creating it and saving it in the sketch folder (remember you may have to close and reopen the project to see the new file). You can declare these variables in your main sketch if you'd like, but we will keep with the newer style. The following shows the contents of our arduino_secrets.h file:

```
#define SECRET_SSID "myssid"
#define SECRET_PASS "mypassword
```

To use the file in the main sketch, simply include it as follows:

```
// Secrets are stored in a separate file
#include "arduino_secrets.h"
```

Next, we will need a single constant and some variables. For this example, we will use a delay to wait for the WiFi to connect. Some examples use a polling loop, but a delay works just as well. The following shows the constant and variables we need. The last variable stores the result from checking the status of the connection:

```
// Constants
#define WIFI_WAIT 5000 // 5 seconds

// Global variables
char ssid[] = SECRET_SSID;     // Your network SSID (name)
char pass[] = SECRET_PASS;     // Your network password
int status = WL_IDLE_STATUS;  // The WiFi radio's status
```

To make things a bit easier and to keep our `setup()` function cleaner, we will use a new function to set up the WiFi, and we'll name it `setupWiFi()`.

The first things we need to do are some error checking. We need to check to make sure the WiFi module exists, and it is always a good idea to check its firmware version (for those that permit firmware updates). If there is no WiFi module, we will return `false`, and the connection is not attempted. Once the connection is made, we will return `true`.

If you detect the firmware needs to be upgraded, we can still continue, but you may want to update the firmware at some point. Fortunately, the Arduino IDE has an example sketch to do that. Simply open *File* ➤ *Examples* ➤ *WiFiNINA* ➤ *Tools* ➤ *FirmwareUpdater* and follow the instructions.

Next, we use a loop to attempt our connection. As you likely know, sometimes it can take some time to connect to WiFi. So a loop will help us retry the connection until we get a connection by simply calling the `begin()` function for the WiFi library passing in the SSID and password. We then use a delay to wait for the connection to stabilize and then check the status at the top of the loop.

Once the connection is made via a success code from the `begin()` function, we then get our local IP address with the `localIP()` function of the WiFi library. That's it! Listing 17-1 shows the completed function.

Listing 17-1. Basic Setup WiFi Function (Arduino)

```
bool setupWiFi() {
  int status = WL_IDLE_STATUS;

  // First, check for the WiFi module:
  if (WiFi.status() == WL_NO_MODULE) {
    Serial.println("ERROR: No WiFi detected.");
    return false;
  }
```

```
  // Next, check firmware for latest version
  String firmware_vers = WiFi.firmwareVersion();
  if (firmware_vers < WIFI_FIRMWARE_LATEST_VERSION) {
    Serial.println("WARNING: You should upgrade the firmware");
  }

  // Attempt to connect to Wifi network:
  while (status != WL_CONNECTED) {
    Serial.print("Connecting to WiFi...");
    status = WiFi.begin(ssid, pass);
    // Wait for connection: set according to your environment
    delay(WIFI_WAIT);
  }
  Serial.println("connected.");
  // print your WiFi shield's IP address:
  IPAddress ip = WiFi.localIP();
  Serial.print("IP Address: ");
  Serial.println(ip);
  return true;
}
```

If you look at the WiFi examples, you will find examples of other operations you can do such as finding the radio status of the connection, printing the MAC address, and so on. If you need those features, check the WiFi example sketches for how to write those functions.

Using the `setupWiFi()` function is easy. We simply check to see if it returns true, and if it does, we proceed. If not, we print an error and stop. Listing 17-2 shows the complete sketch for this example.

Listing 17-2. WiFi_Example Sketch

```
#include <SPI.h>
#include <WiFiNINA.h>       // Arduino Uno, MKR-boards, etc.
//#include <WiFi101.h>      // Arduino WiFi 101 shield
//#include <WiFi.h>         // (retired) Arduino WiFi shields

// Secrets are stored in a separate file
#include "arduino_secrets.h"

// Constants
#define WIFI_WAIT 5000 // 5 seconds

// Global variables
char ssid[] = SECRET_SSID;     // Your network SSID (name)
char pass[] = SECRET_PASS;     // Your network password
int status = WL_IDLE_STATUS;  // The WiFi radio's status

void setup() {
  Serial.begin(9600);
  while(!Serial);
  // Setup the WiFi
  if (!setupWiFi()) {
    Serial.println("ERROR: Cannot setup wifi. Halting.");
    while (true);
  }
  Serial.println("Congrats! Your Arduino is connected to the
  Internet.");
}

void loop() {
}

bool setupWiFi() {
  int status = WL_IDLE_STATUS;
```

```
  // Attempt to connect to Wifi network:
  while (status != WL_CONNECTED) {
    Serial.print("Connecting to WiFi...");
    status = WiFi.begin(ssid, pass);
    // Wait for connection: set according to your environment
    delay(WIFI_WAIT);
  }
  Serial.println("connected.");
  // print your WiFi shield's IP address:
  IPAddress ip = WiFi.localIP();
  Serial.print("IP Address: ");
  Serial.println(ip);
  return true;
}
```

When you execute this sketch, you should see the connection succeed and your local IP address displayed as shown in the following:

```
Connecting to WiFi...connected.
IP Address: 192.168.NNN.NNN
Congrats! Your Arduino is connected to the Internet.
```

Raspberry Pi

Fortunately, unless you are using an old Raspberry Pi board, our Raspberry Pi has everything we need to connect to the Internet. And, if you are using the Raspberry Pi to write your code, you've already connected it to the Internet via WiFi or an Ethernet connection.

In the event that you have not connected your Raspberry Pi to the Internet, you can revisit Chapter 4 to learn how to do that or see the "Desktop" section in the Raspberry Pi documentation (`www.raspberrypi.org/documentation/configuration/wireless/`) for a tutorial on how to set up your WiFi.

OK, now that we've got an idea of how to connect our boards to the Internet, let's learn how to use IFTTT in the most basic of operations: triggering a test SMS message.

Basic Operation

The If This Then That (IFTTT) cloud service is a routing service that allows you to initiate an action based on some event. In IFTTT, we create *applets* that contain an event condition or trigger, which can be one of hundreds of services such as a simple event notice (we use Webhooks) "this," and then connect that to an action with another service "that." Hence, if "this" happens, we execute (initiate) "that". Cool, eh?

Tip For a complete tutorial on what IFTTT is and what you can do with it, see `https://ifttt.com/explore/welcome_to_ifttt`.

IFTTT is one of the easiest cloud services to get started with as it has a very simple, guided setup that makes creating your first applet very easy. While IFTTT has a free account option, you will be limited to only three applets and fewer choices for more advanced development, but for getting started, this is all you need.

Let's see a complete walk-through for setting up an applet for our first example - detecting when someone successfully enters a secret knock. We'll just do the IFTTT portion here and see how to modify our secret knock project later.

We will be using two services. The first will allow us to send a request to trigger the event, and the second is an SMS message to our phone. The first service is called Webhooks and requires us to create an account to use it. The Webhooks account is also free. The second service we will use is called ClickSend, and we need a different account for that. The ClickSend service is not free, but they provide all new accounts with a $6.00 initial credit.

At about $0.05 per SMS message, we won't need to add any funds and can treat ClickSend as "free" for now. Thus, we will need three accounts: one for IFTTT, another for Webhooks, and a third for ClickSend.

Using IFTTT

In this section, we will see a walk-through all of the steps to create our first applet and create the new accounts along the way. For future applets, you can follow the same process but do not need to create the accounts (if you are using the same services; new services may require additional accounts).

Briefly, we will create our IFTTT account and then create our first applet. When we set up the trigger (the "if this"), we will create an account with Webhooks, configure a condition for the trigger, and then set up the service to execute when the trigger fires. At this point, we will need to create an account for ClickSend. Let's get started.

First, we will navigate to `www.ifttt.com` and create an account in IFTTT. We click Get started and then click *sign up* as shown in Figure 17-2. Notice the image shows several options for logging in including using an Apple, Google, or Facebook account. I like to keep my accounts separate, so if you want to do the same or don't have an account in one of the other services, click the *sign up* link at the bottom shown by the arrow.

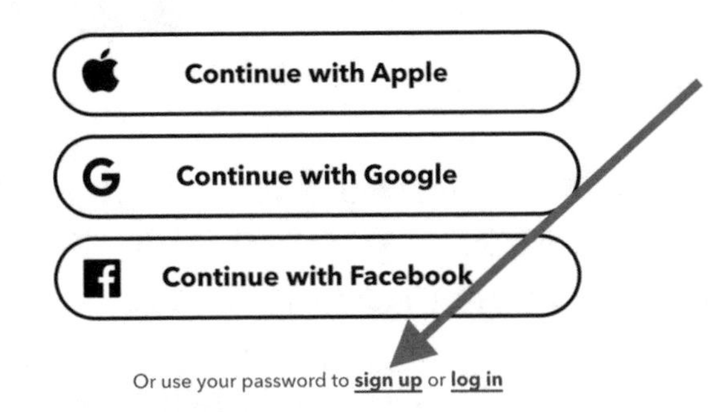

Figure 17-2. *Create a new IFTTT account*

On the next screen, fill in your email address and password and then click the *Sign up* button as shown in Figure 17-3.

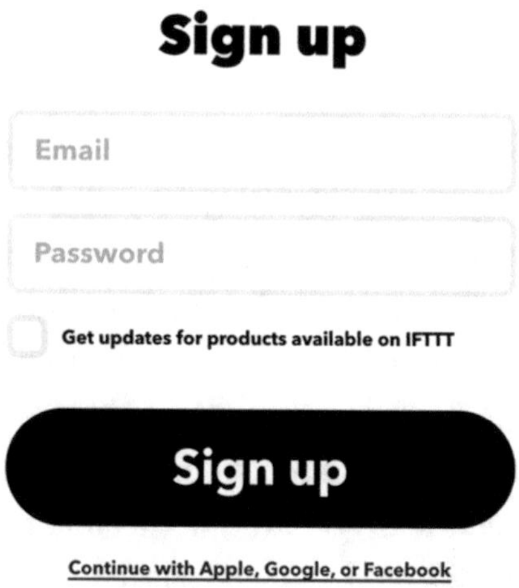

Sign up
Email
Password
Get updates for products available on IFTTT
Sign up
Continue with Apple, Google, or Facebook

Figure 17-3. *Sign up for an IFTTT account*

Once you have a new IFTTT account, you will return to your main screen. To create your first applet, click the Create button as shown in Figure 17-4.

My Applets
Explore
Developers
Create

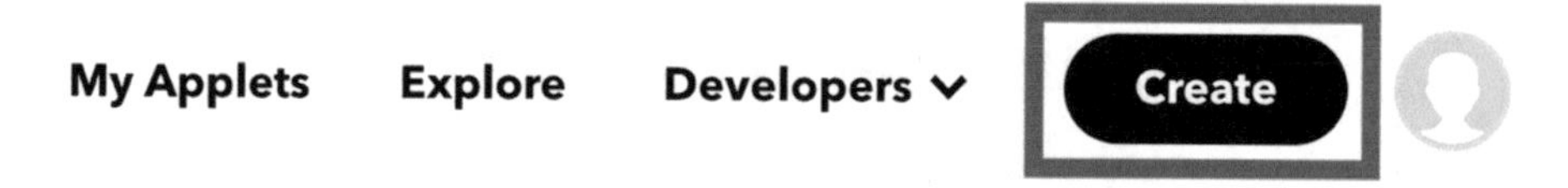

Figure 17-4. *IFTTT home page*

The next screen will start the setup for the applet. Figure 17-5 shows the applet setup screen. Notice the graphics show the "if this" and "then that" sections known as the trigger and service. Click the *Add* button as shown in Figure 17-5 to add the trigger. Notice the "if this" is black, while the "then that" section is gray. We must set up the trigger before we can set up the service.

Figure 17-5. Add a trigger for IFTTT

Recall we want to use Webhooks. We can search for Webhooks by typing `Webhooks` into the text box (capitalization doesn't matter) and then select Webhooks by clicking the Webhooks icon as shown in Figure 17-6.

Figure 17-6. Choose the Webhooks service

Next, we need to choose a trigger. This involves selecting a trigger and naming it. During this process, we will need to create a new account for Webhooks. To proceed, click the Webhooks box as shown in Figure 17-7.

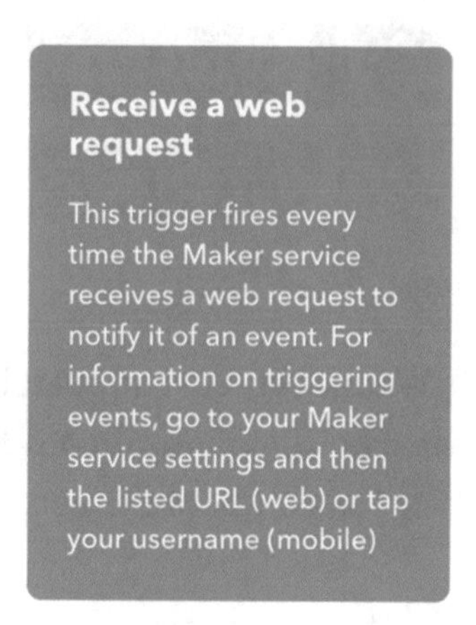

Figure 17-7. *Choose Webhooks for your trigger*

When you click Webhooks, it will take you to a new screen where you can connect the Webhooks service to your account. Click Connect as shown in Figure 17-8.

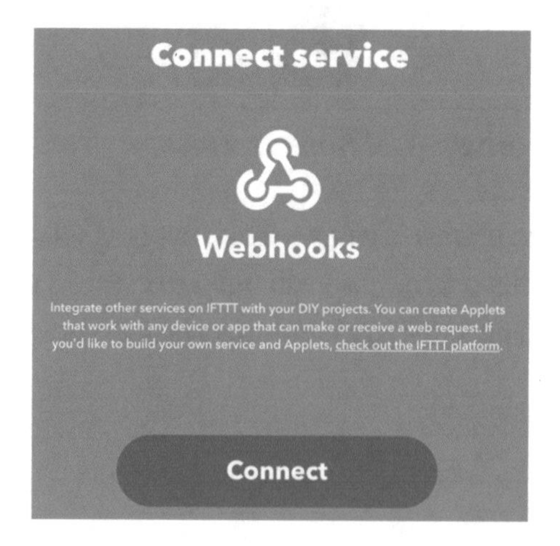

Figure 17-8. *Connect to Webhooks*

Next, create your Webhooks account using your email address and a password. Once you create the account, you will be redirected back to IFTTT where you can name your trigger. For our example, we will use `secret_knock_accepted` as shown in Figure 17-9. When ready, click the *Create trigger* button.

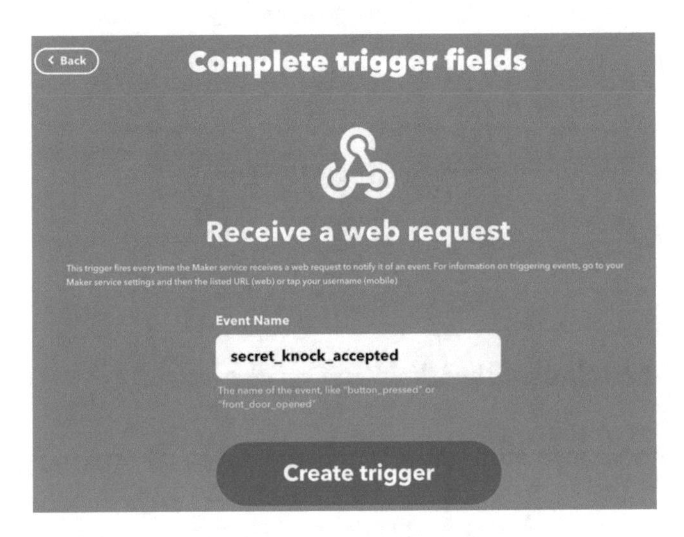

Figure 17-9. *Name the trigger*

When you click the Create trigger button, you will be returned to the main applet creation screen as shown in Figure 17-10. Now, we're ready for the "then that" segment. Click the *Add* button to choose a service to initiate when the trigger fires.

Figure 17-10. *Add a service for the applet*

Recall the service we want to use is named ClickSend. You can find it quickly by typing in SMS and then choose the ClickSend service as shown in Figure 17-11. This (or any search) may return more than one entry. If you have an Android phone, you may want to explore the Android SMS option.

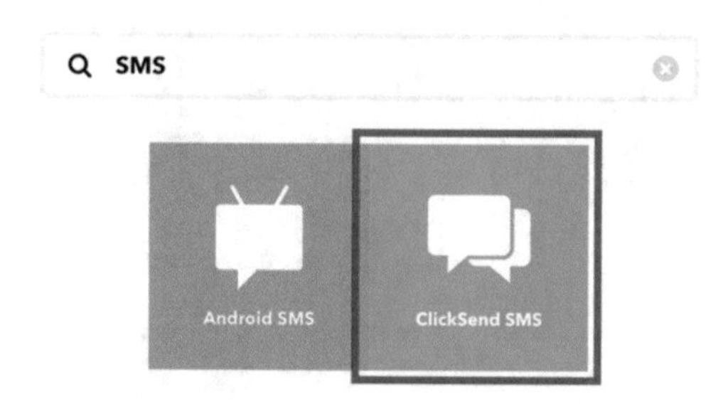

Figure 17-11. *Choose the ClickSend service*

Next, choose an action (there is only the one) as shown in Figure 17-12.

Figure 17-12. *Choose an action for the ClickSend service*

Next, click *Connect* as shown in Figure 17-13.

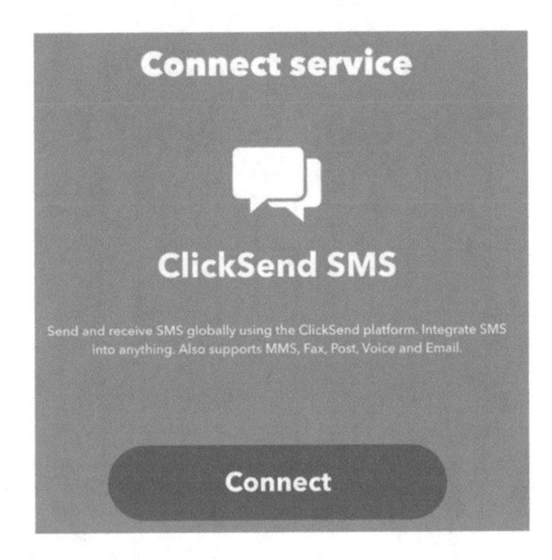

Figure 17-13. *Connect to ClickSend*

This will request you to create an account on ClickSend. Go ahead and do that now. When complete, you will be asked to verify your account with a code emailed to your account. Once that code is entered, you will be sent to your ClickSend home page as shown in Figure 17-14.

***Figure 17-14.** ClickSend home page*

You can log out and go back to the IFTTT page, then click Connect again, log in to ClickSend, and then click Allow if the popup appears as shown in Figure 17-15.

***Figure 17-15.** Allow access (IFTTT)*

At this point, we can complete the Send SMS action as shown in Figure 17-16. Here, we add our mobile number in the *To* box, our name or number in the *From* box, and our message formatted with the date and time the event occurred in the *Message* box.

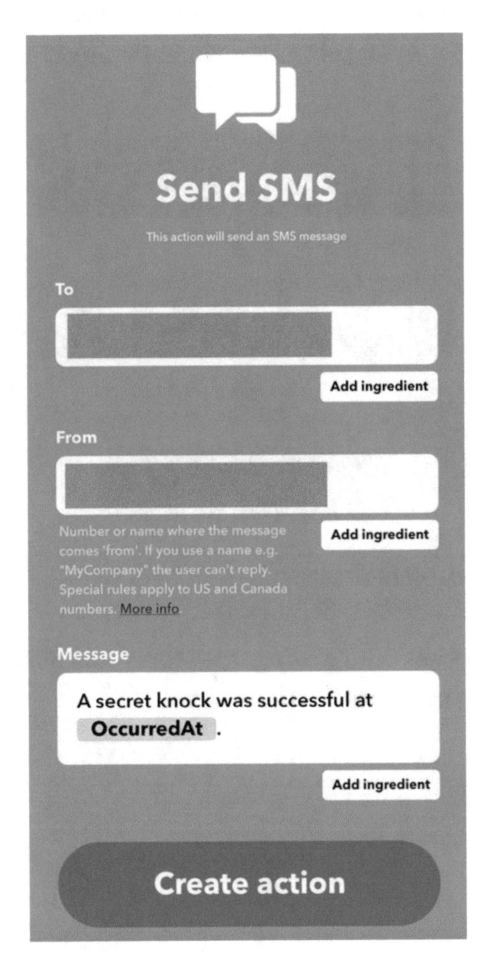

Figure 17-16. *Send SMS action*

You can use *Add ingredient* to choose the data you want to add. For this example, we choose the *OccurredAt* ingredient in the message to capture the date and time when the trigger was fired. When done, click *Create action* to continue.

You will return to the applet creation page where you can click *Continue* to complete the applet as shown in Figure 17-17.

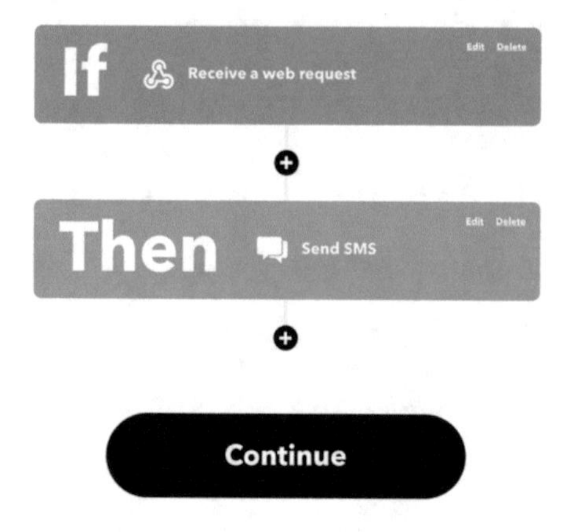

Figure 17-17. *Complete the applet*

Next, you can review your settings and then click *Finish* to complete the applet as shown in Figure 17-18.

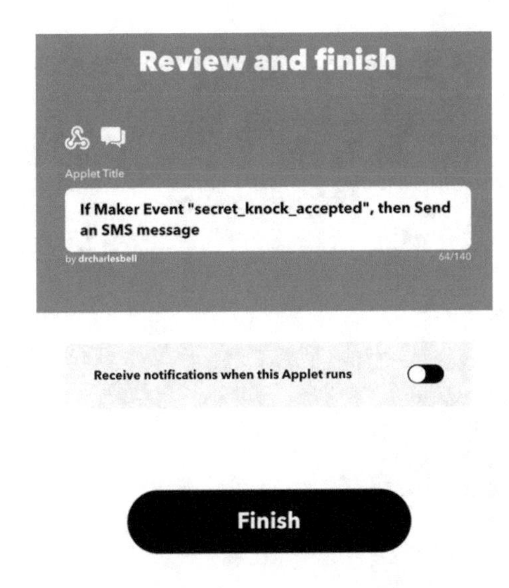

Figure 17-18. *Review and finish applet creation*

This will return you to your IFTTT home page. You should now see your applet as shown in Figure 17-19. Notice you only get three free applets, and the page reminds you of this at the top.

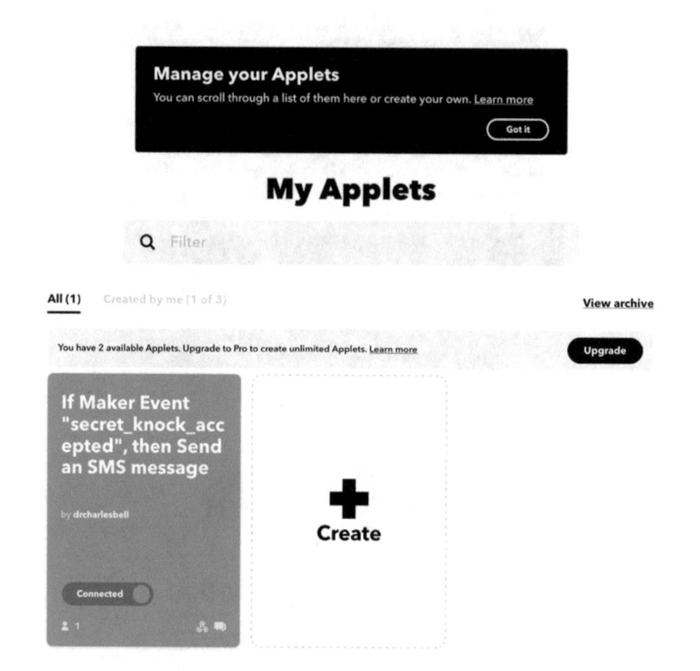

Figure 17-19. IFTTT home page with the new applet

You may think we're done, but we're not. We will need some special credentials created on Webhooks that you will need to trigger the event. Click your applet and then click Documentation as shown in Figure 17-20.

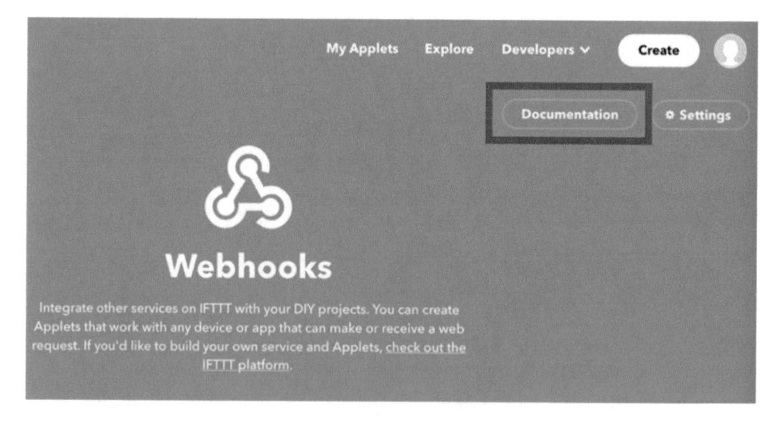

Figure 17-20. Getting Webhooks credentials

This will display your credentials for Webhooks. Figure 17-21 shows an example of the credentials page.

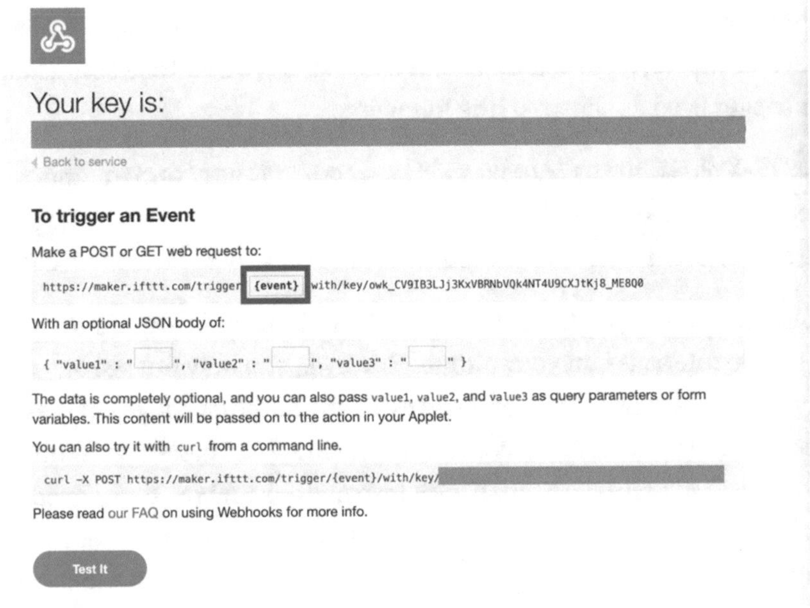

Figure 17-21. *Webhooks credentials page*

There are several things of interest on this page that has been redacted. At the top is the key you will need to place in your code to access the webhook. In fact, you will use the key to send an HTTP request to the Webhooks service, and the key is uniquely associated with your trigger.

Next is a URL you can use to test your webhook. Simply click {event} and replace it with your trigger name. This will complete the URL for you. For example, you will see something like the following appear on the page. You can copy that URL and paste it into your browser, which will trigger the event:

```
https://maker.ifttt.com/trigger/secret_knock_detected/with/key/
XXXXXXX
```

When you do that, you will see a simple message appear like the following:

```
Congratulations! You've fired the secret_knock_detected event
```

You can also copy the curl command and paste it into a terminal to fire the event. When you execute the curl command, you will see confirmation your trigger fired as shown in the following:

```
$ curl -X POST https://maker.ifttt.com/trigger/secret_knock_
accepted/with/key/XXXXXXXX
Congratulations! You've fired the secret_knock_accepted event
```

At this point, if you tried both the URL and the curl command, there will be two messages on your phone. However, you may not see any right away. This is because ClickSend will evaluate your first couple of messages. Once they are approved, you will start receiving SMS messages. If there is a problem, you will get an email with instructions on how to proceed. You may also experience a delay with the first few messages, but after that it should work as expected. Once they are sent, you should see a message on your phone like Figure 17-22.

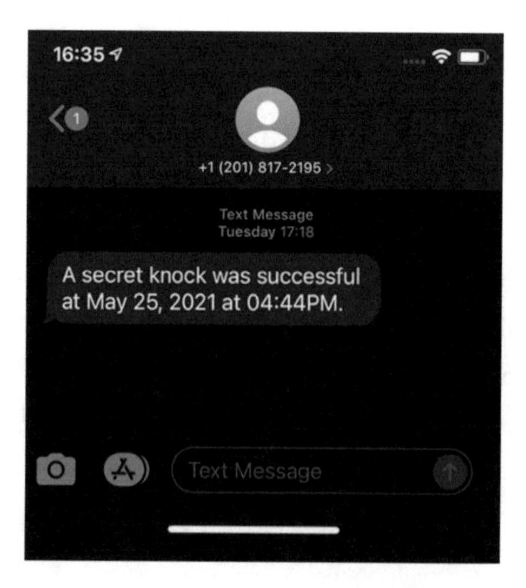

Figure 17-22. *SMS from our secret_knock_accepted test*

Now, if you sign into the ClickSend website (`www.clicksend.com`), you will see the cost of the SMS messages deducted from your initial allowance. If you plan to run your project for a long time, you may want to add more money to the account so you don't miss any messages.

Now that we know how to create an applet in IFTTT and use Webhooks to link to ClickSend to send an SMS when triggered, we can incorporate that into our IoT projects.

Example Projects

Let's see how to apply what we learned to two of our example projects to complete the IoT portion for each. Recall we are going to use the secret knock example from Chapter 13 and the weather project from Chapter 8. But first, let's see what software we need for our platforms.

For the Arduino, you need to add one new software library named ArduinoHttpClient. To install the `ArduinoHttpClient` software library, open the Library Manager from the Arduino IDE menu (*Sketch* ➤ *Include Library* ➤ *Library Manager...*). Then search for `arduinohttp` and install the latest version of the ArduinoHttpClient library as shown in Figure 17-23.

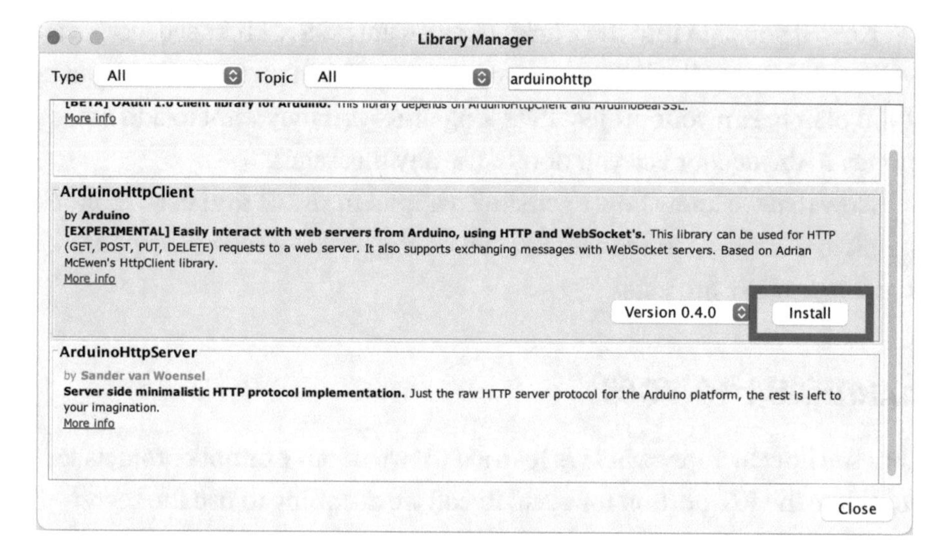

Figure 17-23. *Installing the ArduinoHttpClient library (Arduino)*

For the Raspberry Pi and Python, we need the Python requests library. You can install that with the following command:

```
$ pip3 install requests
```

Now that our platforms are prepared, let's see the code to connect our project to IFTTT.

Example 1: Secret Knock Alert

In this example, we will use the secret knock project from Chapter 13. Recall this project uses Grove modules. Refer to Table 13-1 for the components you will need and Figure 13-7 for how to assemble the project components. Once that's done, we can modify the code. If you haven't written the code for this project, you should review the code explanations in Chapter 13 prior to attempting this example. But first, let's set up IFTTT.

IFTTT Setup

Since we've already set up the applet in IFTTT for the secret knock example, we need only add the code to the project to trigger the event in the applet. To do so, we need to issue an HTTP POST[2] method via a command. As you see, this is very easy in Python, but a bit more work for the Arduino. Let's start with the Arduino version.

Arduino

The best way to start is to copy the Arduino project folder from Chapter 13 and rename it to `secret_knock_iot`. To do so, simply copy the `secret_knock` folder and rename it. You then must open the new folder and rename the `secret_knock.ino` file to `secret_knock_iot.ino`. Once that is done, open the new project in the Arduino IDE. You should see the exact same code as we had in Chapter 13. At this point, you may want to set up the hardware as described in Chapter 13.

There are a number of things we must do to add the code. First, we must add the WiFi connection code. Then we can add code to send the POST method command. Refer to the "Networking: Connecting Your Board to the Internet" section to learn how to do that. It is recommended that you add that code first and then test it before adding the rest of the code. Once you've confirmed the WiFi can connect, you can proceed.

Since the modifications to the Arduino code are nontrivial, we will create a new class to wrap our functionality, make the code easier to use from the main sketch, and allow us to reuse the class in other projects.

Recall we must create the files with the project closed. We will name the class `Webhook` so we need to create two files - `Webhook.h` and `Webhook.cpp` as shown in the following:

```
Secret_knock_iot % touch Webhook.h
Secret_knock_iot % touch Webhook.cpp
```

[2] `www.w3schools.com/tags/ref_httpmethods.asp`

Now, reopen the project and click the *Webhook.h* tab. Here, we will add the class header, which has only one function named `sendTrigger()`, which will accept three values as strings. IFTTT allows you to specify three key values to pass to your trigger using the keys `value1`, `value2`, and `value3`. You cannot change these names, unfortunately. We will include them since we will need them in other projects.

Aside from that, we will need to include the new library and make a couple of strings: one for the URL for IFTTT (`server_name`) and another containing the key for our webhook. Note we don't need the entire string, just the portion starting with the first slash. The following shows those strings. Note that the key string is named `resource_str` and redacted and your string will be much longer. The only other variable is the port, which is port 80 (same as http):

```
const char *server_name {"maker.ifttt.com"};
const char *resource_str {"/trigger/secret_knock_accepted/with/
key/XXX"};
int port {80};
```

Listing 17-3 shows the completed header file.

Listing 17-3. Webhook Header File (Arduino)

```
#include <Arduino.h>
#include <WiFi.h>
#include <ArduinoHttpClient.h>

class Webhook {
public:
  Webhook();
  void sendTrigger(const char *value1="", const char
  *value2="", const char *value3="");
```

```
private:
  const char *server_name {"maker.ifttt.com"};
  const char *resource_str {"/trigger/secret_knock_accepted/
  with/key/XXX"};
  int port = 80;

  WiFiClient wifi;
  HttpClient *client;
};
```

Now, let's complete the constructor and the `sendTrigger()` function. The constructor simply creates an instance of the Arduino `HttpClient` class. Click the Webhook.cpp tab and create the function as shown in Listing 17-4.

Listing 17-4. Webhook Code File (Arduino)

```
#include "Webhook.h"

Webhook::Webhook() {
  client = new HttpClient(wifi, server_name, port);
}

void Webhook::sendTrigger(const char *value1, const char
*value2, const char *value3) {
  Serial.println("Sending webhook trigger to IFTTT.");
  // IFTTT Maker Webhook is limited to (3) key/value pairs
  String postData = "value1=" + String(value1) + "&value2=" +
                    String(value2) + "&value3=" +
                    String(value3);

  String contentType = "application/x-www-form-urlencoded";
  client->post(resource_str, contentType, postData);
```

```
  // read the status code and body of the response
  int statusCode = client->responseStatusCode();
  String response = client->responseBody();

  Serial.print("Status code: ");
  Serial.println(statusCode);
  Serial.print("Response: ");
  Serial.println(response);
}
```

OK, there's a lot going on in there. Notice the first thing we do aside from diagnostic messages is create a string that contains the three values for the three variables (called ingredients in IFTTT). We then use another string for the content type of the message we will send. We then pass those strings along with our resource string to the `client->post()` function. Next, we wait for a response from the server and display that in the serial monitor.

To use this in our project, we add a new `#include` for the `Webhook` header, then locate the place in the code where the secret knock is accepted, and create a new instance of the `Webhook` class and call the `sendTrigger()` as shown in Listing 17-5. The new lines of code are shown in bold.

Listing 17-5. Adding Webhook to the Main Sketch (Arduino – secret_knock_iot)

```
...
    // If knock accepted, turn on green LED, or red LED if
       failure
    if (knockSensor->validateSecret()) {
      Serial.println("Secret knock accepted.");
      digitalWrite(GREEN_LED, HIGH);
      delay(3000);
      digitalWrite(GREEN_LED, LOW);
```

```
        Serial.println("Sending trigger to Webhook.");
        Webhook *webhook = new Webhook();
        webhook->sendTrigger();
        delete webhook;
    }
...
```

OK, that's it. Now the project is ready to run. Go ahead and try it out. You should get a new SMS when the secret knock is accepted. If you do not, go back and make sure you have the correct key in the resource string.

An excerpt of the output you can expect in the serial monitor is shown in the following:

```
Connecting to WiFi...connected.
IP Address: 192.168.NN.NN
...
Sending webhook trigger to IFTTT.
Status code: 200
Response: Congratulations! You've fired the secret_knock_
accepted event
...
```

Python

The Python version is considerably easier. Once the requests library is installed, you need only copy the Python files (`secret_knock.py`, `knock_sensor.py`) and rename the main script to `secret_knock_iot.py`.

Next, add the import statement for the requests class. We will also use a string to store our resource key. Unlike the Arduino version, we can use the complete key as listed in the Webhooks configuration page. Finally in the secret knock accepted code segment, use the requests class and call the `post()` function passing in the resource string. We can print the result from the function call to see its status as shown in Listing 17-6.

Listing 17-6. Adding Webhook Code to Main Script (Python – secret_knock.py)

```
# Import libraries
import sys
import time
import requests
...
RESOURCE_STR = ("https://maker.ifttt.com/trigger/secret_knock_
accepted/with"
                "/key/XXXXXXXXX")
...
from grovepi import pinMode, digitalRead, digitalWrite
from knock_sensor import KnockSensor
...
           # If knock accepted, turn on green LED, or red LED
             if failure
            if sensor.validate_secret():
                print("Secret knock accepted.")
                digitalWrite(GREEN_LED, HIGH)
                time.sleep(3)
                digitalWrite(GREEN_LED, LOW)
                print("Sending trigger to Webhook.")
                retval = requests.post(RESOURCE_STR)
                print(retval.text)
...
```

That's it! Go ahead and try it out. You should see an SMS momentarily after the secret knock is accepted. An excerpt of the output you will see in the terminal is shown in the following. Notice the response from the server:

```
Sending webhook trigger to IFTTT.
Congratulations! You've fired the secret_knock_accepted event
```

Now, let's look at a second example.

Example 2: Weather Alert

In this example, we will use the weather project from Chapter 8. Recall we use a Qwiic environmental sensor to read the temperature, humidity, and barometric pressure. Refer to Table 8-1 for the components you will need and Figure 8-1 for how to assemble the project components. Once that's done, we can modify the code. If you haven't written the code for this project, you should review the code explanations in Chapter 8 prior to attempting this example.

But first, let's set up IFTTT.

IFTTT Setup

The setup we need for this example is very similar to the secret knock example. In fact, you can follow the same steps we used to set up the secret knock applet. The only difference is we will format the message differently so that we can send the environment data.

Simply visit your home page on IFTTT and then create a new applet. Name it *hows_the_weather* as shown in Figure 17-24.

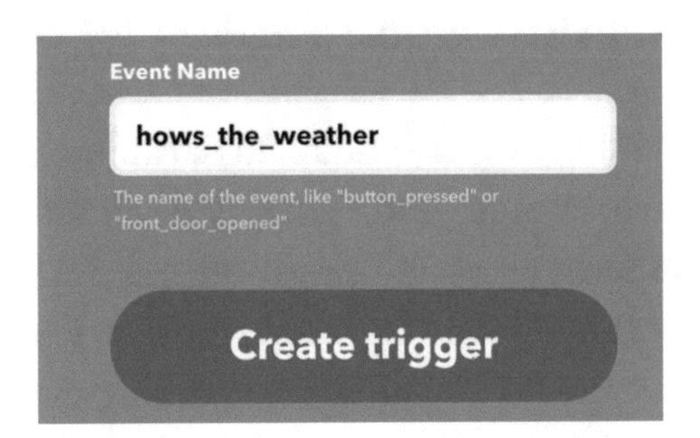

Figure 17-24. *Set up the hows_the_weather applet*

When you set up the ClickSend service, use the following for the message as shown in Figure 17-25.

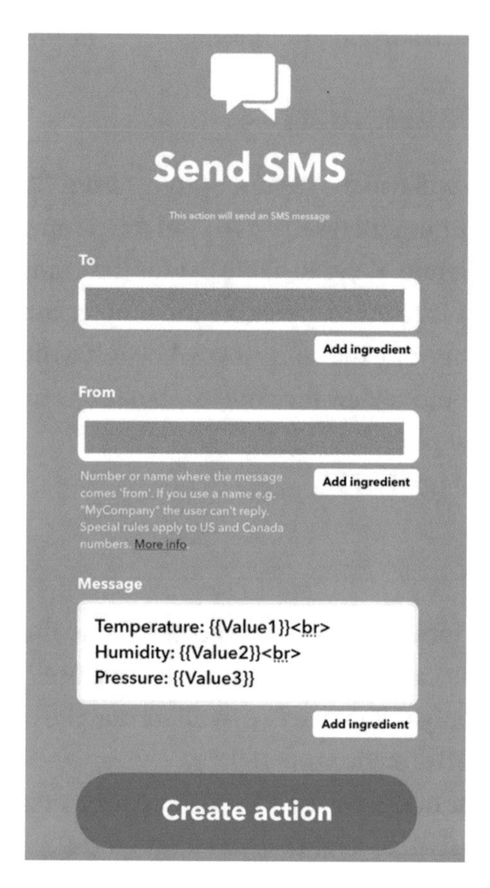

Figure 17-25. *Set up the hows_the_weather ClickSend service*

Notice we used the `{{value1}}` notation to capture the `value1` ingredient as well as the `value2` and `value3`. We will send the temperature, humidity, and barometric pressure using these built-in labels. You can also click *Add ingredient* to automatically format the data item.

Now that we have the IFTTT applet created, we can write the code. Let's start with the Arduino version.

Arduino

The best way to start is to copy the Arduino project folder from Chapter 8 and rename it to weather_iot. To do so, simply copy the weather folder and rename it. You then must open the new folder and rename the weather.ino file to weather_iot.ino. At this point, you may want to set up the hardware as described in Chapter 8.

We will be making the same set of changes that we did with the last example. Specifically, we will add the same WiFi code we used in the last example, and we will use the new Webhook class we created in the last example to send the POST message to invoke the trigger.

Since we are reusing the Webhook class from the last example, all you need to do is copy the Webhook.h and Webhook.cpp files from the secret_knock_iot folder to the weather_iot folder and open the project in the Arduino IDE. You should see the exact same code as we had in Chapter 8.

Rather than step through every line of code we need to change given the WiFi code is the same as the last example, let's focus only on the changes you need to fire the hows_the_weather event. The first thing you need to do is change the resource_str in the Webhook.h file to use the new event name (hows_the_weather) as shown in the following. Your key should be the same as the last example:

```
const char *resource_str {"/trigger/hows_the_weather/with/
key/<KEYHERE>"};
```

In the main sketch, we create a new instance of the Webhook class and then call the sendTrigger() function passing in the temperature, humidity, and barometric pressure values. Listing 17-7 shows the code you will need to use in the loop() function in the main sketch (new code shown in bold).

Listing 17-7. Adding the Webhook to the Main Sketch (Arduino - weather_iot)

```
void loop()
{
  // Read the sensor
  float temperature = bme280.readTempC();
  float humidity = bme280.readFloatHumidity();
  float pressure = bme280.readFloatPressure();

  // Display data to serial monitor
  printDiagnostics(temperature, humidity, pressure);

  // Display data to OLED
  showDataOLED(temperature, humidity, pressure);

  // Send the data to our webhook trigger if data has changed
  if ((old_temperature != temperature) or (old_humidity !=
  humidity) or
      (old_pressure != pressure)) {
    // Save the data
    old_temperature = temperature;
    old_humidity = humidity;
    old_pressure = pressure;
    Serial.println("Sending a weather update.");
    Webhook *webhook = new Webhook();
    String temp_str = String(temperature);
    temp_str += " C";
    String humid_str = String(humidity);
    humid_str += " %rh";
    String press_str = String(pressure);
    press_str += " hPa";
```

```
    webhook->sendTrigger(temp_str.c_str(), humid_str.c_str(),
                         press_str.c_str());
    delete webhook;
  }
  delay(30000);
}
```

Notice there is some more work going on there. We need to convert those floating-point numbers for the data to strings, so we use the `String` class and add the units too. This will make creating the message in the ClickSend easier to format.

OK, that's it. Now the project is ready to run. Go ahead and try it out. You should get a new SMS when the weather data is presented on the micro OLED similar to Figure 17-26. If you do not, go back and make sure you have the correct key in the resource string.

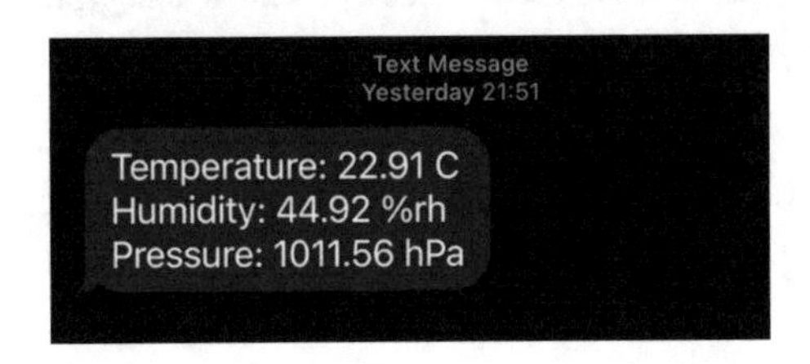

Figure 17-26. *Example SMS from weather_iot*

Python

The Python version is easier but requires a bit more work to format the data values, and we'll modify the code to only trigger the event if one of the data items changes. This will greatly reduce the number of SMS messages sent and shows a nice alternative to lengthening your sample rate. Once the requests library is installed, you need only copy the Python files (`weather.py`, `knock_sensor.py`) and rename the main script to `weather_iot.py`.

Next, add the import statement for the requests class. We will also use a string to store our resource key. Unlike the Arduino version, we can use the complete key as listed in the Webhooks configuration page. Finally in the secret knock accepted code segment, use the requests class and call the `post()` function passing in the resource string. We can print the result from the function call to see its status as shown in Listing 17-8.

Listing 17-8. Adding Webhook Code to Main Script (Python – weather.py)

```
# Import libraries
import sys
import time
import requests
...
RESOURCE_STR = ("https://maker.ifttt.com/trigger/hows_the_
weather/with" "/key/XXXXXXXXX")
...
    # Compensate for reference pressure
    bme280.sea_level_pressure = 1013.25
    while True:
        # Read the sensor data
        temperature = bme280.temperature
        humidity = bme280.humidity
        pressure = bme280.pressure
        # Display diagnostics
        print_diagnostics(temperature, humidity, pressure)
        # Show data on OLED
        show_data_oled(temperature, humidity, pressure)
        # Send data to the Webhook if it has changed
```

```
if old_data['temp'] != temperature or
   old_data['humid'] != humidity or old_data['press']
   != pressure:
    # Save the data
    old_data['temp'] = temperature
    old_data['humid'] = humidity
    old_data['press'] = pressure
    # Trigger the update
    print("Sending a weather update.")
    data = {
        "value1": "{:.2f} C".format(temperature),
        "value2": "{:.2f} %rh".format(humidity),
        "value3": "{:.2f} hPa".format(pressure)
    }
    retval = requests.post(RESOURCE_STR, data)
    print(retval.text)
time.sleep(30)...
```

Notice we create a dictionary for the values read with keys `value1`, `value2`, and `value3`. Similar to how we formatted the values for the Arduino version, we will convert the floating-point values to strings and add the units to the strings.

Notice also how we used three `old_*` variables to store the last values read from the sensor. We can check these values on the next read, and if any one of them has changed, we send the trigger. Otherwise, we skip the trigger. This will help reduce the number of SMS messages to send an SMS only when the data changes. This illustrates how you can create a trigger that can respond to the data rather than the program flow (in this case, when the sensor is read).

That's it! Go ahead and try it out. You should see an SMS momentarily after the weather data is updated.

Once you have both examples working, you have just created your first complete IoT projects! Now your Arduino and Python projects can send you an SMS. How cool is that?

Summary

When you take a typical electronics project such as a weather station, electronic game, home automation, etc. and connect it to the Internet, you've just upped the capabilities of that small project considerably.

We saw two simple examples of this by connecting two of our example projects to the Internet. Each used a simple IoT cloud service to allow us to get information from sensors or an event. Clearly, that same if-this-then-that concept can be applied to some of the other example projects.

For example, you could set up your Simon Says game to send a text to a group of people announcing a winner. Or you could modify the compass project to echo a compass reading to your phone. You are only limited by your imagination!

In this chapter, we learned more about cloud systems and how they can be used in IoT projects. Now that you've seen how easy it is to get started and how little code is needed in your projects, you can begin to modify your own projects. But we've just scratched the surface here. There is so much more that can be done with another simple, free cloud solution.

In the next chapter, we will expand our tour of cloud systems for IoT by looking at one of the most popular free options: ThingSpeak – a popular, easy-to-use, cloud-based IoT data-hosting service from MathWorks. You will learn how to send your data to the cloud and display it using nice, easy-to-use graphics using the previous example projects on both the Arduino and the Raspberry Pi.

CHAPTER 18

Using ThingSpeak

Now that we've built a good foundation of experience working with Qwiic, STEMMA QT, and Grove modules including how to write code to use the sensors, respond to inputs (e.g., buttons), and display data as well as how to use IFTTT to create a simple event-driven IoT solution, it's time to take our IoT skills to a new level.

Thus far, we haven't discussed how to use the data generated from our IoT projects mainly because we haven't covered how to store data locally on our host boards. While doing so is fairly easy for the Raspberry Pi, doing so for the Arduino is harder and more involved since we'd need to add either an onboard chip (EEPROM) or secure digital (SD) card. Due to the limited size of these options, you will encounter issues you need to resolve such as how much data you want to store and for how long.

While those are things that can be solved, the bigger question is, What are you going to do with the data? Would you want to see how the data changes over time, how one sensor data compares to another, how often a value changes, or more basic statistics like min, max, and average values? All of these things require processing power that your host board is unlikely to have (Raspberry Pi excluded).

Furthermore, you may want to see the data presented in one or more graphs that you can use for a pictorial representation. The best way to do this is to take advantage of IoT cloud services. Not only can you store the data easily, but you can also perform analysis on the data and present it in one of several graphics.

C. Bell, *Beginning IoT Projects*, https://doi.org/10.1007/978-1-4842-7234-3_18

In fact, you can store your data in the cloud using a popular, easy-to-use, cloud-based IoT data-hosting service from MathWorks called ThingSpeak (`www.thingspeak.com`). We will see how to take several of the example projects from this book and connect them to ThingSpeak to see how we can gain more insights about the data. We will see examples for both the Arduino and the Raspberry Pi.

But first, let's take a brief tour of ThingSpeak and how to get started using it in our projects.

Getting Started

ThingSpeak offers a free account for noncommercial projects that generate fewer than 3 million messages (or data elements) per year or around 8,000 messages per day. Free accounts are also limited to four channels (a channel is equivalent to a project and can save up to eight data items). If you need to store or process more data than that, you can purchase a commercial license in one of four categories, each with specific products, features, and limitations: Standard, Academic, Student, and Home. See `https://thingspeak.com/prices` and click each of the license options to learn more about the features and pricing.

ThingSpeak works by receiving messages from devices that contain the data you want to save or plot. There are libraries available that you can use for certain platforms or programming languages such as Arduino or Python.

However, you can also use a machine-to-machine (M2M) connectivity protocol (called MQTT[1]) or representational state transfer (REST[2]) API designed as a request-response model that communicates over HTTP to send data to or read data from ThingSpeak. Yes, you can even read your data from other devices.

---

[1] `http://mqtt.org/`

[2] `https://en.wikipedia.org/wiki/Representational_state_transfer`

Tip See www.mathworks.com/help/thingspeak/channels-and-charts-api.html for more details about the ThingSpeak MQTT and REST API.

When you want to read or write from/to a ThingSpeak channel, you can either publish MQTT messages, send requests via HTTP to the REST API, or use one of the platform-specific libraries that encapsulate these mechanisms for you. A channel can have up to eight data fields represented as a string or numeric data. You can also process the numeric data using several sophisticated procedures such as summing, average, rounding, and more.

We won't get too far into the details of these protocols. Rather, we will see how to use ThingSpeak as a quick start guide. MathWorks provides a complete set of tutorials, documentation, and examples. So, if you need more information about how ThingSpeak works, check out the documentation at www.mathworks.com/help/thingspeak/.

The first thing we need to do is create an account.

Create an Account in ThingSpeak

To use ThingSpeak, you must first sign up for an account. Fortunately, they provide the option for a free account. In fact, you get a free account to start with and add (purchase) a license later. To create a free account, visit https://thingspeak.com/, click *Get Started For Free*, and then click Create one! as shown in Figure 18-1.

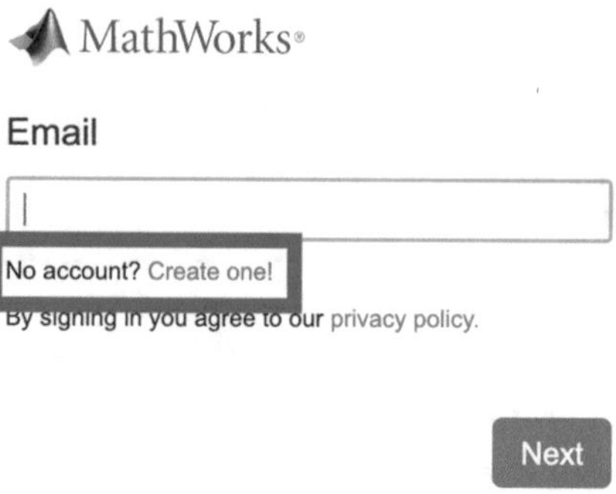

Figure 18-1. *Create a new ThingSpeak/MathWorks account*

On the next page, fill in your email address, location (general geographic), and first and last names and then click *Continue*. You will then be sent a validation email. Open that and follow instructions to verify your email and complete your free account by choosing a password. You may be asked to complete a short questionnaire. Be sure to log in before continuing.

Next, let's create our first channel.

Create a Channel

Once you log in to ThingSpeak, you can create a channel to hold your data. Recall each channel can have up to eight data items (fields). From your login home page, click *New Channel* as shown in Figure 18-2.

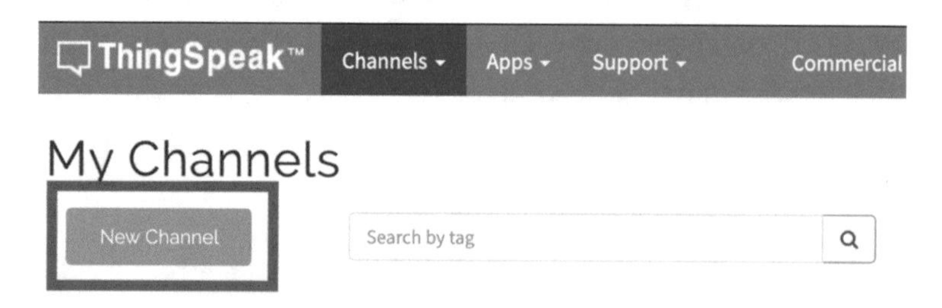

Figure 18-2. *Creating a channel in ThingSpeak*

You will be presented with a really long form that has a lot of fields that you can fill out. Figure 18-3 shows an example of the form.

New Channel

Name

Description

Field 1 Field Label 1

Field 2

Field 3

Field 4

Field 5

Field 6

Field 7

Field 8

Metadata

Tags

(Tags are comma separated)

Link to External Site http://

Link to GitHub https://github.com/

Elevation

Show Channel Location

Latitude 0.0

Longitude 0.0

Show Video

YouTube

Vimeo

Video URL http://

Show Status

Save Channel

Help

Channels store all the data that a ThingSpeak application collects. Each channel includes eight fields that can hold any type of data, plus three fields for location data and one for status data. Once you collect data in a channel, you can use ThingSpeak apps to analyze and visualize it.

Channel Settings

- **Percentage complete:** Calculated based on data entered into the various fields of a channel. Enter the name, description, location, URL, video, and tags to complete your channel.
- **Channel Name:** Enter a unique name for the ThingSpeak channel.
- **Description:** Enter a description of the ThingSpeak channel.
- **Field#:** Check the box to enable the field, and enter a field name. Each ThingSpeak channel can have up to 8 fields.
- **Metadata:** Enter information about channel data, including JSON, XML, or CSV data.
- **Tags:** Enter keywords that identify the channel. Separate tags with commas.
- **Link to External Site:** If you have a website that contains information about your ThingSpeak channel, specify the URL.
- **Show Channel Location:**
  - **Latitude:** Specify the latitude position in decimal degrees. For example, the latitude of the city of London is 51.5072.
  - **Longitude:** Specify the longitude position in decimal degrees. For example, the longitude of the city of London is -0.1275.
  - **Elevation:** Specify the elevation position meters. For example, the elevation of the city of London is 35.052.
- **Video URL:** If you have a YouTube™ or Vimeo® video that displays your channel information, specify the full path of the video URL.
- **Link to GitHub:** If you store your ThingSpeak code on GitHub®, specify the GitHub repository URL.

Using the Channel

You can get data into a channel from a device, website, or another ThingsSpeak channel. You can then visualize data and transform it using ThingSpeak Apps.

See Get Started with ThingSpeak™ for an example of measuring dew point from a weather station that acquires data from an Arduino® device.

Learn More

Figure 18-3. *New Channel form*

At a minimum, you need only name the channel, enter a description (not strictly required but recommended), and then select (tick) one or more fields naming each.

So what are all those channel settings? The following gives a brief overview of each. As you work with ThingSpeak, you may want to start using some of these fields:

- *Percentage complete*: A calculated field based on the completion of the name, description, location, URL, video, and tags in your channel.
- *Channel Name*: Unique name for the channel.
- *Description*: Description of the channel.
- *Field#*: Tick each box to enable the field.
- *Metadata*: Additional data for the channel in JSON, XML, or CSV format.
- *Tags*: A comma-separated list of keywords for searching.
- *Link to External Site*: If you have a website about your project, you can provide the URL here to publish on the channel.
- *Show Channel Location*: Tick this box to include the following fields:
    - *Latitude*: Latitude of the sensor(s) for the project or source of the data
    - *Longitude*: Longitude of the sensor(s) for the project or source of the data
    - *Elevation*: Elevation in meters for use with projects affected by elevation
- *Video URL*: If you have a video associated with your project, you can provide the URL here to be published on the channel.
- *Link to GitHub*: If your project is hosted in GitHub, you can provide the URL to be published on the channel.

Wow, that's a lot of stuff for free! As you will see, this isn't a simple toy or severely limited product. You can accomplish quite a lot with these settings. Notice there are places to put links to video, website, and GitHub. This is because channels can be either private (only your login or API key as we will see can access) or public. Making a channel public allows you to share the data with anyone, and thus those URL fields may be handy to document your project. Cool.

Now, let's create a practice channel that we will use in the next section to see how to write data (sometimes called upload) to ThingSpeak. Use the following parameters for the fields on the New Channel form.

- *Name*: practice_channel
- *Description*: Testing ThingSpeak connection from Arduino and Python
- *Field 1*: RandInt

Enter the values as shown and then click *Save Channel* to complete the process. Now we are ready to test writing some data.

How to Add ThingSpeak to Your Projects

Once you create your channel, it is time to write some data. There are two pieces of information you will need for most projects: the API key for the channel and for some libraries the channel number (the integer value shown on the channel page). There are libraries available for many platforms, and on some platforms, there may be several ways (libraries or techniques) to write data to a ThingSpeak channel.

You can find the API key on the channel page by clicking the *API Keys* tab. When you create a new channel, you will have one write and one read API key. You can add more keys if you need them so that you can use one key per device, location, customer, etc. Figure 18-4 shows the API Keys tab for the channel created previously.

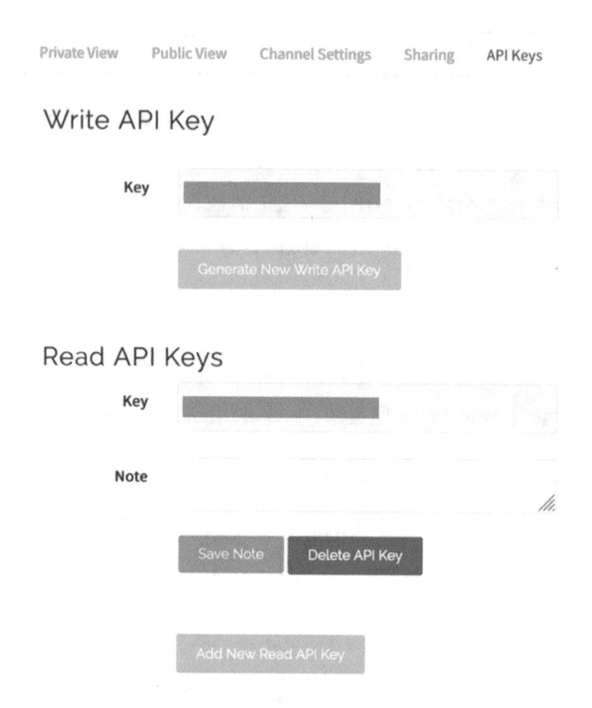

Figure 18-4. *API keys for the practice channel*

Notice I masked out the keys. If you make your channel public, do not share the write key with anyone you don't want to allow to write to your channel. You can create new keys by clicking the *Generate New Write API Key* or *Add New Read API Key* button. You can delete read keys by clicking the *Delete API Key* button.

We use the key in our code to allow the device to connect to and write data to the channel. So we typically copy this string from the channel page and paste it into our code as a string. Recall we may use a library that encapsulates the HTTP or MQTT mechanism or, in the case of the Raspberry Pi Python library, we use a Python library and the HTTP protocol. We will see both in the upcoming sample projects for Arduino and Raspberry Pi.

Now that you understand the basics of writing data to ThingSpeak, let's take a look at how to do it in more detail for the Arduino. This is followed by an example for the Raspberry Pi.

Using ThingSpeak with the Arduino

This project is a very simple sketch to learn how to connect and write data to a ThingSpeak channel. For the data, we will be generating a random integer and send that to the channel. While this won't necessarily give you anything meaningful, we keep things simple so we can see the mechanics of how to interact with ThingSpeak.

The hardware we will use is a WiFi-enabled Arduino board. You don't need any components beyond the host board itself. If you have not read through Chapter 17 and implemented the examples to connect your Arduino to your WiFi, you should review that chapter before continuing. We will see the code needed for a WiFi connection but without explanation.

Now, let's see how we set up the software and the sketch for the project.

Configuring the Arduino IDE

To write data to the ThingSpeak channel, we need to install the ThingSpeak software library. In the Arduino IDE, choose *Sketch* ➤ *Include Library* ➤ *Manage Libraries....* Enter `ThingSpeak` in the search box and then click the *Install* button as shown in Figure 18-5. Once again, click *Install* to install the library.

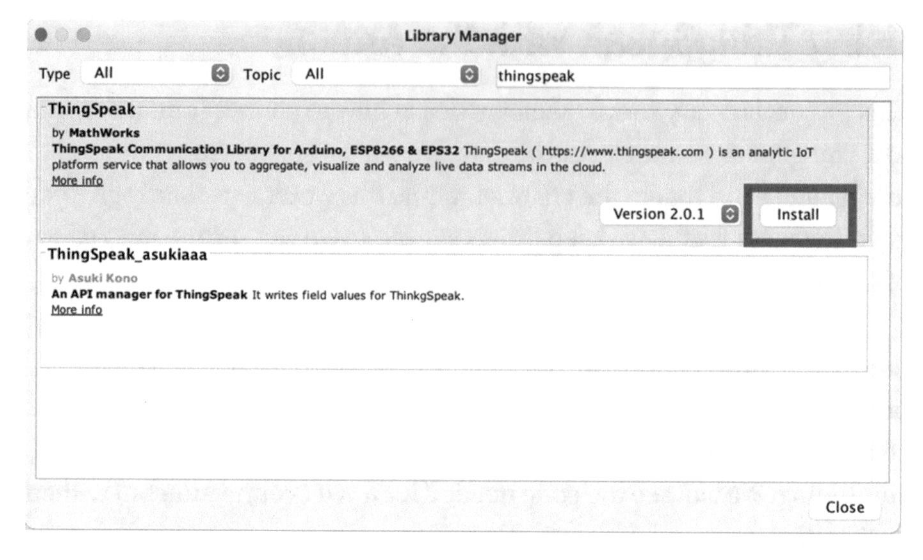

Figure 18-5. *Installing the ThingSpeak library*

Write the Sketch

Now that you have the necessary library installed, open a new Arduino project and name it `arduino_thingspeak.ino`. Recall we need to add the WiFi data to our sketch. We will also store our API key and other critical data in a separate header (`secrets.h`) file, which will be part of the sketch and saved in the same folder. To add a header file, click the small down arrow button to the right of the sketch and select *New Tab* as shown in Figure 18-6. In the prompt, enter `secrets.h` and press *Enter*. This will open a new tab. Click that tab to open the file.

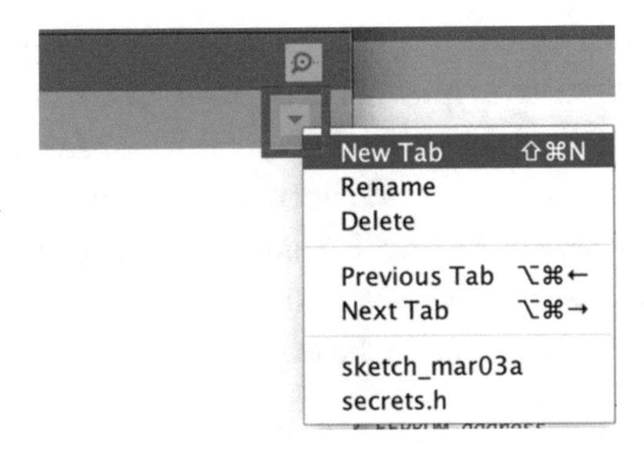

Figure 18-6. *Add a new tab*

We will place the WiFi and our ThingSpeak channel data in this file. Use the #define directive to create new strings that we will use in the main sketch. The following shows the lines and data you need for the file. Type these in and save the file:

```
#define SECRET_SSID "YOUR_SSID"                    // SSID
#define SECRET_PASS "SSID_PASS"                    // WiFi Password
#define SECRET_CH_ID 0000000000                    // Channel number
#define SECRET_WRITE_APIKEY "ABCDEFGHIJKLMNOP"  // Write API Key
```

These include the SSID and password for your WiFi as well as the write API key from your ThingSpeak channel and the channel Id. You can find the channel Id on the channel page in ThingSpeak as shown in Figure 18-7.

ThingSpeak™ Channels Apps Support

practice_channel

Channel ID:
Author:
Access: Private

Testing ThingSpeak connection from Arduino and Python

Private View Public View Channel Settings Sharing API Keys Data Import / Export

Figure 18-7. *Finding the channel Id (ThingSpeak)*

We can also place the `setupWiFi()` function we learned in Chapter 17 in this file. This allows us to move all of the related statements for setting up and using the WiFi to the secrets.h file, which we can later copy to other projects making it really easy to add not only WiFi but also our ThingSpeak credentials to any project. Listing 18-1 shows the complete code for the secrets.h file. Since we've seen most of the code already, you can read through the code and move on to the next part.

Listing 18-1. Secrets Header File for WiFi and ThingSpeak (Arduino)

```
#include <WiFi.h>

#define SECRET_SSID "SSID_GOES_HERE"              // SSID
#define SECRET_PASS "SSID_PASS_HERE"              // WiFi Password
#define SECRET_CH_ID 0000000000000000             // Channel number
#define SECRET_WRITE_APIKEY "API_WRITE_KEY_HERE"   // Write
                                                      API Key
#define WIFI_WAIT 5000 // 5 seconds

char ssid[] = SECRET_SSID;    // your network SSID (name)
char pass[] = SECRET_PASS;    // your network password
WiFiClient  client;

bool setupWiFi() {
  int status = WL_IDLE_STATUS;

  // Attempt to connect to Wifi network:
  while (status != WL_CONNECTED) {
    Serial.print("Connecting to WiFi...");
    status = WiFi.begin(ssid, pass);
    // Wait for connection: set according to your environment
    delay(WIFI_WAIT);
  }
```

```
  Serial.println("connected.");
  // print your WiFi's IP address:
  IPAddress ip = WiFi.localIP();
  Serial.print("IP Address: ");
  Serial.println(ip);
  return true;
}
```

Now, return to the main sketch tab. Begin the sketch with the following includes. You need the `ThingSpeak.h` and the `secrets.h` file we just created:

```
#include "ThingSpeak.h"
#include <WiFiNINA.h>
#include "secrets.h"
```

Next, we add a variable to store the pin number for the sensor. Notice we use those `#defines` we stored in the `secrets.h` file:

```
unsigned long myChannelNumber = SECRET_CH_ID;
const char * myWriteAPIKey = SECRET_WRITE_APIKEY;
```

Next you define your ThingSpeak API key and feed ID:

```
char ThingSpeakKey[] = "<YOUR_KEY_HERE>";
#define FEED_NUMBER <YOUR_FEED_HERE>
```

Now we are ready to write the `setup()` function. Since we aren't using any sensors or modules, we only need to set up the WiFi and connect like we did in Chapter 17. We also need to instantiate the ThingSpeak class and initialize the random library by calling `randomSeed()` (shown in bold). Listing 18-2 shows a sample `setup()` function you can use. We will be using the `setupWiFi()` function seen in Chapter 17 and defined in the `secrets.h` file.

Listing 18-2. Setup Function for ThingSpeak Example (Arduino)

```
void setup() {
  Serial.begin(115200);
  while(!Serial);

  Serial.println("Welcome to the ThingSpeak Arduino
  demonstration!");

  // Setup the WiFi
  if (!setupWiFi()) {
    Serial.println("ERROR: Cannot setup wifi. Halting.");
    while (true);
  }

  // Initialize ThingSpeak
  ThingSpeak.begin(client);

  // Set the random seed
  randomSeed(analogRead(0));
}
```

Finally, the `loop()` function contains the code to generate a random integer and send the data to our ThingSpeak channel. To do so, we first call the `setField()` function for the ThingSpeak library to set each field we want to update (field numbers start at 1). There is no correlation with the field names; rather, we reference the field by number in the order they were defined. So you must remember the order in which the fields were listed when you created your channel. For example, if you used temperature, humidity, and pressure (from top to bottom), temperature would be field 1, humidity field 2, and pressure field 3.

We then use the `writeFields()` function to send the data to ThingSpeak. We can check the result of that call to ensure the code returned is 200, which means success (Ok). Listing 18-3 shows the `loop()` function.

Listing 18-3. Loop Function for ThingSpeak Example (Arduino)

```
void loop() {
  // Generate a random number from 1 to 30
  int randNumber = random(30) + 1;
  Serial.print("Random number generated: ");
  Serial.println(randNumber);

  // Write the data to ThingSpeak
  // Set the fields with the values
  ThingSpeak.setField(1, randNumber);

  // Write to the ThingSpeak channel
  int res = ThingSpeak.writeFields(myChannelNumber,
  myWriteAPIKey);
  if (res == 200) {
    Serial.println("Channel update successful.");
  } else {
    Serial.print("Problem updating channel. HTTP error code ");
    Serial.println(res);
  }

  delay(30000);
}
```

Notice we display the actual result if it does not return a code of 200. Notice also we add a sleep (`delay()`) at the end to sleep for 30 seconds. We do this because the ThingSpeak free account is limited to update once every 15 seconds.

Now that you understand the flow and contents of the sketch, you can complete the missing pieces and start testing. Listing 18-4 shows the complete sketch for this project.

Listing 18-4. Arduino-Based ThingSpeak Channel Write Example

```
#include "ThingSpeak.h"
#include "secrets.h"

// Global Variables
unsigned long myChannelNumber = SECRET_CH_ID;
const char * myWriteAPIKey = SECRET_WRITE_APIKEY;

void setup() {
  Serial.begin(115200);
  while(!Serial);

  Serial.println("Welcome to the ThingSpeak Arduino
  demonstration!");

  // Setup the WiFi
  if (!setupWiFi()) {
    Serial.println("ERROR: Cannot setup wifi. Halting.");
    while (true);
  }

  // Initialize ThingSpeak
  ThingSpeak.begin(client);

  // Set the random seed
  randomSeed(analogRead(0));
}

void loop() {
  // Generate a random number from 1 to 30
  int randNumber = random(30) + 1;
  Serial.print("Random number generated: ");
  Serial.println(randNumber);
```

```
  // Write the data to ThingSpeak
  // Set the fields with the values
  ThingSpeak.setField(1, randNumber);

  // Write to the ThingSpeak channel
  int res = ThingSpeak.writeFields(myChannelNumber,
  myWriteAPIKey);
  if (res == 200) {
    Serial.println("Channel update successful.");
  } else {
    Serial.print("Problem updating channel. HTTP error code ");
    Serial.println(res);
  }

  delay(30000);
}
```

Note Be sure to substitute your API key and channel number in the `secrets.h` file. Failure to do so will result in compilation errors.

Take some time to make sure you have all the code entered correctly and that the sketch compiles without errors. Once you reach this stage, you can upload the sketch and try it out.

Testing the Sketch

To test the sketch, be sure the code compiles and you have your hardware set up correctly. Once you have a sketch that compiles, upload it to your Arduino MKR1000 and launch a serial monitor. The following shows an example of the output you should see:

```
Welcome to the ThingSpeak Arduino demonstration!
Random number generated: 1
Channel update successful.
```

```
Random number generated: 3
Channel update successful.
Random number generated: 17
Channel update successful.
Random number generated: 13
Channel update successful.
Random number generated: 15
Channel update successful.
...
```

Did you see similar output? If you did not, check the return code as displayed in the serial monitor. You should be seeing a return code of 200 (meaning success). If the return code is a single digit (1, 2, 3, etc.), you are likely encountering issues connecting to ThingSpeak. If this occurs, connect your laptop to the same network cable, and try to access ThingSpeak.

If the connection is very slow, you could encounter a situation in which you get an error code other than 200 on every other attempt or every N attempts. If this is the case, you can increase the timeout in the `loop()` function to delay processing further. This may help for some very slow connections, but it is not a cure for a bad or intermittent connection.

Let the sketch run for about 3 minutes before you visit ThingSpeak. Once the sketch has run for some time, navigate to ThingSpeak, log in, and click your channel page and then click the *Private View* tab. We use the private view because channels are private by default. You should see results similar to those shown in Figure 18-8.

Channel Stats
Created: 2 days ago
Last entry: 3 minutes ago
Entries: 35

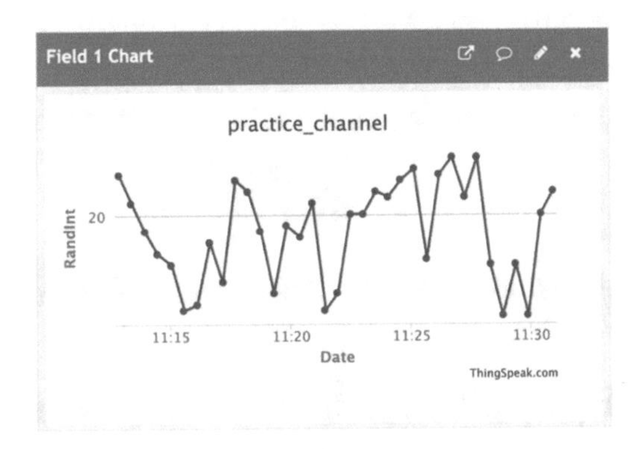

Figure 18-8. *Example channel data (Arduino)*

Notice there isn't much to learn here other than it's an interesting view of our data over time, which is a good default view for most data.

We used the private view because that is how the channel is configured by default. If you want to make the view public, you can by clicking the *Sharing* tab and then selecting either sharing with everyone or sharing to a specific set of people by email address. Keep in mind making the view public to everyone means anyone who browses the ThingSpeak website. Thus, you likely only want to share it with certain people. When you use the Select User option, you type in each email address one at a time and click *Add User*. Your intended recipients will receive an email with an invitation to create a ThingSpeak account. Once they log in, they can click Channels ➤ *Channels shared with me* to see your shared view.

Now, let's look at the Python version of the example. But first, if you have implemented the Arduino example in this section, you will need to reset the data in the channel because we're going to use the same channel. This isn't strictly necessary, but helpful if you want to ensure your Python version is working correctly.

To clear the data in the channel, open the channel home page and then click Channel Settings and scroll down to the bottom to locate the Clear Channel button. Click the button and then reply to the confirmation. Figure 18-9 shows the clear data section of the channel settings page.

Want to clear all feed data from this Channel?

Figure 18-9. *Clear data for the ThingSpeak channel*

Using ThingSpeak with the Raspberry Pi

Once again, the code for this project is easy to learn. We will learn how to connect and write data to a ThingSpeak channel. For the data, we will be generating a random integer and send that to the channel.

The hardware we will use is simply a Raspberry Pi 4B or similar board. No additional hardware is needed, but you will need to connect your Raspberry Pi to your network either via WiFi or Ethernet.

Configuring the Raspberry Pi

To write data to the ThingSpeak channel, we need to ensure we have the request Python library installed. You should already have this installed, but it doesn't hurt to run the install again. The following shows the command you should run on the Raspberry Pi:

```
% pip3 install request
```

That's it! Now, let's write the code. As you will see, it uses a different mechanism for uploading data to ThingSpeak. This is because we do not have a class to encapsulate the functionality, but we will write one!

Write the Code

We will be using a POST message with the Python request library for this version of the example ThingSpeak demonstration. This requires a few more lines of code than the Arduino version, so we will create a class for it that we can reuse.

You could write a similar class for the Arduino (and I encourage you to do so as an exercise), but it won't save you much since we have a ThingSpeak library we're already using on the Arduino.

The class file we will create is named `thingspeak.py` and will contain a class named ThingSpeak. For most of the examples where you will use the class, we need only a constructor and a function to write (upload) data to ThingSpeak. To make it a bit more tolerant of networking issues, we will also build a retry loop into the upload procedure.

Note If you want to read data from ThingSpeak, you can add that function to this class extending its use to other projects.

Let's begin with the imports and constants. We need to import the `http.client`, `time`, and `urllib` libraries as shown in the following. We will use only one constant: a value for the maximum number of retries. The idea is upload will retry up to `MAX_RETRIES` times before aborting. This will help when the Raspberry Pi is connected to a slow or intermittent network:

```
import http.client
import time
import urllib

MAX_RETRIES = 10
...
```

For the constructor, we will accept the API key and user-customized maximum retries with a default of MAX_RETRIES. Since this code is run once and we need to build a proper POST message with a header and given the header doesn't change, we will create that in the constructor too.

For the upload() function, we will require a Python dictionary that includes each of the keys and their values. We have to add the API key, but we can do that easily. In the function, we will create a loop that contains a try...except block for calling the network functions we will use. Specifically, we open a connection to the ThingSpeak server, issue the POST request, and then wait for a status code. One important step is using the urllib class to parse the Python dictionary we pass in with the keys and values. We need to do this to reformat it for the request.post() function. We then test the code to ensure the upload worked.

If we encounter a problem with any of the network functions, we sleep for 5 seconds and then try the commands again. We will do this up to MAX_RETRIES or until the operation succeeds.

Listing 18-5 shows the complete code for this class. Take some time to read through it so that you familiarize yourself with how it works.

Listing 18-5. The ThingSpeak Class (Python)

```
# Import libraries
import http.client
import time
import urllib

MAX_RETRIES = 10

class ThingSpeak:

    def __init__(self, key, num_retries=MAX_RETRIES):
        self.api_key = key
        # Create the header
        self.headers = {
```

```
            "Content-type": "application/x-www-form-urlencoded",
            'Accept': "text/plain"
        }
        self.max_retries = num_retries

    def upload(self, param_dict):
        param_dict.update({'key': self.api_key})
        # Setup the data to send in a JSON (dictionary)
        params = urllib.parse.urlencode(param_dict)
        retry = 0
        while retry <= self.max_retries:
            try:
                # Create a connection over HTTP
                conn = http.client.HTTPConnection("api.
                thingspeak.com:80")
                data = None
                # Execute the post (or update) request to
                  upload the data
                conn.request("POST", "/update", params, self.
                headers)
                # Check response from server (200 is success)
                response = conn.getresponse()
                # Display response (should be 200)
                if response.status != 200:
                    print("Response: {0} {1}".format(response.
                    status, response.reason))
                    # Read the data for diagnostics
                    data = response.read()
                print("Channel update successful.")
                conn.close()
                retry = self.max_retries + 1  # stop the loop
```

```
        except Exception as err:
           print("WARNING: ThingSpeak connection failed: {0}, "
                 "data: {1}".format(err, data))
            if retry <= self.max_retries:
                print("Retrying in 5 seconds. [{}]".
                format(retry+1))
                time.sleep(5)
                retry = retry + 1
            else:
                retry = self.max_retries + 1
                print("WARNING: Cannot send data. Exceeded
                retries.")
```

Next is the code for the main script. We will use the new class to upload the random number we generate. We will name the main script `thingspeak_python.py`. If you are following along, open a new file now with that name. Be sure to place it in the same folder as the `thingspeak.py` module.

Listing 18-6 shows the complete code for the script for this project. It follows a now familiar pattern where we create a `main()` function and call it from a `try...except` block to catch a *CTRL+C* key sequence. The code is very simple. All you need to do is put your API key in the constant and run it.

Listing 18-6. Complete Code for the thingspeak_python.py Script

```
# Import libraries
import random
import sys
import time
from thingspeak import ThingSpeak

# API KEY
THINGSPEAK_APIKEY = 'YOUR_WRITE_API_KEY_HERE'
```

```
def main():
    """main"""
    print("Welcome to the ThingSpeak Raspberry Pi
    demonstration!")
    print("Press CTRL+C to stop.")
    thing_speak = ThingSpeak(THINGSPEAK_APIKEY)
    while True:
        # Generate a random integer
        rand_int = random.randint(1, 20)
        print("Random number generated: {}".format(rand_int))
        thing_speak.upload({'field1': rand_int})
        # Sleep for 30 seconds
        time.sleep(30)

if __name__ == '__main__':
    try:
        main()
    except (KeyboardInterrupt, SystemExit) as err:
        print("\nbye!\n")
sys.exit(0)
```

Notice the dictionary we used to pass the data to the `upload()` function. Here, we used `field1` as the key for the channel field. As it turns out, we must use `field1`, `field2`, etc. for the field key names regardless of how we may name them in the channel. While this may be a little strange, you should get in the habit of listing the fields in the dictionary in the order they appear in the channel setup.

Note Be sure to substitute your API key in the location marked. Failure to do so will result in runtime errors.

Now that you have all the code entered, let's test the script and see if it works.

Testing the Script

To run the script, enter the following command. Let the script run for several iterations before using *Ctrl+C* to break the main loop. Listing 18-7 shows an example of the output you should see. You may see retry attempts if your network drops or you lose connectivity.

Listing 18-7. Sample Output for the Example (Python)

```
$ python3 ./thingspeak_python.py
Welcome to the ThingSpeak Raspberry Pi demonstration!
Press CTRL+C to stop.
Random number generated: 1
Channel update successful.
Random number generated: 3
Channel update successful.
Random number generated: 17
WARNING: ThingSpeak connection failed: [Errno -3] Temporary
failure in name resolution, data: None
Retrying in 5 seconds. [1]
WARNING: ThingSpeak connection failed: [Errno -3] Temporary
failure in name resolution, data: None
Retrying in 5 seconds. [2]
WARNING: ThingSpeak connection failed: [Errno -3] Temporary
failure in name resolution, data: None
Retrying in 5 seconds. [3]
Channel update successful.
Random number generated: 13
Channel update successful.
Random number generated: 15
Channel update successful.
Random number generated: 9
Channel update successful.
```

```
Random number generated: 19
Channel update successful.

bye!
```

If the connection is very slow, you could encounter a situation in which you get an error code other than 200 every other attempt or every N attempts. If this is the case, you can increase the timeout in the `loop()` function to delay processing further. This may help for some very slow connections, but it is not a cure for a bad or intermittent connection.

Let the sketch run for about 3 minutes before you visit ThingSpeak. Once the sketch has run for some time, navigate to ThingSpeak, log in, and click your channel page and then click the *Private View* tab. We use the private view because channels are private by default. You should see results similar to those shown in Figure 18-10.

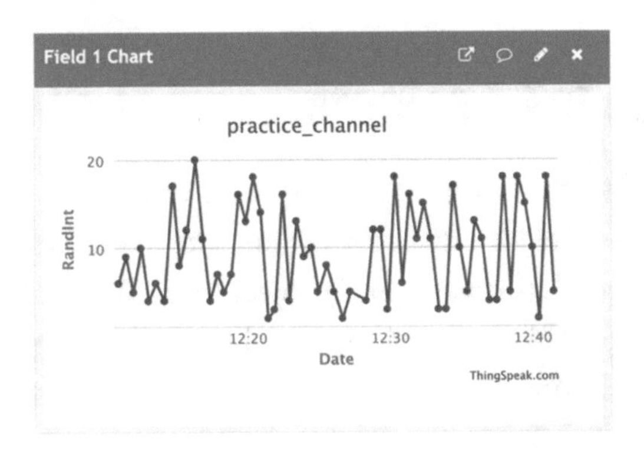

***Figure 18-10.** Example channel data (Python)*

If you do not see similar data, go back and check the return codes as discussed in the last project. You should see return codes of 200 (success). Check and correct any errors in network connectivity or syntax or logic errors in your script until it runs successfully for several iterations (all samples stored return code 200).

If you see similar data, congratulations! You now know how to generate data and save it to the cloud using two different platforms.

Now, let's turn our attention to several of our previous example projects retooling them to upload their data to ThingSpeak.

Note ThingSpeak free accounts are limited to four channels. If you plan to implement all of the example projects in this chapter, you may need to delete one or more channels or upgrade your account to a paid subscription. To delete a channel, navigate to your home page, then click the *Channel Settings* tab, scroll down to the bottom, and click *Delete Channel*.

Example IoT Projects

This section includes three of the projects from previous chapters that we will update to send data to ThingSpeak for visualization. The projects include the "How's the weather?" project from Chapter 8, the "digital gardener" project from Chapter 9, and the "monitoring your environment" project from Chapter 15. If you have not implemented these projects, you may want to do so before attempting the following examples.

Each example presents details at a high level, and much of the details for the original project are omitted for brevity. Rather, we will see details on the channel to create and how to prepare and modify the source files and then a demonstration of executing the project.

The hardware for each should be the same you used in the previous projects. The only difference is you will need to substitute an Arduino with WiFi capabilities and your Raspberry Pi 4B or similar for the Arduino and Python versions.

Example 1: IoT Weather Station

Let's take the project from Chapter 8 and make it a true IoT project. We will add the WiFi code and ThingSpeak code we saw in the previous section to send our data to the cloud. Let's begin with creating a new ThingSpeak channel.

The hardware for this project is the same as used in Chapter 8. Refer to the "Hardware Required" section to assemble the components needed. Read through both the Arduino and Python versions for any differences or things you need to consider.

Create the ThingSpeak Channel

The data for this project include three items: temperature in Celsius, relative humidity, and barometric pressure. We will create a channel that has three fields named for each of these data items.

Log in to your ThingSpeak account and click *New Channel*. We will name the channel *Weather IoT*. Use the information shown in Figure 18-11 to complete the form and then click *Save Channel* at the bottom of the form. Or you can press *Enter*, which will save the channel for you. Note that you will need to tick the checkbox for fields 2 and 3 to get them to accept input.

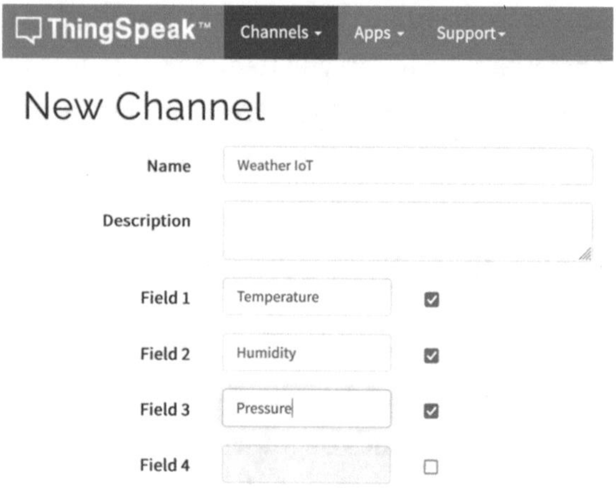

ThingSpeak™ Channels ▾ Apps ▾ Support ▾

New Channel

Name: Weather IoT

Description

Field 1: Temperature ☑

Field 2: Humidity ☑

Field 3: Pressure ☑

Field 4: ☐

Figure 18-11. *Weather IoT channel settings*

Recall we need to remember the order of the fields. Here, we have defined *Temperature, Humidity,* and *Pressure* where they will be referenced as field 1, field 2, and field 3 in our Arduino sketch and `field1`, `field2`, and `field3`, respectively, in our Python code.

Now that we have the channel created, go to the *API Keys* tab and record the API key. You will need this information in the next step. Figure 18-12 shows which key you will need.

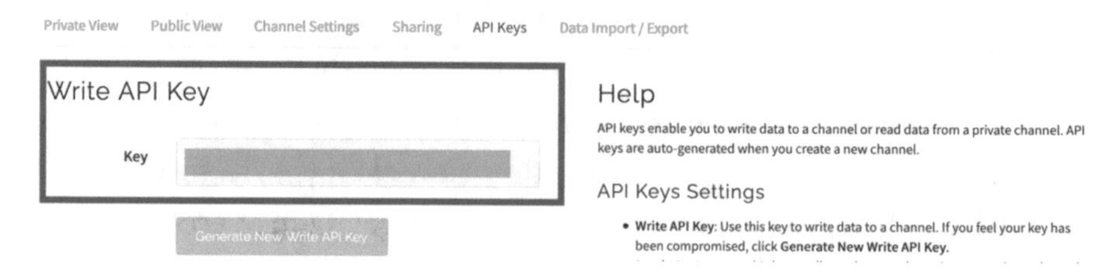

Figure 18-12. *Weather IoT channel API key*

Now we are ready to modify our code.

Prepare the Project Files

For this project, you should create a new project folder and copy the code files from Chapter 8 renaming them as follows. For the Arduino version, rename the project folder weather_iot and the Arduino sketch weather_iot. For the Python version, rename the project script to weather_iot.py.

We also need to copy our helper code files for ThingSpeak (and WiFi for the Arduino) as follows. For the Arduino version, copy the secrets.h file from the thingspeak_arduino project to your new project folder (weather_iot). For the Python version, copy the thingspeak.py file to the same folder as your weather_iot.py script. When you are done, you should have a folder structure like the following (includes the previous examples) with the "new" files shown in bold. Note that Ch18 is the main folder. You can name it whatever you want:

```
Ch18
  |
  +--- thingspeak_arduino
  | +--- thingspeak_arduino.ino
  | +--- secrets.h
  +--- weather_iot.py
  +--- weather_iot
  | +--- weather_iot.ino
  | +--- secrets.h
  +--- thingspeak.py
  +--- thingspeak_python.py
```

With that administrative work done, we can add the preliminary code.

Update the Project Code

In this section, we will modify the project code to add the WiFi (Arduino) and ThingSpeak code to turn our offline project into an IoT project. Let's start with the Arduino version and then visit the Python version.

Arduino

Recall we need only import the `secrets.h` file and add a few variables. We also need to call the `setupWiFi()` function. Listing 18-8 shows an excerpt of the code with the new lines added. The rest of the code from Chapter 8 remains the same. We will update the `loop()` function in the next section.

Listing 18-8. Updates to the Weather IoT Main Sketch (Arduino)

```
#include <Wire.h>
#include <SFE_MicroOLED.h>
#include "SparkFunBME280.h"
#include "ThingSpeak.h"
#include "secrets.h"

...

unsigned long myChannelNumber = SECRET_CH_ID;
const char * myWriteAPIKey = SECRET_WRITE_APIKEY;

...

void setup()
{
  Serial.begin(115200);
  Serial.println("How's the weather?");
  Serial.println("-----------------");
  Wire.begin();
```

```
  if (!setupBME280()) {
    Serial.println("The sensor did not respond. Please check
    wiring.");
    while(1); //Freeze
  }
  if (!setupOLED()) {
    Serial.println("ERROR: OLED not found!");
    while(1);
  }
  bme280.setReferencePressure(SEALEVEL_REFERENCE);
  delay(3000);

  // Setup the WiFi
  if (!setupWiFi()) {
    Serial.println("ERROR: Cannot setup wifi. Halting.");
    while (true);
  }

  // Initialize ThingSpeak
  ThingSpeak.begin(client);
}
```

Don't forget to initialize the ThingSpeak library at the end of the code with a call to the begin() function. The client variable is defined in the secrets.h file (in case you're wondering where it came from).

Next, we need to set up the fields and then call the writeFields() function from the ThingSpeak Arduino library. Easy, eh? Listing 18-9 shows the completed loop() function with the new code highlighted.

Listing 18-9. Updates to the Weather IoT Loop Function (Arduino)

```
...
void loop()
{
  // Read the sensor
  float temperature = bme280.readTempC();
  float humidity = bme280.readFloatHumidity();
  float pressure = bme280.readFloatPressure();

  // Display data to serial monitor
  printDiagnostics(temperature, humidity, pressure);

  // Display data to OLED
  showDataOLED(temperature, humidity, pressure);

  // Write the data to ThingSpeak
  // Set the fields with the values
  ThingSpeak.setField(1, temperature);
  ThingSpeak.setField(2, humidity);
  ThingSpeak.setField(3, pressure);

  // Write to the ThingSpeak channel
  int res = ThingSpeak.writeFields(myChannelNumber,
  myWriteAPIKey);
  if (res == 200) {
    Serial.println("Channel update successful.");
  } else {
    Serial.print("Problem updating channel. HTTP error code ");
    Serial.println(res);
  }

  delay(30000);
}
```

Notice the `delay()` at the bottom of the loop. Recall we need to slow our data writes down to allow for the ThingSpeak free account time restrictions. We choose 30 seconds as the delay.

Raspberry Pi

Recall we need only add the import statement and API key. Listing 18-10 shows an excerpt of the code with the new lines added. The rest of the code from Chapter 8 remains the same. We will update the `main()` function to add the ThingSpeak code in the next section.

Listing 18-10. Updates to the Weather IoT Main Script (Python)

```
# Import libraries
import sys
import time
import board
import busio
import adafruit_bme280
import qwiic
from thingspeak import ThingSpeak

# API KEY
THINGSPEAK_APIKEY = 'YOUR_WRITE_API_KEY_HERE'
...
```

Next, we need to declare an instance of our `ThingSpeak` class from the `thingspeak.py` library module and then, after reading the data, form a Python dictionary and pass that to our `thing_speak.upload()` function call. Listing 18-11 shows the function with changes in bold. Notice we sleep for 30 seconds at the end like we did in the Arduino version. The rest of the code for this version remains the same as we had in Chapter 8.

Listing 18-11. Updates to the Weather IoT Main Function (Python)

```
...
def main():
    """main"""
    print("\nHow's the Weather?")
    print("------------------")
    if not setup_oled():
        print("ERROR: The OLED module is not found. "
              "Please check your connections!")
        sys.exit(1)

    # Compensate for reference pressure
    bme280.sea_level_pressure = 1013.25
    thing_speak = ThingSpeak(THINGSPEAK_APIKEY)
    while True:
        # Read the sensor data
        temperature = bme280.temperature
        humidity = bme280.humidity
        pressure = bme280.pressure
        # Display diagnostics
        print_diagnostics(temperature, humidity, pressure)
        # Show data on OLED
        show_data_oled(temperature, humidity, pressure)

        # Send data to ThingSpeak channel
        data = {
            'field1': temperature,
            'field2': humidity,
            'field3': pressure
        }
        thing_speak.upload(data)
```

```
        # Sleep for 30 seconds before next reading
        time.sleep(30)
...
```

That's it. We're ready to execute the project. We will need to let it run for a few minutes so we can get some data. If you're running the project in a controlled environment where the values do not change, you may not notice much variation. As an exercise, consider altering the environment to stimulate changes in the data. Don't use flame or touch the electronics in any way while they are running.

Execute and Visualize the Data

At this point, you can set up the hardware and run the project. Let it run for about 20 minutes and then visit your ThingSpeak channel page. You should see your data in the channel private view similar to Figure 18-13.

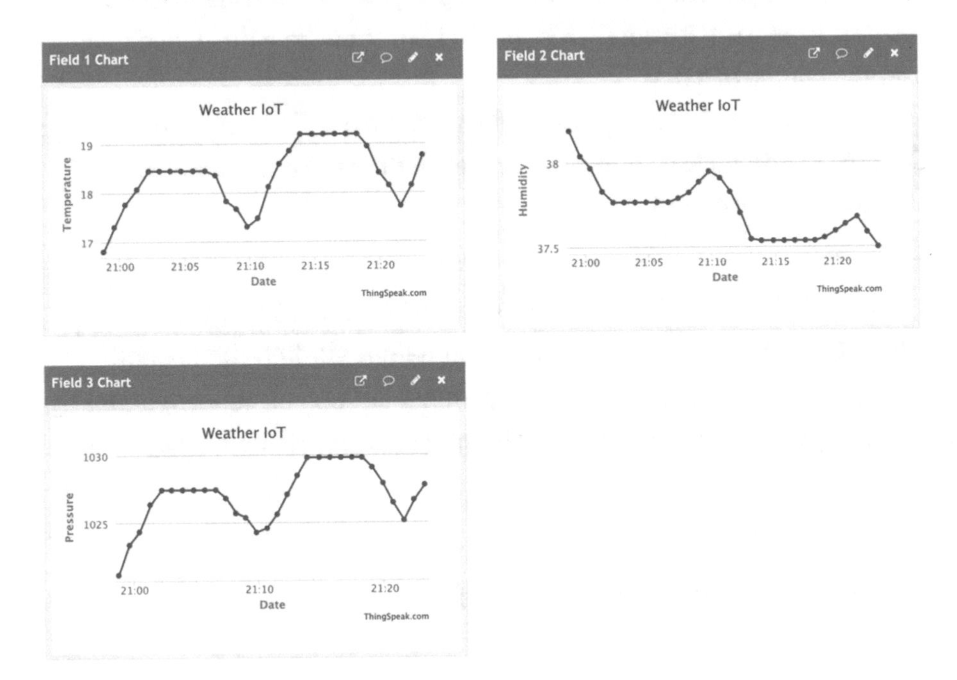

Figure 18-13. *Example results (Weather IoT example)*

While the line graphs are nice in that they show you data changed over time, they are more than pretty displays. In fact, you can hover your mouse over any data point and see the raw data as shown in Figure 18-14. Nice!

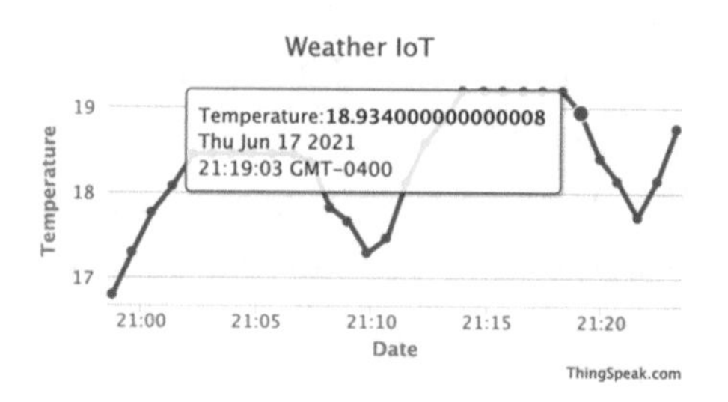

Figure 18-14. *Chart data details (ThingSpeak)*

Once again, you may not see a lot of variances in the data if you run it in a controlled environment. For better results in a controlled environment, you should consider changing the sample rate from 30 seconds to every 4–6 hours. This should help show how the data changes over the course of a day.

You could even combine this code with the IFTTT example from Chapter 17 and have it send you a text each time the data changes. Go ahead and experiment!

Note If you choose to run both versions of the project, you may want to delete the data in the channel before starting the second version. Recall we can delete the channel data by clicking *Channel Settings* for the channel on your home page and clicking *Clear Channel* at the bottom.

Now, let's look at the next example project.

Example 2: IoT Digital Gardener

Let's take the project from Chapter 9 and make it a true IoT project. We will add the WiFi code and ThingSpeak code we saw in the previous section to send our data to the cloud. Let's begin with creating a new ThingSpeak channel.

The hardware for this project is the same as used in Chapter 9. Refer to the "Hardware Required" section to assemble the components needed. Read through both the Arduino and Python versions for any differences or things you need to consider.

Create the ThingSpeak Channel

The data for this project is a bit different than others. Recall we can have one or more soil sensors and we generate a moisture evaluation for each. For this example, we will capture the raw data rather than the label (category). Thus, we will need one channel with one field for each sensor. We'll keep it simple and use only two sensors, but we will write the code to allow for more. We will create a channel that has two fields named for each of the sensors. We'll use the name of the plant(s) they monitor for clarity.

Log in to your ThingSpeak account and click *New Channel*. We will name the channel *IoT Gardener*. Use the information shown in Figure 18-15 to complete the form and then click *Save Channel* at the bottom of the form. Or you can press *Enter*, which will save the channel for you. Note that you will need to tick the checkbox for fields 2 and 3 to get them to accept input.

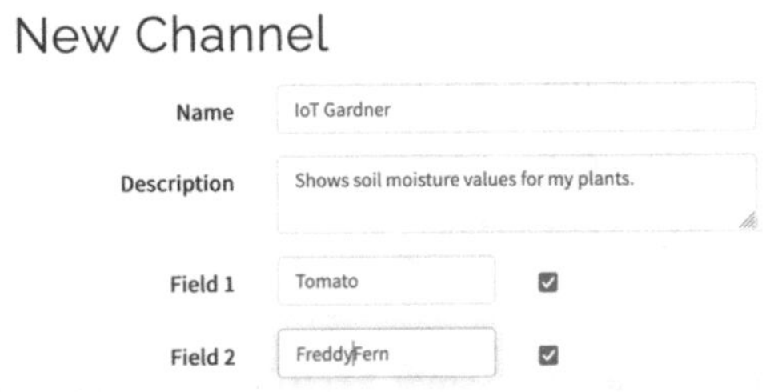

New Channel

Name: IoT Gardner

Description: Shows soil moisture values for my plants.

Field 1: Tomato

Field 2: FreddyFern

Figure 18-15. *IoT Gardener channel settings*

Recall we need to remember the order of the fields. Here, we have defined *Tomato* and *FreddyFern* where they will be referenced as field 1 and field 2 in our Arduino sketch and `field1` and `field2`, respectively, in our Python code.

Now that we have the channel created, go to the *API Keys* tab and record the API key. You will need this information in the next step. Now we are ready to modify our code.

Prepare the Project Files

For this project, you should create a new project folder and copy the code files from Chapter 9 renaming them as follows. For the Arduino version, rename the project folder `gardener_iot`[3] and the Arduino sketch `gardener_iot`. For the Python version, rename the project script to `gardener_iot.py`.

We also need to copy our helper code files for ThingSpeak (and WiFi for the Arduino) as follows. For the Arduino version, copy the `secrets.h` file from the `thingspeak_arduino` project to your new project folder (`gardener_iot`). For the Python version, copy the `thingspeak.py` file to the same folder as your `gardener_iot.py` script. When you are done, you should have a folder structure like the following (includes the previous examples) with the "new" files shown in bold. Note that Ch18 is the main folder. You can name it whatever you want:

```
Ch18
  |
  +--- gardener_iot
  | +--- QwiicSoilMoisture.h
  | +--- gardener_iot.ino
  | +--- QwiicSoilMoisture.cpp
```

[3] Yes, the channel is named IoT Gardener, yet the files are named `gardener_iot*`. That is intentional.

```
| +--- secrets.h
+--- gardener_iot.py
+--- soil_moisture.py
...
```

With that administrative work done, we can add the preliminary code.

Update the Project Code

In this section, we will modify the project code to add the WiFi (Arduino) and ThingSpeak code to turn our offline project into an IoT project. Let's start with the Arduino version and then visit the Python version.

Arduino

Recall we need only import the `secrets.h` file and add a few variables. We also need to call the `setupWiFi()` function. Listing 18-12 shows an excerpt of the code with the new lines added. The rest of the code from Chapter 9 remains the same. We will update the `loop()` function in the next section.

Listing 18-12. Updates to the IoT Gardener Main Sketch (Arduino)

```
#include <Wire.h>
#include <SerLCD.h>
#include <SparkFun_I2C_Mux_Arduino_Library.h>
#include "QwiicSoilMoisture.h"
#include "ThingSpeak.h"
#include "secrets.h"

// Constants
#define NUMBER_OF_SENSORS 2  // Set number of sensors here
#define DRY_THRESHOLD 250    // Low threshold for dry soil
#define WET_THRESHOLD 400    // High threshold for wet soil
```

```
// Global Variables
QWIICMUX myMux;    // Mux
SerLCD lcd;        // Serial LCD

// Create pointer to an array of pointers to the sensor class
QwiicSoilMoisture **soilMoistureSensors;

unsigned long myChannelNumber = SECRET_CH_ID;
const char * myWriteAPIKey = SECRET_WRITE_APIKEY;

void setup()
{
  Serial.begin(9600);
  Serial.println("Digital Gardener!");

  Wire.begin();

  // Now, setup the LCD
  lcd.begin(Wire);
  lcd.clear();                      // Clear the display
  lcd.setBacklight(255, 255, 255); // Set backlight to bright
                                       white
  lcd.setContrast(5);               // Set contrast. 0<- for
                                       higher contrast
  //          01234567890123456789   - Max characters we can display
  lcd.print("Digital Gardener!");

  // Create set of pointers for instantiated soil moisture
     classes
  soilMoistureSensors = new QwiicSoilMoisture *[NUMBER_OF_
  SENSORS];

  // Instantiate the instances of the class
```

```
for (int x = 0; x < NUMBER_OF_SENSORS; x++)
  soilMoistureSensors[x] = new QwiicSoilMoisture(Wire);

if (myMux.begin() == false) {
  Serial.println("ERROR: Mux not detected. Freezing...");
  while (1);
}

// Initialize all the sensors
bool initSuccess = true;

for (byte x = 0; x < NUMBER_OF_SENSORS; x++) {
  myMux.setPort(x);
  if (!soilMoistureSensors[x]->begin()) {
    Serial.print("Sensor ");
    Serial.print(x);
    Serial.println(" did not initialize! Check wiring?");
    initSuccess = false;
  }
}

if (initSuccess == false) {
  Serial.print("Freezing...");
  while (1);
}
Serial.println("Mux ready...");

// Setup the WiFi
if (!setupWiFi()) {
  Serial.println("ERROR: Cannot setup wifi. Halting.");
  while (true);
}
```

```
  // Initialize ThingSpeak
  ThingSpeak.begin(client);

  delay(3000);
}
```

Don't forget to initialize the ThingSpeak library at the end of the code with a call to the begin() function. The client variable is defined in the secrets.h file (in case you're wondering where it came from).

Next, we need to set up the fields and then call the writeFields() function from the ThingSpeak Arduino library. Listing 18-13 shows the completed loop() function with the new code highlighted.

Listing 18-13. Updates to the IoT Gardener Loop Function (Arduino)

```
...
void loop()
{
  // Read the sensor
  float temperature = bme280.readTempC();
  float humidity = bme280.readFloatHumidity();
  float pressure = bme280.readFloatPressure();

  // Display data to serial monitor
  printDiagnostics(temperature, humidity, pressure);

  // Display data to OLED
  showDataOLED(temperature, humidity, pressure);

  // Write the data to ThingSpeak
  // Set the fields with the values
  ThingSpeak.setField(1, temperature);
  ThingSpeak.setField(2, humidity);
  ThingSpeak.setField(3, pressure);
```

```
  // Write to the ThingSpeak channel
  int res = ThingSpeak.writeFields(myChannelNumber,
  myWriteAPIKey);
  if (res == 200) {
    Serial.println("Channel update successful.");
  } else {
    Serial.print("Problem updating channel. HTTP error code ");
    Serial.println(res);
  }

  delay(30000);
}
```

Notice the delay() at the bottom of the loop. Recall we need to slow our data writes down to allow for the ThingSpeak free account time restrictions. We choose 30 seconds as the delay.

Raspberry Pi

Recall we need only add the import statement and API key. Listing 18-14 shows an excerpt of the code with the new lines added. The rest of the code from Chapter 9 remains the same. We will update the main() function to add the ThingSpeak code in the next section.

Listing 18-14. Updates to the IoT Gardener Main Script (Python)

```
# Import libraries
import time
import sys
import qwiic_serlcd

from soil_moisture import SoilMoisture
from thingspeak import ThingSpeak
```

```
# API KEY
THINGSPEAK_APIKEY = 'YOUR_WRITE_API_KEY_HERE'
...
```

Next, we need to declare an instance of our ThingSpeak class from the thingspeak.py library module and then, after reading the data, form a Python dictionary and pass that to our thing_speak.upload() function call. Listing 18-15 shows the function with changes in bold. Notice we sleep for 30 seconds at the end like we did in the Arduino version. The rest of the code for this version remains the same as we had in Chapter 9.

Listing 18-15. Updates to the IoT Gardener Main Function (Python)

```
...
def main():
    """Main function to run the digital gardener example."""
    lcd = qwiic_serlcd.QwiicSerlcd()
    soil_moisture_sensor = SoilMoisture(NUMBER_OF_SENSORS)

    # Use the serial LCD
    print("\nDigital Gardener!")
    if not lcd.connected:
        print("The Qwiic SerLCD device isn't connected to "
              "the system. Please check your connection",
              file=sys.stderr)
        sys.exit(1)

    lcd.setBacklight(255, 255, 255) # set backlight to bright
                                       white
    lcd.setContrast(5)               # set contrast
    lcd.clearScreen()                # clear the screen
    lcd.print("Digital Gardener!")
    lcd.setCursor(0, 1)
    lcd.print("Getting ready")
```

```
for i in range(0, 5):
    lcd.print(".")
    time.sleep(2) # wait sec for system messages to
    complete

thing_speak = ThingSpeak(THINGSPEAK_APIKEY)
while True:
    lcd.clearScreen()
    data = {}
    for i in range(0, NUMBER_OF_SENSORS):
        value, voltage = soil_moisture_sensor.read_
        sensor(i)
        if value > WET_THRESHOLD:
            condition = "Too WET!"
        elif value < DRY_THRESHOLD:
            condition = "Too DRY!"
        else:
            condition = "Ok"
        msg = "#{0}: {1:5} {2}".format(i, value, condition)
        data.update({'field{0:1}'.format(i): value})
        print(msg)
        lcd.setCursor(0, i)
        lcd.print(msg)
        time.sleep(0.5)

    # Send data to ThingSpeak channel
    thing_speak.upload(data)

    # Sleep for 30 seconds
    time.sleep(30)
...
```

Notice we build a dictionary for the data inside the loop that loops over the set of sensors. Only once all sensors are read do we upload the data to ThingSpeak. Take some time to review this code until you're comfortable how it works.

That's it. We're ready to execute the project. We will need to let it run for a few minutes so we can get some data. If you're running the project in a controlled environment where the values do not change, you may not notice much variation. As an exercise, consider altering the environment to stimulate changes in the data. Don't use flame or touch the electronics in any way while they are running.

Execute and Visualize the Data

At this point, you can set up the hardware and run the project. Let it run for about 20 minutes and then visit your ThingSpeak channel page. You should see your data in the channel private view like Figure 18-16.

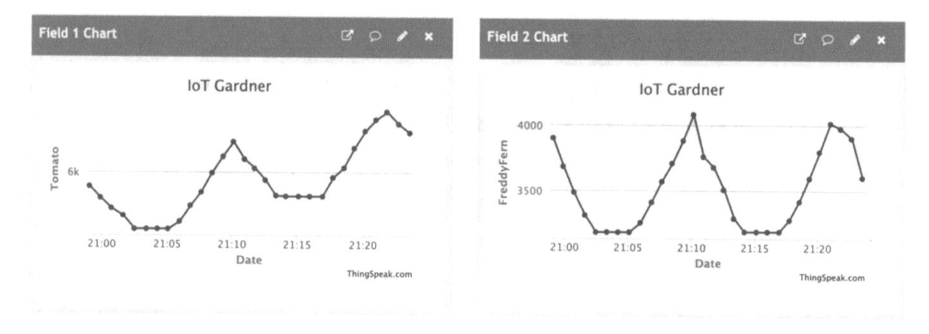

Figure 18-16. *Example results (IoT Gardener example)*

Once again, you may not see a lot of variances in the data if you run it in a controlled environment. For better results in a controlled environment, you should consider changing the sample rate from 30 seconds to every 4–6 hours. This should help show how the data changes over the course of a day.

There is more you can do here inside ThingSpeak to enhance your visuals. The easiest customization is simply changing the chart setup. If you click the little pencil icon on one of your charts, you will open the settings dialog for the chart. Here, you can change a number of labels such as the chart title and axis, change the color, and more.

Let's give it a go. Figure 18-17 shows the dialog for the tomato plant data. Notice I changed the title only for demonstration purposes. Once you have the data entered, click *Save* to make the changes. The chart will refresh with the new options.

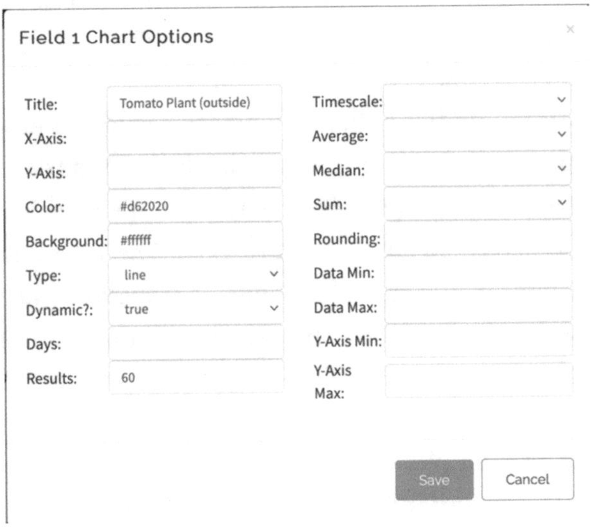

Figure 18-17. *Modifying a chart in ThingSpeak*

Let's see what this looks like. I also changed the title for the fern data. Figure 18-18 shows the results.

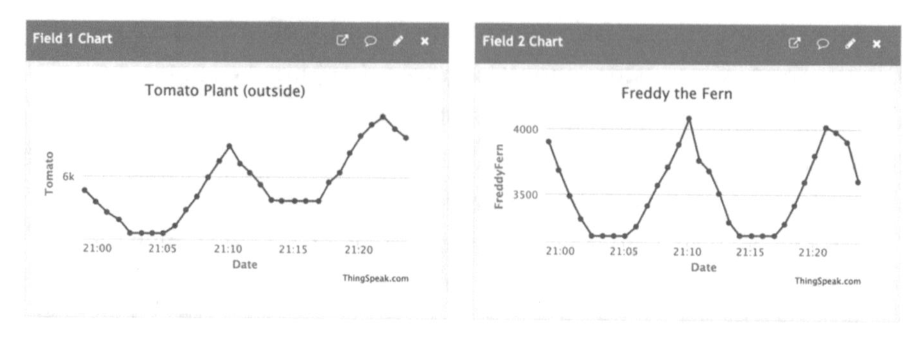

***Figure 18-18.** Example results with titles (IoT Gardener example)*

You can experiment with the other options. The average, median, and sum allow you to change the view to those parameters giving you a lot of control over how the data is presented.

You could also make use of the MatLab connections and widgets to build a sophisticated view and perform analysis on the data. While I encourage you to give that a go, it is a bit out of scope for this book.

The good news is there are a few widgets we can use to make our data more informative. You can add a gauge that shows an analog face, a numerical display that can show the data in only numeric form, or an indicator light that you can set to illuminate under certain conditions. I encourage you to experiment with these and see what you can create.

Let's look at an example. We are going to add two gauges to our private view. We will add a gauge configured with a "red zone" for the dry value range. What this will display is the last value read for the field, and this permits us to see if the soil is dry at a glance. The default line graphs show us how the values changed over time, but the more important question we will likely ask is if the soil is dry.

Start by clicking the *Add Widgets* button on the private view as shown in Figure 18-19.

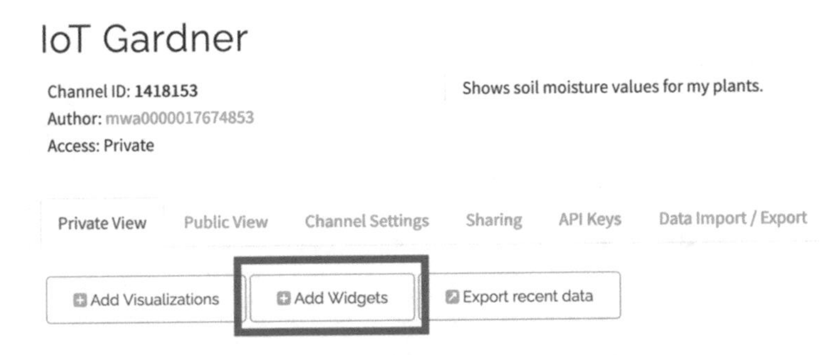

Figure 18-19. *Adding a widget (ThingSpeak)*

Next, select the gauge widget from the list as shown in Figure 18-20 and click *Next*.

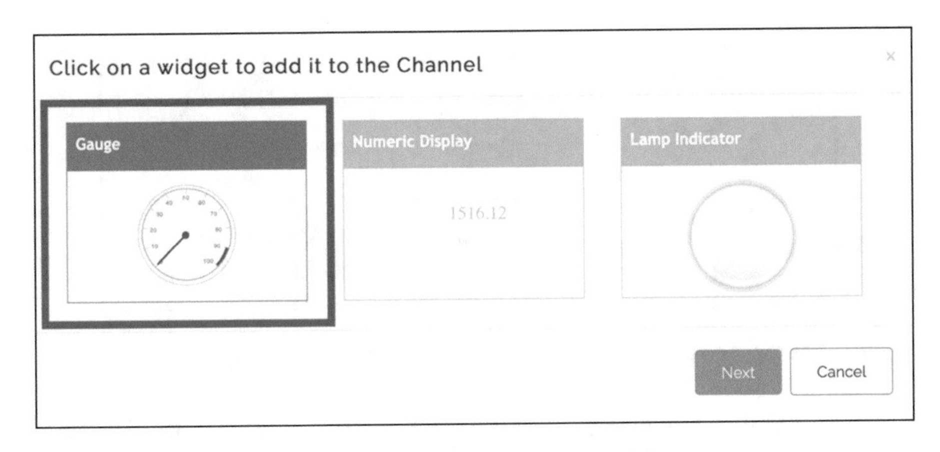

Figure 18-20. *Select graph widget (ThingSpeak)*

Next, you will see a dialog where you can configure the gauge. Figure 18-21 shows the settings used to create a dry soil gauge. Notice I set the title, selected the field (tomato plant is field1), and set min and max for the data range, the interval to 2000, and the dry range from 0 to 4000. Once you have the data entered, click *Save*. It may take a few moments for the graph to refresh.

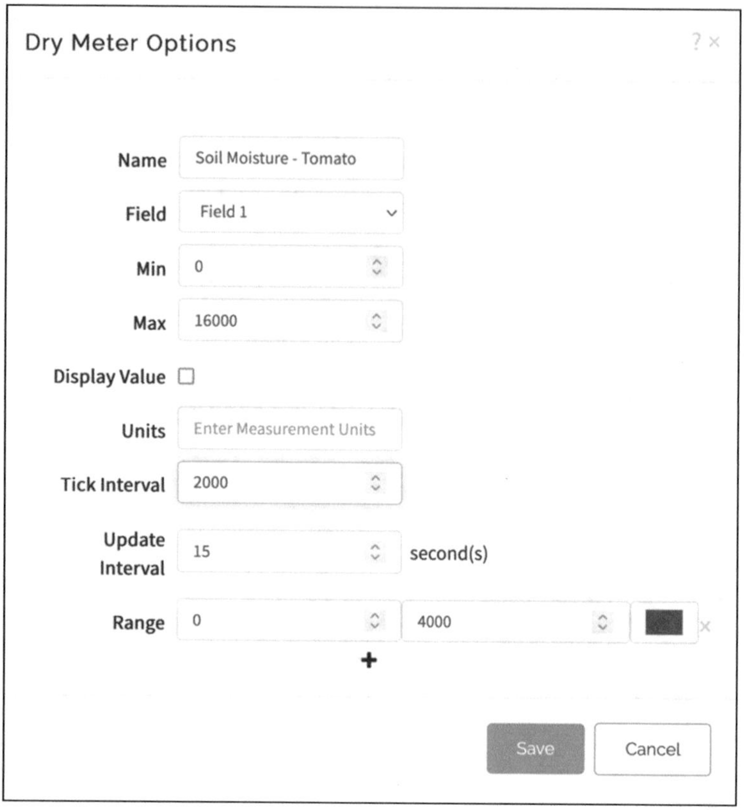

Dry Meter Options

Name	Soil Moisture - Tomato		
Field	Field 1		
Min	0		
Max	16000		
Display Value	☐		
Units	Enter Measurement Units		
Tick Interval	2000		
Update Interval	15	second(s)	
Range	0	4000	

Save Cancel

Figure 18-21. *Creating the dry soil gauge (ThingSpeak)*

If you're following along, go ahead and create a second gauge for the fern (field2). When you're done, and ThingSpeak refreshes the view, you should see four widgets like those in Figure 18-22.

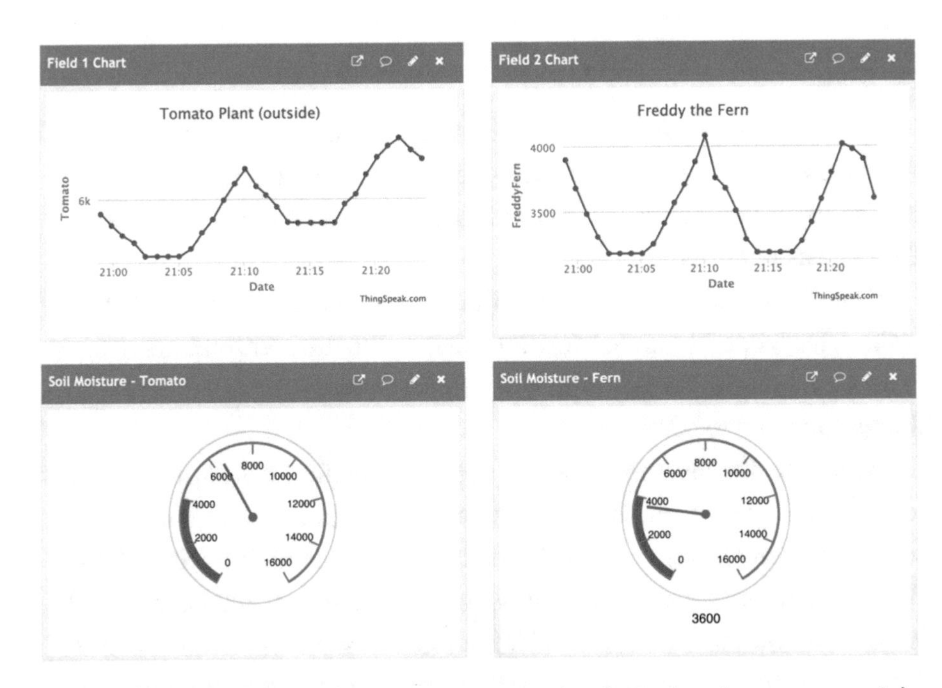

***Figure 18-22.** Example results with gauges (IoT Gardener example)*

Notice we can now see the data changes over time as well as the current soil moisture status for each plant. Cool!

Also notice the difference in the chart and gauge for the fern. Here, the line graph doesn't tell the whole story, but looking at the gauge tells us that soil is dry and needs watering. Go ahead and try this same concept with the indicator widget.

Note If you choose to run both versions of the project, you may want to delete the data in the channel before starting the second version. Recall we can delete the channel data by clicking *Channel Settings* for the channel on your home page and clicking *Clear Channel* at the bottom.

Now, let's look at the next example project.

Example 3: IoT Environment Monitor

Let's take the project from Chapter 15 and make it a true IoT project. We will add the WiFi code and ThingSpeak code we saw in the previous section to send our data to the cloud. Let's begin with creating a new ThingSpeak channel.

The hardware for this project is the same as used in Chapter 15. Refer to the "Hardware Required" section to assemble the components needed. Read through both the Arduino and Python versions for any differences or things you need to consider.

More specifically, this project requires a specific set of Arduino boards for the Arduino version. The programming technique used for one of the sensors requires using the AVR-based boards, which include the original Uno and others. Once we add the WiFi class, the sketch will become too large to run on the Uno and similar boards. Thus, we will need to use an Arduino Mega or Mega 2560 to build this project on the Arduino platform.

Create the ThingSpeak Channel

The data for this project is like both previous examples. We have numerical data as well as categorized data. Like the last project, we will capture the raw data rather than the label (category). The data generated includes the temperature, barometric pressure, dust concentration, and air quality. So we will need one channel with one field for each sensor or four fields in all.

Log in to your ThingSpeak account and click *New Channel*. We will name the channel *IoT Environment Monitor*. Use the information shown in Figure 18-23 to complete the form and then click *Save Channel* at the bottom of the form. Or you can press *Enter*, which will save the channel for you. Note that you will need to tick the checkbox for fields 2–4 to get them to accept input.

New Channel

Name	IoT Environment Monitor	
Description	An indoor environment monitor that monitors temperature, barometric pressure, dust, and air quality.	
Field 1	Temperature	☑
Field 2	Pressure	☑
Field 3	Dust Concentration	☑
Field 4	Air Quality	☑

Figure 18-23. *IoT Environment Monitor channel settings*

Recall we need to remember the order of the fields. Here, we have defined *Temperature, Pressure, Dust Concentration,* and *Air Quality* where they will be referenced as field 1, field 2, field 3, and field 4 in our Arduino sketch and `field1`, `field2`, `field3`, and `field4`, respectively, in our Python code.

Now that we have the channel created, go to the *API Keys* tab and record the API key. You will need this information in the next step. Now we are ready to modify our code.

Prepare the Project Files

For this project, you should create a new project folder and copy the code files from Chapter 15 renaming them as follows. For the Arduino version, rename the project folder `environment_iot`[4] and the Arduino sketch `environment_iot`. For the Python version, rename the project script to `environment_iot.py`.

[4] Once again, the channel is named IoT Environment Monitor, yet the files are named `environment_iot*`.

We also need to copy our helper code files for ThingSpeak (and WiFi for the Arduino) as follows. For the Arduino version, copy the `secrets.h` file from the `thingspeak_arduino` project to your new project folder (`environment_iot`). For the Python version, copy the `thingspeak.py` file to the same folder as your `environment_iot.py` script. When you are done, you should have a folder structure like the following (includes the previous examples) with the "new" files shown in bold. Note that Ch18 is the main folder. You can name it whatever you want:

```
Ch18
 |
 +--- air_monitor.py
 +--- environment_iot.py
 +--- environment_iot
 | +--- AirMonitor.h
 | +--- AirMonitor.cpp
 | +--- environment_iot.ino
 | +--- secrets.h
 ...
```

With that administrative work done, we can add the preliminary code.

Caution Attempting to run the project on an Uno or Leonardo board may result in unstable execution or compilation errors. This version of the project requires an Arduino AVR board with more memory than the Uno. It is suggested you use the Arduino Mega or Mega 2560.

Update the Project Code

In this section, we will modify the project code to add the WiFi (Arduino) and ThingSpeak code to turn our offline project into an IoT project. Let's start with the Arduino version and then visit the Python version.

Arduino

Recall we need only import the `secrets.h` file and add a few variables. We also need to call the `setupWiFi()` function. Listing 18-16 shows an excerpt of the code with the new lines added. The rest of the code from Chapter 15 remains the same. We will update the `loop()` function in the next section.

Listing 18-16. Updates to the IoT Environment Monitor Main Sketch (Arduino)

```
#include <Arduino.h>
#include <U8x8lib.h>
#include <Wire.h>
#include "AirMonitor.h"
#include "ThingSpeak.h"
#include "secrets.h"

...

unsigned long myChannelNumber = SECRET_CH_ID;
const char * myWriteAPIKey = SECRET_WRITE_APIKEY;
...

void setup() {
  // Setup buzzer
  pinMode(BUZZER_PIN, OUTPUT);
  Serial.begin(115200);
  while (!Serial);
  // Setup OLED
  oled = new U8X8_SSD1306_128X64_NONAME_HW_I2C(U8X8_PIN_NONE);
  oled->begin();
  oled->setFont(u8x8_font_chroma48medium8_r);
```

```
  Serial.println("Welcome to the Environment Monitor!");
  Serial.print("Starting....");
  oled->drawString(0, 1, "Environment");
  oled->drawString(0, 2, "Monitor");
  oled->drawString(0, 4, "Starting...");
  airQuality = new AirMonitor();
  delay(3000);
  Serial.println("done.");
  oled->drawString(11, 4, "done.");

  // Setup the WiFi
  if (!setupWiFi()) {
    Serial.println("ERROR: Cannot setup wifi. Halting.");
    while (true);
  }

  // Initialize ThingSpeak
  ThingSpeak.begin(client);

  beep();
  delay(3000);
  oled->clear();
}
```

Don't forget to initialize the ThingSpeak library at the end of the code with a call to the begin() function. The client variable is defined in the secrets.h file (in case you're wondering where it came from).

Next, we need to set up the fields and then call the writeFields() function from the ThingSpeak Arduino library. Listing 18-17 shows the completed loop() function with the new code highlighted.

Listing 18-17. Updates to the IoT Environment Monitor Loop Function (Arduino)

```
...
void loop(void) {
  if (airQuality->readData()) {
    // Retrieve the data
    float tempC = airQuality->getTemperature();
    float pressure = airQuality->getPressure();
    float dust = airQuality->getDust();
    AirMonitor::air_quality air = airQuality->getAirQuality();

    oled->drawString(0, 0, "ENVIRONMENT DATA");
    oled->drawString(0, 3, "Temp: ");
    oled->drawString(5, 3, String(tempC, 2).c_str());
    oled->drawString(11, 3, "C    ");
    oled->drawString(0, 4, "Pres: ");
    oled->drawString(5, 4, String(pressure, 1).c_str());
    oled->drawString(14, 4, "Pa");
    oled->drawString(0, 5, "Dust: ");
    if (dust == 0.0) {
      oled->drawString(5, 5, "--          ");
    } else {
      oled->drawString(5, 5, String(dust, 2).c_str());
      oled->drawString(10, 5, "%     ");
    }
    oled->drawString(0, 6, "AirQ: ");
    switch (air) {
      case AirMonitor::air_quality::ERROR_POOR:
      case AirMonitor::air_quality::POOR:
        oled->drawString(5, 6, "POOR");
        break;
```

```
      case AirMonitor::air_quality::FAIR:
        oled->drawString(5, 6, "FAIR");
        break;
      case AirMonitor::air_quality::GOOD:
        oled->drawString(5, 6, "GOOD");
        break;
      default:
        oled->drawString(5, 6, "--         ");
    }

    // Write the data to ThingSpeak
    // Set the fields with the values
    ThingSpeak.setField(1, tempC);
    ThingSpeak.setField(2, pressure);
    ThingSpeak.setField(3, dust);
    ThingSpeak.setField(4, air);

    // Write to the ThingSpeak channel
    int res = ThingSpeak.writeFields(myChannelNumber,
    myWriteAPIKey);
    if (res == 200) {
      Serial.println("Channel update successful.");
    } else {
      Serial.print("Problem updating channel. HTTP error code ");
      Serial.println(res);
    }

    // Check for environmental quality
    if ((dust > MAX_DUST) or (tempC > MAX_TEMP) or
        (air == AirMonitor::air_quality::POOR) or
        (air == AirMonitor::air_quality::ERROR_POOR)) {
      for (int x = 0; x < WARNING_BEEPS; x++) {
```

```
        oled->drawString(3, 7, "ENV NOT OK");
        beep(250);
        delay(250);
        oled->drawString(3, 7, "          ");
        delay(250);
      }
    }
  } else {
    oled->clear();
    oled->drawString(0, 2, "ERROR! CANNOT");
    oled->drawString(0, 3, "READ DATA");
  }
  delay(SAMPLING_RATE);
}
```

Since the sampling rate was set at 60 seconds, we do not need to modify the delay at the end of the loop for this project.

Raspberry Pi

Recall we need only add the import statement and API key. Listing 18-18 shows an excerpt of the code with the new lines added. The rest of the code from Chapter 15 remains the same. We will update the `main()` function to add the ThingSpeak code in the next section.

Listing 18-18. Updates to the IoT Environment Monitor Main Script (Python)

```
# Import libraries
import sys
import time

from grovepi import pinMode, digitalWrite
import grove_128_64_oled as oled
from air_monitor import AirMonitor, AirQualityEnum
```

```
from thingspeak import ThingSpeak

# API KEY
THINGSPEAK_APIKEY = 'YOUR_WRITE_API_KEY_HERE'
...
```

Next, we need to declare an instance of our ThingSpeak class from the thingspeak.py library module and then, after reading the data, use the existing Python dictionary we created in the class (env_data) and pass that to our thing_speak.upload() function call. Listing 18-19 shows the function with changes in bold. The rest of the code for this version remains the same as we had in Chapter 15.

Listing 18-19. Updates to the IoT Environment Monitor Main Function (Python)

```
...
def main():
    """Main"""
    print("Welcome to the Environment Monitor!")
    # Setup the buzzer
    pinMode(BUZZER_PIN, "OUTPUT")
    # Setup the OLED
    setup_oled()
    # Start the AirMonitor
    air_quality = AirMonitor()
    time.sleep(3)
    oled_write(11, 4, "done")
    beep()
    oled.clearDisplay()

    thing_speak = ThingSpeak(THINGSPEAK_APIKEY)
    while True:
        if air_quality.read_data():
```

```
# Retrieve the data
env_data = air_quality.get_data()

oled_write(0, 0, "ENVIRONMENT DATA")
oled_write(0, 2, "Temp: ")
oled_write(5, 2, "{:3.2f}C".format(env_
data["temperature"]))
oled_write(0, 3, "Pres: ")
oled_write(5, 3, "{:05.2f}hPa".format(env_
data["pressure"]))
oled_write(0, 4, "Dust: ")
if env_data["dust_concentration"] == 0.0:
    oled_write(5, 4, "--          ")
else:
    oled_write(5, 4,
        "{:06.2f}%".format(env_data["dust_
        concentration"]))
oled_write(0, 5, "airQ: ")
if env_data["air_quality"] in {AirQualityEnum.ERROR,
                                AirQualityEnum.POOR}:
    oled_write(5, 5, "POOR")
elif env_data["air_quality"] == AirQualityEnum.FAIR:
    oled_write(5, 5, "FAIR")
elif env_data["air_quality"] == AirQualityEnum.GOOD:
    oled_write(5, 5, "GOOD")
else:
    oled_write(5, 5, "--        ")

# Check for environmental quality
if ((env_data["dust_concentration"] > MAX_DUST) or
        (env_data["temperature"] > MAX_TEMP) or
```

```
                (env_data["air_quality"] == AirQualityEnum.
                POOR) or
                (env_data["air_quality"] == AirQualityEnum.
                ERROR)):
            for i in range(0, WARNING_BEEPS):
                oled_write(3, 7, "ENV NOT OK")
                beep(0.250)
                time.sleep(0.250)
                oled_write(3, 7, "          ")
                time.sleep(0.250)

        # Send data to ThingSpeak channel
        data = {
            'field1': env_data['temperature'],
            'field2': env_data['pressure'],
            'field3': env_data['dust_concentration'],
            'field4': env_data['air_quality']
        }
        thing_speak.upload(data)

    else:
        oled.clearDisplay()
        oled_write(0, 2, "ERROR! CANNOT")
        oled_write(0, 3, "READ DATA")

    time.sleep(SAMPLING_RATE)...
```

That's it. We're ready to execute the project. We will need to let it run for a few minutes so we can get some data. If you're running the project in a controlled environment where the values do not change, you may not notice much variation. As an exercise, consider altering the environment to stimulate changes in the data. Don't use flame or touch the electronics in any way while they are running.

Execute and Visualize the Data

At this point, you can set up the hardware and run the project. Let it run for about 20 minutes and then visit your ThingSpeak channel page. You should see your data in the channel private view similar to Figure 18-24.

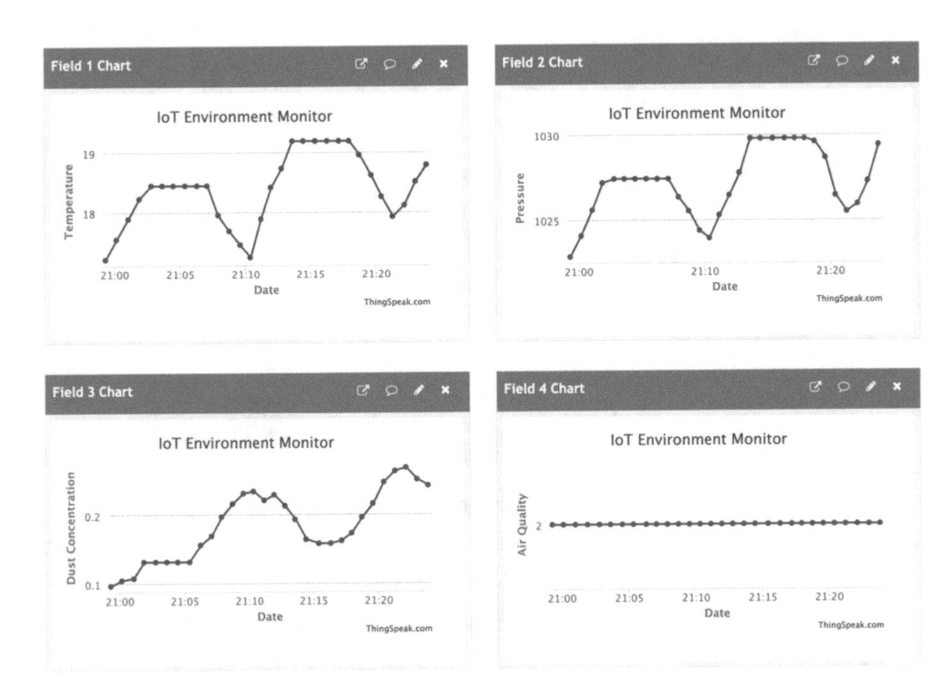

***Figure 18-24.** Example results (IoT Environment Monitor example)*

Once again, you may not see a lot of variances in the data if you run it in a controlled environment. For better results in a controlled environment, you should consider changing the sample rate from 30 seconds to every 4–6 hours. This should help show how the data changes over the course of a day.

However, notice the air quality line graph. That's not telling us anything, is it? What if we created an indicator widget for that data that changes color when the air quality gets poor?

Let's do that. Go ahead and click *Add Widgets* and then select the indicator and fill in the settings as shown in Figure 18-25 and then click *Create*. Notice I set the indicator to turn on only if the air quality (field 4) reaches 3 or greater.

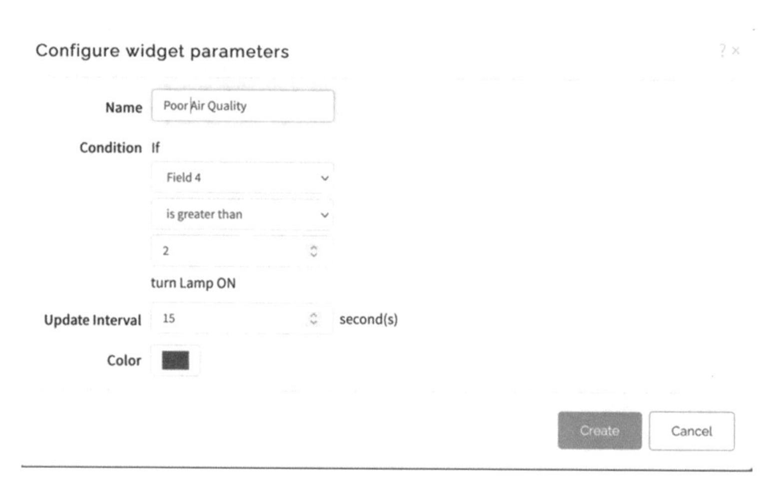

Figure 18-25. *Creating an indicator (IoT Environment Monitor example)*

When air quality is good, the indicator is dim as shown in Figure 18-26.

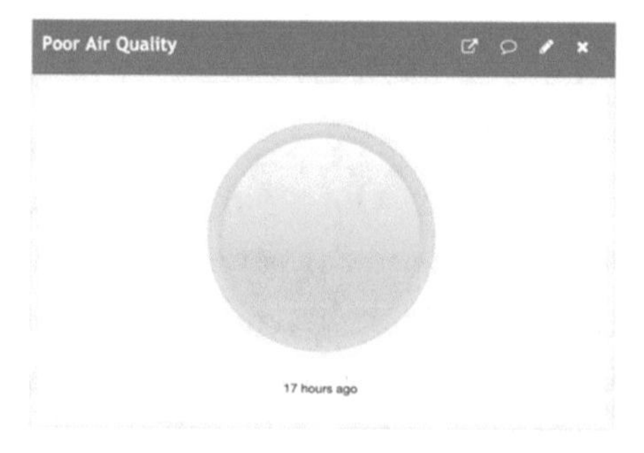

Figure 18-26. *Indicator off (IoT Environment Monitor example)*

Should the data reach a value of 3 to indicate poor air quality, the indicator will turn on as shown in Figure 18-27. This shows us how we can use the data to show thresholds reached. It can be used for high thresholds or low thresholds in which case you may want to choose a less alarming color such as green or so on.

Figure 18-27. *Indicator on (IoT Environment Monitor example)*

Now, we can take this a step further and create an array of indicators for the air quality. For example, we can create one for good air quality, another for poor, and another for bad. Figure 18-28 shows an example of the indicators. Note that you can drag and drop the widgets on the view to rearrange them. Nice! Note: The indicator colors are green for good, yellow for poor, and red for bad quality.

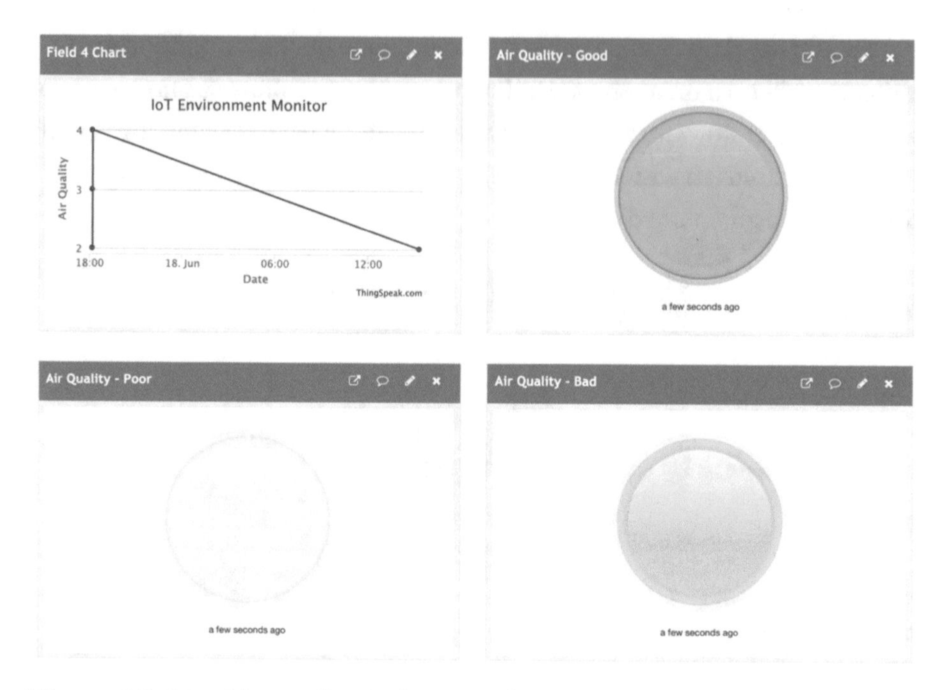

Figure 18-28. *Air quality indicators (IoT Environment Monitor example)*

Now I can see at a glance what the air quality is at the moment of last data read. Very nice!

Note If you choose to run both versions of the project, you may want to delete the data in the channel before starting the second version. Recall we can delete the channel data by clicking *Channel Settings* for the channel on your home page and clicking *Clear Channel* at the bottom.

Summary

If you have implemented all of the projects in this book, congratulations! You are now ready to tackle your own IoT projects. If you're still working on the examples, keep at it until you've learned everything you need to know to build your own IoT projects.

Our journey in learning how to build IoT projects for the Arduino and Python has concluded with a dive into how to use ThingSpeak to satisfy the needs of your IoT project for storing and displaying your data. In this chapter, we learned how to get started with ThingSpeak from creating our account to creating channels to storing our data and even some insights into how to modify the visualizations. Together with the knowledge you gained in this chapter and the previous chapters, you now have the skills to complete your own IoT projects.

In fact, you can now put down this book in triumph and start thinking of some really cool ways you can implement what you have learned. Perhaps you want to monitor events and data in your house, workshop, or garage. Or perhaps you want to design a more complex project that monitors sound, movement, and ambient temperature changes (like a home security system). Or maybe you want to revisit the example project chapters and implement the suggestions at the end of each chapter. All that and more is possible with what you have learned in this book. Good luck, and happy IoT projects!

Appendix

This appendix presents a list of the hardware required to complete the projects presented in the book. It presents the common hardware needed for all projects and then a section for each component system. While these lists are included in each chapter and discussed in greater detail, listing them here helps to see all of the hardware used in the book as a set and helps when planning to purchase the components you do not already own.

General Hardware List

The following are the hardware you should obtain in order to complete the projects in this book. Table A-1 lists the hardware common to all component systems.

***Table A-1.** General Hardware Needed*

Component	URL	Qty	Cost
Arduino MKR 1010 WiFi	www.sparkfun.com/products/15251	1	$35.95
Raspberry Pi 3B or later	www.sparkfun.com/categories/233 www.adafruit.com/category/176	1	$35.00+
SparkFun RedBoard Qwiic (Arduino Uno or compatible)	www.sparkfun.com/products/15123	1	$19.95

(*continued*)

C. Bell, *Beginning IoT Projects*, https://doi.org/10.1007/978-1-4842-7234-3

Table A-1. (*continued*)

Component	URL	Qty	Cost
Qwiic pHAT for Raspberry Pi	www.sparkfun.com/products/15945	1	$5.95
Grove Base Shield V2.0 for Arduino	www.seeedstudio.com/Base-Shield-V2.html	1	$4.45
GrovePi+	www.sparkfun.com/products/15945	1	$5.95

You will also need the host-to-computer cables associated with the specific Arduino board you choose. Similarly, you will also need a keyboard, mouse, power supply, and monitor for your Raspberry Pi (see Chapter 4 for more details).

Note If you plan to implement only one of the component system example projects, you do not need both host adapter boards listed – just choose the one for your chosen platform.

Consolidated Hardware Lists

This section presents a table that lists the hardware needed to complete the projects in this book broken down by component system.

Qwiic Component System

The components needed for the Qwiic and STEMMA QT examples in this book are listed in Table A-2 organized by chapter for quick reference.

Table A-2. Qwiic and STEMMA QT Components Needed

Chapter	Component	URL	Qty	Cost
7	Proximity Sensor Breakout – 20cm, VCNL4040	www.sparkfun.com/products/15177	1	$6.95
	Micro OLED Breakout	www.sparkfun.com/products/14532	1	$16.95
	Qwiic cable (any length can be used)	www.sparkfun.com/products/14427	2	$1.50
8	Environmental Combo Breakout – BME280	www.sparkfun.com/products/15440	1	$14.95
	Micro OLED Breakout	www.sparkfun.com/products/14532	1	$16.95
	Qwiic cable (any length can be used)	www.sparkfun.com/products/14427	2	$1.50
9 (Arduino)	Qwiic Soil Moisture Sensor	www.sparkfun.com/products/17731	1*	$8.50
	20×4 SerLCD – RGB Backlight (Qwiic)	www.sparkfun.com/products/16398	1	$24.95
	Qwiic Mux Breakout – 8 Channel	www.sparkfun.com/products/16398	1	$11.95
	Qwiic cable (any length can be used)	www.sparkfun.com/products/14427	3**	$1.50

(continued)

Table A-2. (*continued*)

Chapter	Component	URL	Qty	Cost
9 (Python)	Soil Moisture Sensor (with Screw Terminals)	www.sparkfun.com/products/13637	1*	$6.95
	20×4 SerLCD – RGB Backlight (Qwiic)	www.sparkfun.com/products/16398	1	$24.95
	Qwiic 12 Bit ADC – 4 Channel	www.sparkfun.com/products/15334	1	$10.50
	Qwiic cable (any length can be used)	www.sparkfun.com/products/14427	3**	$1.50
	Jumper Wires Premium 12" M/M	www.sparkfun.com/products/9387	1	$4.50
10	Adafruit LSM6DS33 6-DoF Accel + Gyro IMU – STEMMA QT / Qwiic	www.adafruit.com/product/4480	1	$5.95
	Micro OLED Breakout	www.sparkfun.com/products/14532	1	$16.95
	Qwiic cable (any length can be used)	www.sparkfun.com/products/14427	2	$1.50
11	SparkFun Triple Axis Magnetometer Breakout – MLX90393	www.sparkfun.com/products/14571	1	$15.95
	Micro OLED Breakout	www.sparkfun.com/products/14532	1	$16.95
	Qwiic cable (any length can be used)	www.sparkfun.com/products/14427	2	$1.50

(*continued*)

Table A-2. (*continued*)

Chapter	Component	URL	Qty	Cost
14	SparkFun Qwiic TMP102	www.sparkfun.com/products/16304	1	$6.50
16	SparkFun Qwiic LED Button Breakout	www.sparkfun.com/products/15931	4	$3.10
	LED Tactile Button – White	www.sparkfun.com/products/10439	1	$2.10
	LED Tactile Button – Green	www.sparkfun.com/products/10440	1	$2.10
	LED Tactile Button – Red	www.sparkfun.com/products/10442	1	$2.10
	LED Tactile Button – Blue	www.sparkfun.com/products/10443	1	$2.10
	Qwiic cable	www.sparkfun.com/products/14426	4	$0.95

Grove Component System

The components needed for the Grove examples in this book are listed in Table A-3 organized by chapter for quick reference. Note that Chapters 14 and 16 require Qwiic components as shown in Table A-2.

Table A-3. Grove Components Needed

Chapter	Component	URL	Qty	Cost
13	Grove Sound Sensor	www.seeedstudio.com/Grove-Sound-Sensor-Based-on-LM386-amplifier-Arduino-Compatible.html	1	$4.90
	Grove Red LED	www.seeedstudio.com/Grove-Red-LED.html	1	$1.90
	Grove Green LED	www.sparkfun.com/products/14532	1	$1.90
	Grove Button	www.seeedstudio.com/buttons-c-928/Grove-Button.html	1	$1.90
	Grove cables (any length can be used)	Included with each preceding module	5	
14	Grove Light Sensor	wiki.seeedstudio.com/Grove-Light_Sensor	1	$2.90
	Grove Chainable RGB LED	www.seeedstudio.com/Grove-Chainable-RGB-Led-V2-0.html	1	$5.99
	Grove LCD RGB Backlight	www.seeedstudio.com/Grove-LCD-RGB-Backlight.html	1	$11.90
	Grove Qwiic Hub	www.seeedstudio.com/Grove-Qwiic-Hub-p-4531.html	1	$1.90
	Grove cables (any length can be used)	Included with each preceding module	4	
	Qwiic cable	Included with the Qwiic Hub	1	

(continued)

Table A-3. (*continued*)

Chapter	Component	URL	Qty	Cost
15	Grove OLED 0.96 v1.3	seeedstudio.com/Grove-OLED-Display-0-96.html	1	$16.40
	Grove Buzzer	seeedstudio.com/Grove-Buzzer.html	1	$2.10
	Grove I2C High Accuracy Temperature Sensor (MCP9808)	seeedstudio.com/Grove-I2C-High-Accuracy-Temperature-Sensor-MCP9808.html	1	$5.20
	Grove Temperature and Barometer Sensor (BMP280)	seeedstudio.com/Grove-Barometer-Sensor-BMP280.html	1	$9.80
	Grove Air Quality Sensor	www.seeedstudio.com/Grove-Air-Quality-Sensor-v1-3-Arduino-Compatible.html	1	$10.90
	Grove Dust Sensor	www.seeedstudio.com/Grove-Dust-Sensor-PPD42NS.html	1	$12.70
	Grove cables (any length can be used)	Included with each preceding module	6	

(*continued*)

Table A-3. (*continued*)

Chapter	Component	URL	Qty	Cost
16	Grove Dual Button	www.seeedstudio.com/Grove-Dual-Button-p-4529.html	1	$2.20
	Grove Buzzer	www.seeedstudio.com/Grove-Buzzer.html	1	$1.90
	Grove LCD RGB Backlight	www.seeedstudio.com/Grove-LCD-RGB-Backlight.html	1	$11.90
	Grove Qwiic Hub	www.seeedstudio.com/Grove-Qwiic-Hub-p-4531.html	1	$1.90
	Grove cables (any length can be used but longer may be best)	Included with each preceding Grove module	3	
	Grove Female Breakout (Python version only)	www.seeedstudio.com/Grove-4-pin-Female-Jumper-to-Grove-4-pin-Conversion-Cable-5-PCs-per-PAck.html	1	$3.90

Index

A

B

C. Bell, *Beginning IoT Projects*, https://doi.org/10.1007/978-1-4842-7234-3

E

J

K

L

M

N

O

P

S

T

U, V

W, X, Y, Z